Templated Fabrication of Graphene-Based Materials for Energy Applications

Templated Fabrication of Graphene-Based Materials for Energy Applications

Edited by

Chunnian He, Naiqin Zhao, and Junwei Sha

WILEY-VCH

Editors

Prof. Chunnian He
Tianjin University
Materials Science and Engineering
Peiyang Park Campus
No.135 Yaguan Road
Haihe Education Park
300350 Tianjin
China

Prof. Naiqin Zhao
Tianjin University
Materials Science and Engineering
Peiyang Park Campus
No.135 Yaguan Road
Haihe Education Park
300350 Tianjin
China

Assoc. Prof. Junwei Sha
Tianjin University
Materials Science and Engineering
Peiyang Park Campus
No.135 Yaguan Road
Haihe Education Park
300350 Tianjin
China

Cover Image: © Neon_dust/Shutterstock

Library of Congress Card No.: applied for

British Library Cataloguing-in-Publication Data
A catalogue record for this book is available from the British Library.

Bibliographic information published by the Deutsche Nationalbibliothek
The Deutsche Nationalbibliothek lists this publication in the Deutsche Nationalbibliografie; detailed bibliographic data are available on the Internet at <http://dnb.d-nb.de>.

Print ISBN: 978-3-527-34600-4
ePDF ISBN: 978-3-527-82209-6
ePub ISBN: 978-3-527-34625-7
oBook ISBN: 978-3-527-82208-9

Typesetting Straive, Chennai, India
Printing and Binding CPI Group (UK) Ltd, Croydon, CR0 4YY

Printed on acid-free paper

C092508_290622

Contents

Preface

With carbon neutrality being raised as the high-priority mission for human society, there is an urgent need for technological development in related fields. In particular, the demand for energy storage and conversion applications, represented by batteries, is increasing rapidly. Moreover, their development of energy-related applications greatly boosted the requirements for new materials.

Among various new materials, graphene is undoubtedly the most popular one. Since its discovery, graphene has become a star material due to its excellent mechanical, electrical, and chemical properties. However, in energy-related fields such as batteries, supercapacitors, and electrocatalysis, the demand for materials has a different focus. How to manufacture and improve the graphene-based materials to meet different needs is a question worth exploring. Among the many strategies to prepare graphene-based materials, the template method is one of the most popular methods. The advantage of the template method is that it can effectively regulate the microstructure of graphene. Also, such a method can introduce heteroatoms or other phases in graphene by the interaction between the template and precursors during the preparation process. There are more and more researchers recognizing the benefits of the template method for graphene-based materials production. There has been a rapid growth in research in this area and many promising applications have emerged. Therefore, we think it is necessary to summarize and review the development in this field, which is the main reason why we have written this book.

The framework of this book can be broadly divided into three parts. Firstly, we will start with a basic introduction to graphene-based materials (Chapters 1 and 2); the second part is the frontier of template methods for the preparation of graphene-based materials (Chapters 3–5); the third part is the research progress of graphene-based materials in different energy-related applications (Chapter 6–10).

Chapter 1 mainly introduces the basic knowledge of graphene, including its history and physical properties. The purpose of this chapter is to give the reader a background for the following chapters. Chapter 2 will give readers a grasp of the current synthesis strategies for graphene. To this regard, the classification of graphene preparations is described and some typical researches are introduced in this chapter. Chapters 3 to 5 will focus on a brief overview of different kinds of template methods for graphene production. The study of porous metals for graphene preparation is presented in Chapter 3. Nanoporous graphene shows excellent physics and electrochemical performance in the fields of energy storage and conversion due to its high-quality and unique interconnected structure. Chapter 3 presents an overview of the recent research about the nanoporous graphene-based materials using nanoporous metal as the substrates. Then, Chapter 4 focuses on how to prepare graphene in large quantities, in particular. Considering the cost of graphene preparation with the potential for a large number of applications, substantial efforts have been devoted to developing a facile and versatile method, and several low-cost template methods will be reviewed in this part. In Chapter 5, the strategy of powder metallurgy and additive manufacturing procedures to prepare graphene materials is highlighted, which is one of the current research interests of our group. Subsequent chapters will discuss the various applications of graphene-based materials, such as lithium-ion batteries (Chapter 6), lithium-metal batteries (Chapter 7), lithium-sulfur batteries (Chapter 8), supercapacitors (Chapter 9), electrocatalysis (Chapter 10), and so on. Chapters 6 to 10 all follow a similar framework of discussion. At first, we will give the background of these fields, such as the basic concepts in energy applications and the physicochemical principles for different devices. Then, the discussion of the current bottlenecks in materials encountered in these applications will be presented. Consequently, we will describe why graphene-based materials are promising in these fields and how graphene should be improved to suit the different requirements. Meanwhile, we will review the specific applications of graphene-based materials prepared by the template methods in these fields and give the properties that can be achieved or the performance in practical cases. At the end of each chapter, we will discuss the current challenges of these graphene-based materials in each energy-related application, as well as possible improvement strategies and directions.

In these chapters, relevant content includes both the authors' studies and the research of others. This content has been reorganized and reviewed to form systematic frameworks. It is my pleasure to write and edit this book on graphene-based materials and their energy applications. It is hoped that the publication of this book will be helpful to researchers in this field and provide guidelines for related researches. Special thanks go to my students, colleagues, and the publisher's editors for their discussions and help.

List of Abbreviations

0D	zero-dimensional
1D	one-dimensional
2D Fe_3O_4@C@PGC	2D porous graphitic carbon nanosheets uniformly embedded with carbon-encapsulated Fe_3O_4 nanoparticles
2D	two-dimensional
3D BMG	three-dimensional bi-functional modular graphene network
3D CG	3D graphene-based hosts with continuous ductlike structure
3D FL-MoS_2@PCNNs	few layers MoS_2 nanosheets anchored on 3D porous carbon nanosheet networks
3D G	3D graphene architectures
3D GF	3D graphene foam
3D GF-FeS_2	cauliflower-like FeS_2 anchored on three-dimensional graphene foams
3D GM	3D graphene monolith
3D GN	three-dimensional graphene foams
3D GNs	3D porous graphene-like networks
3D PG	3D porous graphene
3D Rebar GF	CNTs-reinforced 3D graphene foam
3D S@PGC	sulfur nanoparticles in three-dimensional (3D) porous graphitic carbon (PGC)
3D SnSb@N–PG	SnSb in-plane nanoconfined three-dimensional N-doped porous graphene composite microspheres
3D ZnO@PCCMs	ZnO nanoparticle-confined 3D porous carbon composite microspheres
3D	three-dimensional
3DCu@NG	nanoporous Cu@N-doped graphene
3D-DG	free-standing 3D duct-like graphene
3D-DG@MnO_2	3D nanoporous graphene @MnO_2 composite
ΔG	Gibbs free energy
AFM	atomic force microscopy
AGNRs	armchair graphene nanoribbons
ALD	atomic layer deposition
a-MEGO	KOH-activated microwave-exfoliated graphite oxide

a-MnO_x	amorphous manganese oxide
AMs	anode materials
An	aniline
APCVD	atmospheric-pressure CVD
APS	aminopropyltriethoxysilane
APTMS	3-aminopropyl-trimethoxysilane
ASCs	asymmetric supercapacitors
a-TiO_2	amorphous TiO_2
BET	Brunauer–Emmett–Teller
BF-STEM	bright-field scanning transmission electron microscopy
B-LIG-MSCs	flexible supercapacitors based on boron-doped laser scribing graphene
BN-GAs	nitrogen and boron co-doped monolithic graphene aerogels
C	capacitance
C60	fullerene
carbide@CNS	vertically aligned 2D N-doped carbon nanosheets embedded with uniform nanosized metal carbides
Cat	catecho
CE	coulombic efficiency
CNF	carbon nanofiber
CNFs	carbon nanofibers
CNs	carbon nanosheets
CNTs	carbon nanotubes
CO_2RR	carbon dioxide reduction reaction
Co_3O_4/GS	Co_3O_4/graphene sheets
Co-MOF/3DGN	Co-MOF/three-dimensional graphene network
CoS_2/G	CoS_2/graphene/CoS_2 heterostructure
CoS_2-N-C/3DGN	CoS_2 encapsulated in N-doped carbon/3DGN
CPD	critical point dryer
CS-800A	S-doped carbons synthesized from K_2SO_4 at 800 °C
CS–800B	S-doped carbons synthesized from $Na_2S_2O_3$ at 800 °C
CTAB	cetyltrimethylammonium bromide
CV	cyclic voltammogram
CVD	chemical vapor deposition
DA	dopamine hydrochloride
DFT	density functional theory
DMA	dynamic mechanical analysis
DOS	density-of-states
E	energy density
ECR	electrochemical CO_2 reduction
EDLCs	electrical double-layer capacitors
EELS	electron energy-loss spectroscopy
EIS	electrochemical impedance spectroscopy
EOG	edge-oriented multilayer graphene

Fe@C@PGC	2D porous graphitic carbon nanosheets uniformly embedded with carbon-encapsulated Fe nanoparticles
Fe_2O_3/GS	Fe_2O_3/graphene sheets
FeS_2/rGO	FeS_2 microspheres anchored on rGO
G@MoS_2-C	MoS_2 encapsulated in carbon and coupled on graphene sheets
GAMs	graphene aerogel microlattices
g-C_3N_4	carbon nitride
G-CNF film	carbon nanofiber-stabilized graphene aerogel film
G-Co_3O_4	graphene-embedded Co_3O_4 rose-spheres
GF separator	glass fibers separator
GF	graphene fiber
GNR	graphene nanoribbons
GO	graphene oxide
H-3DRG	heteroatom-doped edge-enriched 3D rivet graphene
HAADF–STEM	high-angle annular dark-field scanning transmission electron microscopy
HAH	hydroxylamine hydrochloride
HER	hydrogen evolution reaction
HG	hydrogenated graphite
HIP	hot isostatic pressing
hnp-G	hierarchical nanoporous graphene
HOPG	highly oriented pyrolytic graphite
HPC-BMS	hierarchical porous carbons with different size of pores (-B, -M, S, meaning big, medium, and small, respectively.)
HPGN	layered porous graphene nanoparticles
HR-TEM	high-resolution TEM
I_D/I_G	The ratio of D-band and G-band in the Raman spectroscopy
ITO	indium tin oxide
K	thermal conductivity
LCGO	liquid crystalline graphene oxide
LDH	layered double hydroxide
LEG	light-emitting diode
LIBs	lithium-ion batteries
LIC	lithium-ion capacitors
LMA	lithium metal anode
LMB	lithium metal battery
LMO	lithium transition-metal oxide
LOG	laterally oriented graphene
LPCVD	low-pressure CVD
LPM-3D Rebar GF	3D rebar GF prepared via a loose powder metallurgy templates method
LSG	laser scribing graphene
MBE	molecular beam epitaxy
MG	MoS_2-reduced graphene oxide

ML	monolayer
MMT	montmorillonite
MnO_2/e-CMG	graphene–MnO_2 composites
MOF	metal–organic framework
MoS_2/G	MoS_2/graphene
MO_x	the transition metal oxides
MS	molten salt
MS_2	metal disulfides
MS_x	transition metal sulfides
NCS-SG	SG on the inner surface of nano-structured Cu Scaffold
NDG/MoS_2/NDG	nitrogen-doped graphene/MoS_2/nitrogen-doped graphene
NG	nitrogen-doped graphene
Ni_3Fe/N-C	Ni_3Fe nanoparticles embedded in porous nitrogen-doped carbon sheets
NiCo-LDH	NiCo-layered double hydroxide
NiO NS/graphene	NiO nanosheets anchored on the graphene through Ni—O—C bonds composites
NiO/GF	NiO microspheres loaded on graphene foam
NiS/GNS	graphene nanosheets decorated with NiS
NiS_2@G	NiS_2@graphene
N-MG-600	N-doped nanoporous graphene synthesized at 600 °C
NO-3DG@CNC	nitrogen and oxygen co-doped 3D nanoporous duct-like graphene@carbon nano-cage hybrid films
NO-3DG@CNCs	N, O co-doped 3D nanoporous graphene@carbon nanocages
NO-3DNG	N, O co-doped nanoporous graphene
NPC	nanoporous Cu
NPs	nanoparticles
NRR	nitrogen reduction reaction
NS-800	3D N, S co-doped nanoporous graphene heated at 800 °C
NSG	nitrogen and sulfur co-doped graphene
NSP750	N, S, P tri-doped 3D nanoporous graphene obtained at 750 °C
NW	nanowire
OER	oxygen evolution reaction
OMFLC	N-doped ordered mesoporous few-layer carbon
ORR	oxygen reduction reaction
PAM	polyacrylamide
PAN	polyacrylonitrile
PANI	polyaniline
PCCF	porous cellular carbon framework
PDA	polydopamine
PECVD	plasma-enhanced chemical vapor deposition
PEDOT	poly(3,4-ethylenedioxythiophene)
PF	phenol-formaldehyde

PG	porous graphite
PIP13-FSI	*N*-methyl-*N*-propylpiperidinium bis(fluorosulfonyl)imide
PMTs	powder metallurgy templates
PP separator	polypropylene separator
PPy	polypyrrole
PPy/PG	PPy-coated porous graphene
PrGO	partially reduced graphene oxide
PTh	polythiophene
PVA	polyvinyl alcohol
PYR14-FSI	*N*-butyl-*N*-methylpyrrolidinium bis(fluorosulfonyl)imide
QD	quantum dot
QHE	quantum Hall effect
rGO	reduced graphene oxide
rGO@p-FeS_2@C	porous FeS_2 nanoparticles anchored on the surface of rGO and coated by carbon layer
rGO@ReS_2@N-C	ReS_2 nanosheets are confined in 2D-honeycombed carbon nanosheets that comprise an rGO inter-layer and a N-doped carbon coating-layer
RHE	reversible hydrogen electrode
RPECVD	remote plasma-enhanced CVD
RT	room temperature
SAED	selected area electron diffraction
SANi-NG	single-atom (SA) Ni doping on nitrogen-doped graphene frameworks
Sb/rGO	Sb-decorated rGO composite
Sb@NS-3DPCMSs	Sb@C nanoconfined nitrogen-sulfur co-doped 3D porous carbon microspheres
SCG	SEI-coated graphene
SEAD	selected area diffraction
SEASA	solvent-evaporation-assisted self-assembly
SEI	solid electrolyte interphase
SEM	scanning electron microscope
SG	stacked graphene
SG	sulfur-doped graphene
SHG	stereoscopic holes over the graphitic surface
SIBs	sodium-ion batteries
SiC	silicon carbide
Sn/NS-CNFs@rGO	ultra-small Sn nanoparticles encapsulated in N, S co-doped CNFs sheathed within rGO scrolls films
Sn@G-PGNWs	Sn nanoparticles encapsulated with graphene shells of about 1 nm
Sn-Co-G	Sn-Co nanoparticles decorated rGO composites
Sn-G	Sn nanoparticles decorated rGO composites
SPS	spark plasma sintering
SSA	specific surface area

STEM	scanning transmission electron microscopy
TBOT	tetrabutyl titanate
TEM	transmission electron microscope
UHVCVD	ultrahigh-vacuum CVD
V	voltage window
ν-ReS_2	vertically aligned ReS_2 nanosheets
ν-ReS_2/rGO	ReS_2 nanosheets vertically aligned on reduced graphene oxide
WGC	wrinkled graphene cage
XPS	X-ray photoelectron spectroscopy
yolk-shell CoS_2@NG	N-doped graphene-coated yolk–shell CoS_2 spheres
YS-γ-Fe_2O_3@G-GS	yolk–shell γ-Fe_2O_3@graphene
ZGNRs	zigzag graphene nanoribbons

1

Graphene-Based Materials: Structure and Properties

Xiaoyang Deng[1,2] *and Yue Li*[1]

[1] *Tianjin University, School of Materials Science and Engineering and Tianjin Key Laboratory of Composites and Functional Materials, 135 Yaguan Rd., Jinnan District, Tianjin 300350, P.R. China*
[2] *Taiyuan University of Technology, Institute of New Carbon Materials, School of Materials Science and Engineering, P.R. China*

1.1 Introduction to Carbon Materials

Carbon materials have played important role in human society due to their extremely widespread applications, as shown in Figure 1.1. For example, while carbon black was used as paint, charcoal was as a heat source according to historical materials as early as the legend of Yao and Shun. Graphite is one of the most important carbon materials with numerous functions, including as electrodes in casting/stamping/pyrometallurgy processes, in powder form as pencil lead and polishing powder, as high-quality blocks in nuclear reactors, and as refractory materials for crucibles and molds. Diamond is the hardest three-dimensional carbon material. The rare, natural form of diamond is usually used in jewelry, while the synthetic form is used in cutting tools, infrared window materials, and abradants. In addition, carbon black is an important reinforcement material for tires in the development of automobiles. Conductive carbon black is important as a conductive additive in electrode production. Carbon fibers have been a star material in their use as additive-reinforcement materials in recent decades. For example, carbon-fiber-reinforced composite materials are widely used in aircraft body parts and aerospace field. In automobiles, carbon-fiber-reinforced carbon materials can be used in brakes production, while the motor-body manufacture is a promising application for carbon-fiber-based materials. In other fields, carbon-fiber-reinforced materials can be used to fabricate fan blades for wind power generation and in sports equipment. Some examples of applications of carbon materials are illustrated, but listing every application is not possible here due to the innumerable fields where they are used.

With the present rapidly rising world population and industrialization, energy demand has led to rigorous energy-sources consumption and environmental

Templated Fabrication of Graphene-Based Materials for Energy Applications, First Edition.
Edited by Chunnian He, Naiqin Zhao, and Junwei Sha.

Figure 1.1 Typical carbon materials, such as charcoals, pencil lead, diamond, and carbon black-reinforced tires. Source: Pixabay.com. Reproduced with permission of Pixabay.

pollution problem. Hence, the development of sustainable and clean energy technologies is a great challenge for human society, and the promising solutions are exploring solar/wind/tidal energies. Although these energy sources are renewable and clean, the highly intermittent feature makes them necessarily be coupled with the energy storage systems. On the other hand, the coming electronic and information age also requires high-performance energy storage equipment. Lithium-ion batteries and supercapacitors are the most widely used energy storage devices in industry and daily life – to name a few, electric vehicles, portable electronic equipment, and power station. Active carbon and graphite are used as active materials in the electrodes of supercapacitors and lithium-ion batteries, respectively; conductive carbon black is a commonly used additive to enhance the electrical conductivity of electrodes, while now carbon nanotube (CNT) and graphene are also used as conductive additives in some products.

Carbon materials are predominantly composed of carbon atoms, and they have diverse structures and properties. There are various methods for the classification of carbon materials according to different bases – e.g. the chemical bonding modes of the carbon–carbon atoms, the production process, micro/nano-texture, and time of appearance. According to the chemical feature of carbon–carbon bonding (Figure 1.2), various families of carbon materials have been defined: C—C bonds based on the sp^3 orbitals for the diamond family materials, C—C bonds based on sp^2 orbitals to construct graphite family, C—C bonds based on sp. orbitals for carbynes, and C—C bonds with hybrid orbitals, such as graphdiyne constructed by

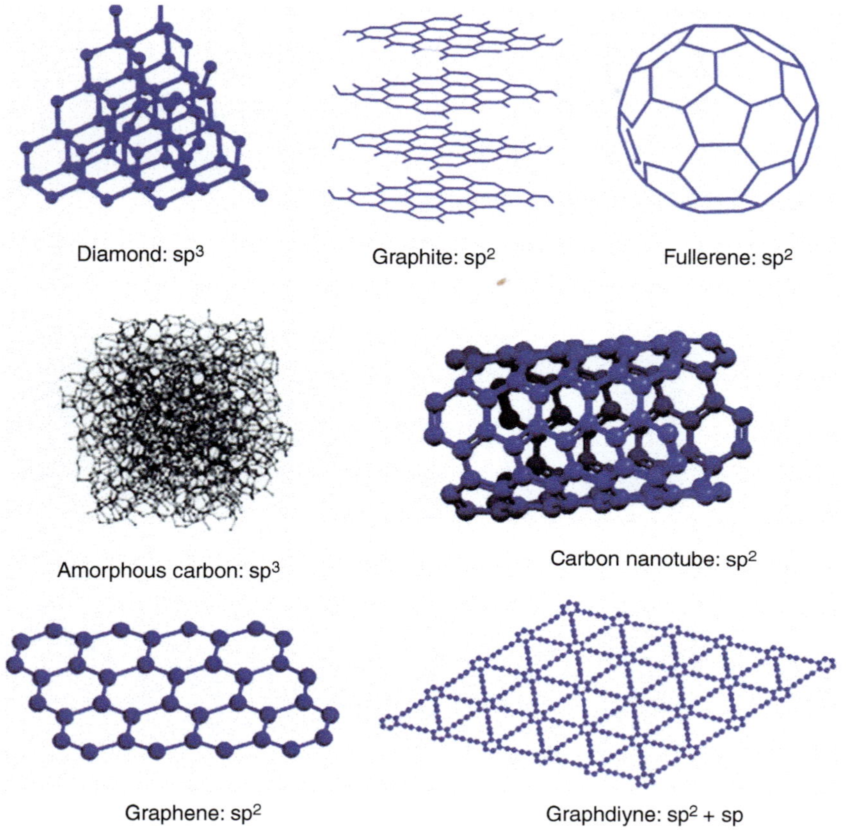

Figure 1.2 Carbon family based on carbon–carbon bonding. Source: Li et al. [1]. Reproduced with permission of Royal Society of Chemistry.

C—C bonds based on sp/sp^2 hybrid orbitals [2–5]. The most commonly used carbon materials in industry belong to the graphite family with a layered stack of carbon hexagons. Such carbon layers have strong anisotropy due to the strong covalent bonding based on sp^2 orbitals in the layers and weak bonding of van der Waals force of π electron clouds between stacked layers [4].

In the nanometer era, there have been many new members in the carbon family since the discovery of carbon nanomaterials. In 1985, zero-dimensional (0D) fullerene (C60) was first discovered with zero dimension by Smalley and coworkers [6, 7]. In 1991, the CNTs, a new one-dimensional (1D) carbon allotrope, were proposed by Iijima [8, 9]. The CNTs with various structures, i.e. thickness (single-, double-, or multi-walled), diameter, and length, have been synthesized and some types are produced on industrial scale [10]. When scaling down the thickness of graphite into nanoscale, graphene can be obtained; in other words, the monolayer graphene is the mono-unit of graphite. In 2004, graphene with two-dimensional (2D) structure was first experimentally evidenced and characterized by Geim and Novoselov, and it quickly became the most widely investigated material [11].

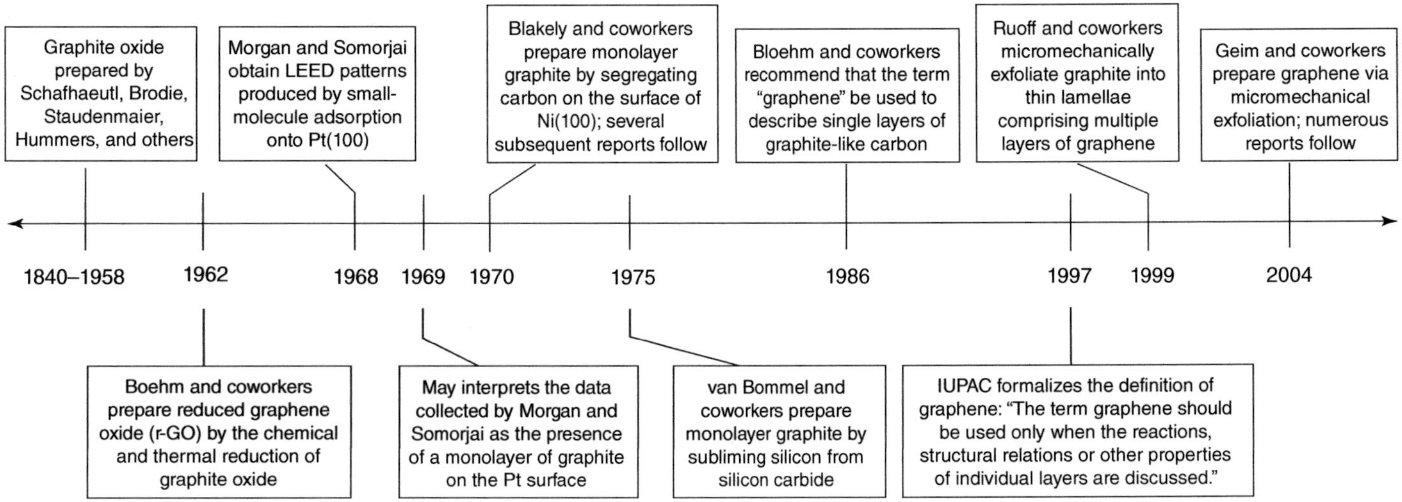

Figure 1.3 Timeline of selected events in the history of the preparation, isolation, and characterization of graphene. Source: Dreyer et al. [14]. Reproduced with permission of John Wiley and Sons.

1.2 History of Graphene

Despite a series of theoretical studies stating otherwise, as early as the 1940s, Wallace had advised that graphite might possess extraordinary electronic characteristics if its layer is isolated [12]. The term "graphene" was first proposed by Boehm et al. in 1986, who recommended that the term "graphene layer" should be used for the isolated graphitic structure constructed by only carbon atoms in one plane, which means that graphene is the most basic structure of graphite [13]. Geim, Novoselov, and coworkers used the mechanical approach to synthesize graphene in 2004, in which the highly ordered pyrolytic graphite surface was pressed against a surface of silicon wafer, and, when removed with a scotch tape, thin flakes of graphene were detected and characterized [11] (Figure 1.3). This was the first time that the existence of graphene was demonstrated through experimental evidence. In 2010, Geim and Novoselov were awarded a Nobel Prize for "groundbreaking experiments regarding the two-dimensional material graphene."

"Graphene," therefore, is defined as "an isolated single layer of carbon hexagons with sp^2-hybridized C—C bonding." In other words, as a new two-dimensional (2D) allotrope of carbon materials graphene has monoatomic thick honeycomb lattice structure (Figure 1.4a) [2, 15–18]. Hence, it can be considered as the basic building block to construct other graphitic materials, e.g. wrapped up into 0D fullerenes, rolled into 1D CNTs, and stacked to 3D graphite (Figure 1.4b) [15]. The distinctive structure and physical properties of graphene make it important for applications in technological field, such as polymer-based nanocomposites, energy storage and conversion devices (e.g. lithium-ion batteries, supercapacitors, air batteries, and fuel cells), flexible electronic and optical devices, and chemical sensors [2, 3, 19–22].

In general, some physical and chemical methods can be used to synthesize graphene, e.g. mechanical exfoliation of graphene layers from pristine graphite; chemical vapor deposition (CVD) of graphene layers on different crystals; thermal decomposition of silicon carbide (SiC); lengthwise unzipping of CNTs; exfoliation

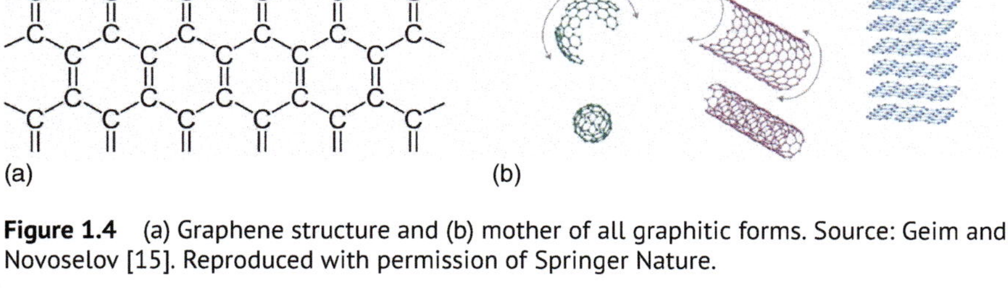

Figure 1.4 (a) Graphene structure and (b) mother of all graphitic forms. Source: Geim and Novoselov [15]. Reproduced with permission of Springer Nature.

of graphite through its intercalation compound; reduction of graphene oxides (GOs) and graphene fluoride; and organic synthesis processes [2, 23, 24]. But some methods, such as mechanical exfoliation, synthesis on SiC, and organic synthesis, exhibit difficulty in scalability and are expensive to produce, which restrict the widespread use of graphene. CVD method is unsuitable for large-scale production due to the high cost and rather low yield. The liquid-phase exfoliation process is so highly scalable and low cost that it can produce graphene in large quantities. Furthermore, although the reduced graphene oxide possesses low quality, it is still suitable for mass production due to the high yield and low cost.

1.3 Structure of Graphene

As shown in Figure 1.5a, the honeycomb lattice of graphene consists of two interpenetrating triangular sublattices, which are designated A and B. The sites of one sublattice (A) are at the centers of triangles (B) with a carbon-to-carbon interatomic length of 1.42 Å. Figure 1.5b shows the first Brillouin zone of graphene containing the high-symmetry points M, K, and K′. Each carbon atom has one *s* orbital and three p orbitals. The single s orbital is tied up with two in-plane p orbitals in the strong covalent bonding of graphene, which do not contribute to the conductivity. The remaining p orbital is oriented perpendicular to the molecular plane. This odd p orbital is hybridized to form π (valence) and π^* (conduction) bands [27]. The π and π^* bands touch at the K and K′ points (called Dirac points). The linear bands derived from crystal symmetry of graphene are a hallmark of graphene. This feature leads to many interesting physical properties, such as Berry's phase, half-integer quantum Hall effect (QHE), and Klein paradox. The energies of

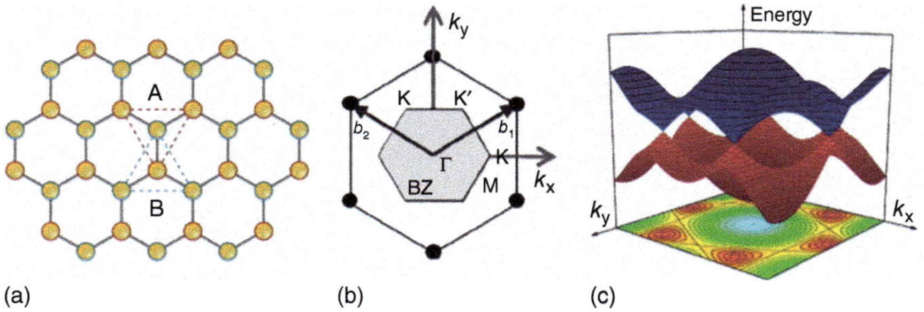

Figure 1.5 Atomic and electronic structures of graphene. (a) Graphene lattice consists of two interpenetrating triangular sublattices, each with different colors. The atoms at the sites of one sub-lattice, (i.e. A) are at the centers of the triangles defined by the other lattice (i.e. B), with a carbon-to-carbon interatomic length of 1.42 Å. Source: Rao et al. [25]. Reproduced with permission of John Wiley and Sons. (b) Reciprocal lattice of graphene. The shaded hexagon is the first Brillouin zone. (c) π–π^* band structure of graphene. The three-dimensional first Brillouin zone is displayed in red and blue for the valence and conduction p bands, respectively, above the planar projection of the valence band. The six Dirac cones are positioned on a hexagonal lattice. Source: Soldano et al. [26]. Reproduced with permission of Elsevier).

the bands lie on the momentum of the charge carriers within the Brillouin zone. The constant energy contours within the linear-band approximation are circles around the K and K′ points [26]. As displayed in Figure 1.5c, the two bands meet each other and then produce cone-shaped valleys in the low-energy regime of the vicinity of the K and K′ points. In this low-energy limit, the dispersion relation of energy–momentum is linear and the carriers are seen as zero-rest mass relativistic particles. In the high-energy limit, the energy–momentum relation is changed and the distorted bands lead to anisotropy, also known as trigonal warping [26]. In addition, with stacked layers on top of each other, the electronic dispersion of graphene is changed. For example, the first-obtained bilayer graphene exhibits its own specific properties [25].

1.4 Properties of Graphene

Intrinsic (undoped) graphene is a semimetal or a semiconductor with zero bandgap. Pristine graphene possesses amazing properties, such as exceedingly high charge carriers (electrons and holes) mobility = 230 000 $cm^2\ V^{-1}\ s^{-1}$ at room temperature and an intriguing thermal conductivity of 4.84×10^3 to $5.30 \times 10^3\ W\ m^{-1}\ K^{-1}$ at room temperature [28]. Furthermore, graphene exhibits many mechanical properties, e.g. the high mechanical stiffness of about 1 TPa [29]. In addition, graphene has an ultra-high surface area of about 2630 $m^2\ g^{-1}$, which is significantly higher than the CNT and graphite counterparts. Besides these, we will now give a detailed account of some properties of graphene.

The existence of massless Dirac quasiparticles in graphene may be verified through the experimental observation, based on the code that the cyclotron mass was dependent on the square root of the electronic density in graphene [30, 31]. Graphene can exhibit an ambipolar electric field effect due to the zero bandgap semiconductor feature. The charge carriers in graphene can be tuned continuously between electrons and holes with concentrations as high as 1013 cm^{-2}, and mobilities of up to 15 000 $cm^2\ V^{-1}\ s^{-1}$ even under ambient conditions (Figure 1.6) [15, 30, 31]. Moreover, the low dependence of mobilities of charge carriers in graphene with temperature suggests that ultrahigh mobility would be accomplished at room temperature [32]. More importantly, when minimizing the influence of impurity scattering, mobilities in excess of 200 000 $cm^2\ V^{-1}\ s^{-1}$ could be achieved in suspended graphene [33]. The mobilities in graphene can still remain high even at a huge carrier concentration (>1012 cm^{-2}) in both electrically and chemically doped devices [34, 35]. Another measure of the electronic quality of graphene is that the QHE can be directly observed even at room temperature, which increases the temperature range for the QHE by a factor of 10 compared with previous reports [36].

Based on the Wiedemann–Franz law, the contribution of electronics is negligible for thermal conductivity in graphene; hence, the thermal conductivity (κ) of graphene is just dependent on phonon transport, namely high-temperature diffusive conduction and low-temperature ballistic conduction [37]. As early as

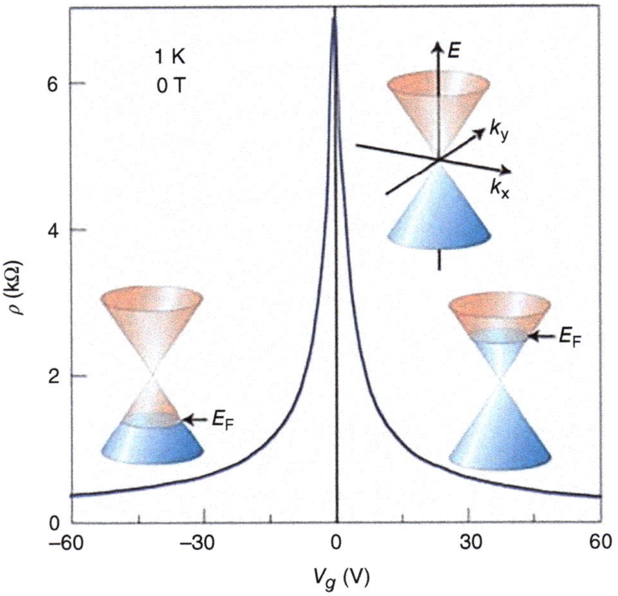

Figure 1.6 Ambipolar electric field effect in single-layer graphene. The rapid decrease in resistivity ρ on adding charge carriers indicates their high mobility (in this case, $\mu \approx 5000\,\text{cm}^2\,\text{V}^{-1}\,\text{s}^{-1}$ and does not noticeably change with increasing temperature to 300 K). Source: Geim and Novoselov [15]. Reproduced with permission of Springer Nature.

2000, the thermal conductivity of suspended monolayer graphene was predicted to be about $6000\,\text{W}\,\text{m}^{-1}\,\text{K}^{-1}$ at room temperature, much higher than that of graphitic carbon [38]. For a suspended monolayer graphene flake obtained from mechanical exfoliation, a high thermal conductivity value of about $5000\,\text{W}\,\text{m}^{-1}\,\text{K}^{-1}$ was observed through an optical measurement according to the shift of Raman G band [28].

The mechanical properties, including Young's modulus and fracture strength of monolayer graphene, have been investigated by atomic force microscopy (AFM) and numerical simulations (e.g. molecular dynamics) [39]. Particularly, the nanoindentation technique was measured using AFM to detect elastic properties and intrinsic breaking strength for free-standing monolayer graphene [29]. The results displayed that Young's modulus and fracture strength of defect-free graphene can reach 1.0 TPa and 130 GPa, respectively. The mechanical properties of some graphene materials, produced by reducing the graphene oxide, were investigated but exhibited decreased properties compared with the defect-free monolayer graphene [40–42].

The measured white light absorbances of one- and two-layered suspended graphene sheets are 2.3% and 4.6%, respectively. According to the experimental observation, the transmittance was linearly decreased based on the number of layers for *n*-layer graphene, and the strict linearity feature has been further demonstrated up to five monolayers [43]. This macroscopic linear dependence between the transmittance and the thickness of graphene films is intimately related to the two-dimensional gapless electronic structure of graphene. However, when the energy of incident photons was lower than 0.5 eV, a deviation from the universal linear behavior could be observed. This deviation can be attributed to the finite temperature and chemical potential shift of the charge-neutrality (Dirac) point induced by doping [44].

1.5 Structure Defects of Graphene

To adapt to different application fields, one of the research focuses on graphene is the targeted control of its key physical and chemical properties. For this purpose, various modification methods are developed. Among them, defect engineering is considered an efficient method that can tailor mechanical, electrical, chemical, and magnetic properties of graphene [45–47]. Structural defects of graphene can be divided into two main categories. The first type is **intrinsic defects**, composed of non-sp^2 orbit hybrid carbon atoms on graphene. The carbon atomic orbital hybrid forms of change are usually caused by the absence of extra carbon atoms in the surrounding carbon six-membered rings. Therefore, obvious holes in non-six-membered carbon rings or even point or line domains can be observed in this graphene sheet at atomic resolution. The second type is **extrinsic defects**, which are generated by foreign (non-carbon) atoms covalently bonded to graphene carbon atoms. The foreign atoms (such as N, O, and B) strongly affect the charge distribution and properties of graphene [48, 49].

Specifically, **graphene intrinsic defects** can be divided into five categories: Stone–Wales defects, linear defects, single vacancies, multiple vacancies, and carbon adatoms defects. **Stone–Wales defect** is caused by simply rotating the C—C bonds to form different carbon polygon combinations (switching between pentagons, hexagons, and heptagons) without adding or removing carbon atoms. Figure 1.7 demonstrates typical Stone–Wales defects formed by rotating a C—C bond by 90° (within four neighboring hexagons that transform to two pentagons and two heptagons: $C_{6,6,6,6} \rightarrow C_{5,5} + C_{7,7}$) [50, 51]. Continuous rotation of C—C bond can be extended to linear defects consisting of paired pentagons and heptagons, as shown in Figure 1.7c. Such defects also require relatively large energy to form (~5 eV), so they can be produced by electron-beam bombardment or rapid cooling at high temperatures [52]. The existence of Stone–Wales defects significantly affects the electrical conductivity of graphene and enhances its electrochemical activity, but its strength decreases significantly [53].

Single vacancies can be considered as the simplest defects, which are formed by a missing carbon atom from continuous planar carbon six-membered ring. Similarly, on the basis of single vacancies, a continued absence of carbon atoms leads to **multiple vacancies**. Figure 1.8 displays the TEM photographs and typical atomic structures of single vacancies and multiple vacancies. The absence of dangling bonds and the steady rotation of C—C bonds make the $V_{2(555\text{-}777)}$ (7 eV) have lower formation energy than $V_{1(5\text{-}9)}$ (7.5 eV) and $V_{2(5\text{-}8\text{-}5)}$ (8 eV), so the probability of the $V_{2(555\text{-}777)}$ defect being observed by TEM is indeed greater [55, 56].

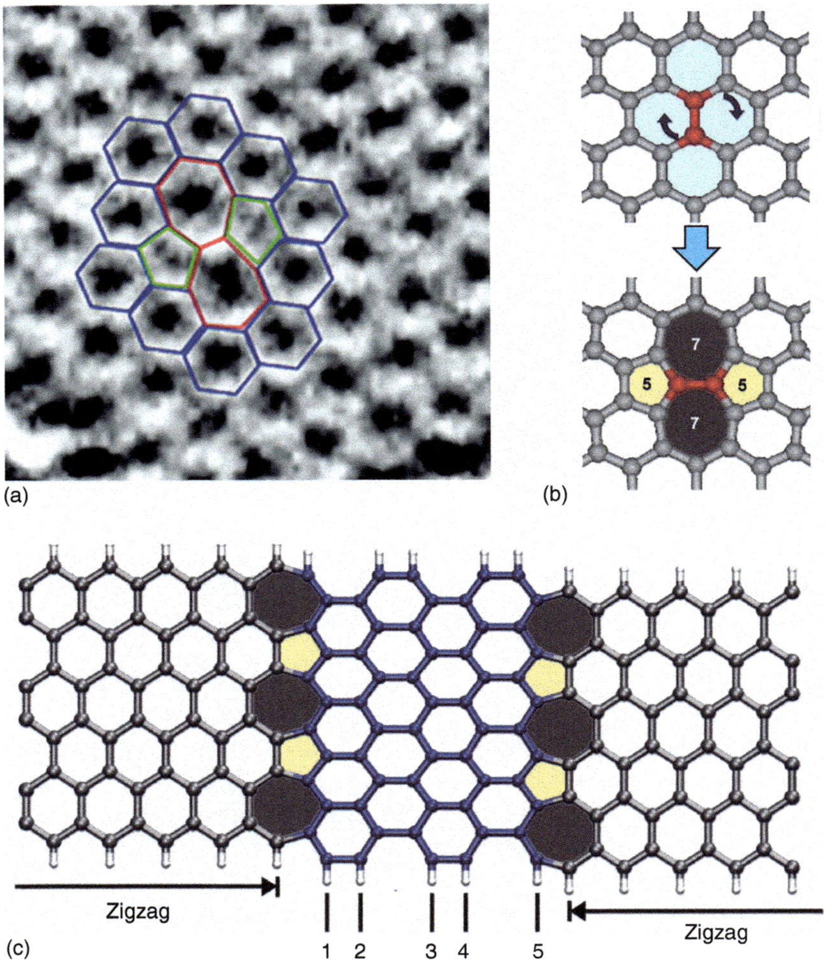

Figure 1.7 TEM image. (Source: Meyer et al. [50]. Reproduced with permission of American Chemical Society) (a) and (b) atomic structure of Stone–Wales defect in graphene by rotating a C—C bond by 90°. Source: Terrones et al. [51] Reproduced with permission of Elsevier. (c) Atomic structure of linear defect with a chain of paired rings (pentagons and heptagons). Source: Terrones et al. [51] Reproduced with permission of Elsevier.

1.5.1 Carbon Adatoms Defects

Free carbon atoms interacting with the desirable planar graphene region may destroy the original planar structure of graphene and cause sp^3-hybridization defects [57]. Figure 1.9 exhibits the spatial arrangement of carbon adatoms defects and introduction positions of free carbon atoms. The existence of carbon adatoms defects undoubtedly destroys the two-dimensional crystal structure of graphene. In particular, some defects (as presented in Figure 1.9b) directly change the orbital sp^2-hybridization type to sp^3-hybridization, which are bound to affect the electrical properties of graphene. Indeed, making such defects manageable is a big challenge for researchers.

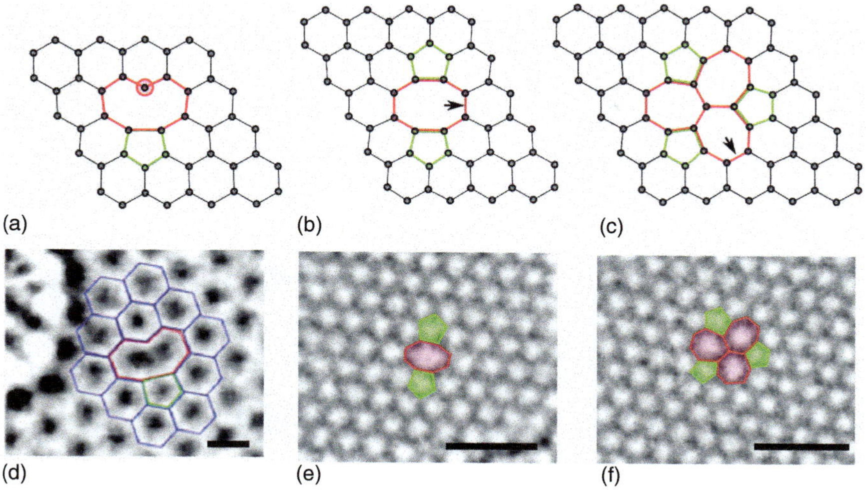

Figure 1.8 Atomic structures and TEM images of single-vacancy $V_{1(5\text{-}9)}$ (a, d), double-vacancy $V_{2(5\text{-}8\text{-}5)}$ (b, e), and $V_{2(555\text{-}777)}$ (c, f) defects in graphene. Source: Banhart et al. [54]. Reproduced with permission of American Chemical Society).

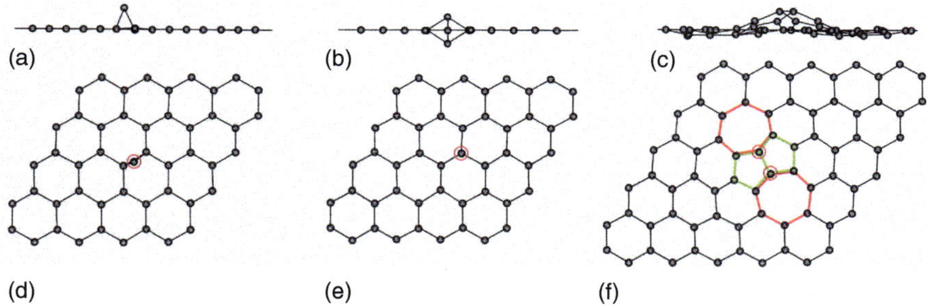

Figure 1.9 Atomic structures of carbon adatoms: (a, d) single adatom in the bridge; (b, e) single adatom in the dumbbell configuration; (c, f) the C_{7557} defect formed by two close adatoms. Source: Banhart et al. [54]. Reproduced with permission of American Chemical Society.

1.5.2 Graphene Extrinsic Defects

Graphene extrinsic defects can be classified as out-of-plane and in-plane heteroatom-introduced defects according to the position of introduction. Under the conditions of CVD or strong oxidation, metal atoms or oxygen-containing functional groups are inevitably introduced on the surface of graphene, resulting in **out-of-plane heteroatom-introduced defects**. Figure 1.10 demonstrates the structure diagram of graphene with out-of-plane defects introduced by metal and oxygen adatoms. The metal adatom can cause obvious migration movement on the graphene layer, and the different metal adatoms can effectively regulate the electronic and magnetic properties of graphene sheets [57, 59]. The graphene-containing oxygen functional group prepared by strong oxidant has obvious better hydrophilicity than graphene. Based on the particularity of this

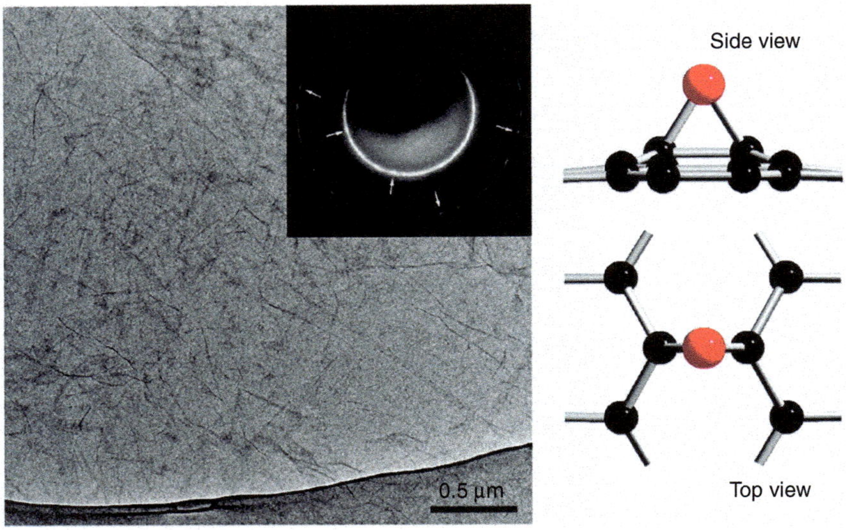

Figure 1.10 TEM image and corresponding atomic structure of GO films. The inset is the diffraction pattern from these GO films. Source: Mkhoyan et al. [58]. Reproduced with permission of American Chemical Society.

material and its broad application prospect, this material is named as graphene oxide (GO). Actually, the oxygen adatoms are difficult to be completely removed in the subsequent reduction process, and the relative content of graphene defects does not change or even increase, which is because when oxygen atoms are removed, carbon atoms are removed at the same time to form cavities, inducing intrinsic defects [58]. **In-plane introduced defects**: Doping by foreign atoms, such as nitrogen and boron, can form three chemical bonds and thus can replace the carbon atoms in graphene, which constitute the **in-plane introduced defects** in graphene. B and N doping affect the charge distribution on the carbon atoms in the sublattice of the substituted dopant. As foreign atoms, N and B atoms also have their own unique properties, which also change the properties of graphene. Moreover, the B- and N-doped graphene exhibit higher electrical conductivity than the pristine graphene [48]. Therefore, researchers are committed to developing and studying heteroatom-doped graphene with excellent catalytic activity and electrical conductivity [60–62].

1.6 Different Dimensional Graphene

According to the different dimensions of graphene nanostructures discovered, graphene can be divided into **0-dimensional (0D) graphene clusters, one-dimensional (1D) graphene nanoribbons, two-dimensional (2D) graphene sheets, and three-dimensional (3D) graphene architectures.** The lateral dimensions of graphene materials vary from nanoscale to microscale to macroscopic, which may affect permeability thresholds, bandgaps, and many other properties and behaviors. When lateral dimensions of graphene materials are extremely limited into nanoscale, three incongruous atoms in the edges would be formed. As this chain-like structure exhibits a cyclic sequence with a larger aspect ratio, 1D sp^2 hybridized carbon crystal graphene nanoribbon was formed. However, if the chains present dimensionless sexual or no periodicity, this structure can be seen as a 0D graphene cluster [51]. Graphene clusters have a large number of dangling bonds of edge atoms, so the structure is quite unstable. The diversity of possible types of graphene clusters significantly hinders the theoretical exploration of the regularities of changes in their properties that vary with many factors (such as size, shape, temperature, and charge). According to this situation, few types of graphene clusters were simulated [63, 64].

Graphene nanoribbons usually exhibit two different possible edge geometries, so-called zigzag and armchair, as shown in Figure 1.11a, which exhibit very different electronic properties due to their contrasting boundary conditions. Early tight-binding calculations suggested that the structure of narrow bands depended on their orientation. These results show the zigzag graphene nanoribbons (ZGNRs) are always metallic (top of Figure 1.11b), and the armchair graphene nanoribbons (AGNRs) show alternate metal structure and gap electronic structure according to the width (bottom of Figure 1.11b). However, for ZGNRs, first-principle calculations verify that the flat band of the high density-of-states (DOS) at zero energy calculated by simple tight-binding calculations is unstable relative to spin splitting.

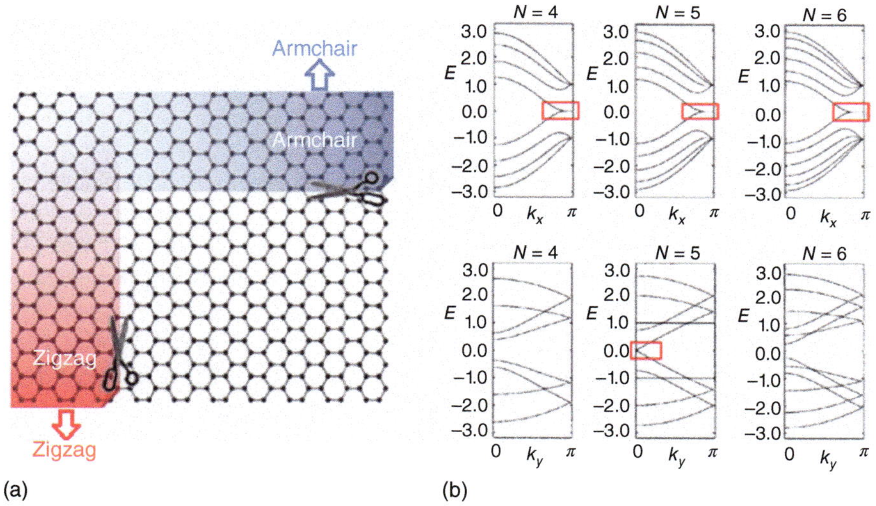

Figure 1.11 (a) The schematic representation of cutting the two-dimensional graphene sheet to obtain ZGNRs and AGNRs [65]. Source: Dutta and Pati [65]. Reproduced with permission of Royal Society of Chemistry. (b) Electronic band structure obtained by tight-binding calculations for various graphene nanoribbon widths. The width is measured by N carbon dimer lines ($N = 4, 5, 6$). Top row: ZGNRs have a metallic state at $E = 0$. Bottom row: AGNRs exhibit alternating metallic and semiconducting behaviors with changing value of N. Source: Nakada et al. [66]. Reproduced with permission of Royal Society of Chemistry.

Because the edge of the narrow ZGNRs is very close to charge neutral, electron conductance tends to be in a spin direction. Using the electric field in-plane to modify the natural energy distribution of spin in the band, semi-metallicity was proposed [67]. At present, the preparation methods of graphene nanoribbons mainly include lithographic-patterning methods (etching graphene) [68], chemical methods (cutting CNTs or other graphitic precursors) [68, 69], CVD methods [70, 71], and organic synthesis [72]. As a two-dimensional variant of graphene, GNRs have room-temperature QHE and higher electron migration. Due to their excellent performance, GNRs demonstrate more flexible and controllable performance than graphene and have a broader application prospect, which presents great development potential in the **fields of polymer modification, lithium-ion battery, fuel cell, and other energy storage materials** and electronic devices.

1.6.1 3D Graphene Architectures (3DG)

In practical applications, 2D graphene sheets are prone to stacking and agglomeration, which significantly reduces specific surface area and electrical conductivity compared with intrinsic graphene. 3D graphene architectures, as 3D derivatives of graphene, not only maintain good electronic conductivity, heat stability, oil-absorption, lightweight, and biocompatible properties inherited from graphene, but also process stable structure and high-surface utilization. Therefore, 3DG has great development prospects in the fields of electronic devices, environmental cleanup, shock damping, biomedical sciences, and nanocomposites.

As usual, 3DG with few defects can be directly prepared on a porous metal template (e.g. Ni, Cu foam). Qin et al. synthesized the 3D porous graphene (3DPG) by CVD method using porous Cu as template and catalysts [73]. The 3DPG exhibits large specific surface area ($1160\,m^2\,g^{-1}$) and desirable electrochemical performance. Furthermore, the obtained 3DPG can be used to make composite (e.g. 3DPG/MnO_2) for electrochemical applications [74]. To obtain 3DG with a large number of defects and functional groups, the common strategies to achieve macroscopic preparation of 3DG are the facile and versatile graphene oxide (GO)-based chemical synthesis methods [75]. However, the GO is usually produced by using strong oxidizing reagents, which are not suitable for environmental considerations. As a consequence, environment-friendly, mass-produced, and cost-effective salt template methods based on soluble organic matter have become considerable strategies to synthesize 3DG [76]. The 3DG was formed by the space limitation of self-assembled salt template during the pyrolysis of organics [77]. By adjusting the type and combination of salt templates, 3DG with special porous structure can be designed to adapt to various electrochemical applications [78]. In addition, through the selection of appropriate organics, heteroatom-doped 3DG materials have been synthesized, endowing them special physical and chemical properties [79, 80].

In recent years, utilizing 3D printing technology to prepare 3DG has also received considerable attention from researchers. Ink-based 3D printed 3DG, GO, or chemically modified graphene is often used as precursors or as additives for printable inks [81]. Jiang et al. reported a facile ion-induced gelation method to directly print 3D graphene aerogel microlattices (GAMs) from GO-based ink with trace additives [82]. Due to the rich porous structure and fast electronic transmission, 3D GAMs exhibited admirable electrochemical performances as electrode material for supercapacitors. Sha et al. used Ni powder and sucrose powder as precursors, commercial CO_2 lasers for 3D printing in atmosphere conditions at room temperature, and obtained 3D graphene foam (3DGF) with high porosity (~99.3%), low density (~0.015 g cm^{-3}), high quality, good damping performance (~0.06), and electrical conductivity (~8.7 S cm^{-1}) [83]. This strategy combines 3D printing technology and template powder metallurgy technology, and exhibits the characteristics of simplicity, efficiency, and strong designability.

1.7 Graphene Composites

To utilize its admirable properties, graphene was used as a typical modified phase to prepare functional graphene–based composites, which have shown excellent application potential in many technical fields, such as electronics, biomedicine, machinery, and aerospace. At present, the energy storage researches of functional graphene–based materials mainly focus on **graphene/conductive polymer composites** and **graphene/inorganic composites**.

1.7.1 Graphene/Conductive Polymer Composites

Conductive polymers, such as polyaniline (PANI), polyacrylonitrile (PAN), polypyrrole (PPy), poly(3,4-ethylenedioxythiophene) (PEDOT), are commonly used to prepare graphene/conductive polymer composites [84–86]. Conductive polymers have the advantages of high energy storage capacity, good electrical conductivity, and low cost. However, the poor mechanical strength performance restricts their electrochemical stability and further application [87]. Therefore, during preparation, graphene is not only required to maintain decentralized structure, but also for effective interface interaction with polymer to achieve the synergistic effect between polymer and graphene.

Up to now, the main methods of preparing graphene/conductive polymer nanocomposites include in situ polymerization, electronic deposition, and hydrothermal synthesis. Zhang et al. prepared 3D PPy-coated porous graphene (PPy/PG) composites by in situ polymerization [88]. The PPy/PG displays a 3D hierarchical porous structure, which can provide a conductive trap and a stable skeleton to lithium polysulfides, thereby improving the cycling performance. Hu et al. combined electrostatic self-assembly approach and hydrothermal reduction method for synthesizing continuous 3D rGO/PANI composite [89]. 3D rGO/PANI possesses interconnected 3D macroporous networks, and the PANI nanofibers are tightly wrapped inside graphene, which can improve the conductivity and the structural stability of PANI effectively. Moreover, self-assembled 3D rGO/PANI also shows good flexibility, which can be fabricated to a flexible storage device.

1.7.2 Graphene/Inorganic Composites

The insertion of inorganic materials can reduce the interaction between graphene sheets, while the combination of graphene with specific inorganic phase can avoid single material deficiencies and a wide application prospect in the field of energy storage [90, 91]. Inorganic materials, mainly including metal and metallic compounds, Si-based materials, and P-based materials, are widely studied materials compound with graphene [90, 92–94].

Chemical reduction is a common method of preparing metal/graphene composites by chemically reducing some metal precursors on the surface of graphene. Lv et al. utilized nano-zeolites as template, $Pt(NH_3)_4Cl_2$ as Pt sources, and developed a novel in situ confined growth route for constructing Pt nanoparticles@graphene nanobox composites [95]. Due to the good dispersion of ultrafine Pt nanoparticles in 3D hollow box graphene skeleton, Pt nanoparticles@graphene nanobox exhibited excellent electrocatalytic activity for the ORR. Qin et al. synthesized SnSb-in-plane nanoconfined 3D N-doped porous graphene network (3D SnSb@N-PG) by a simple spray drying followed by annealing strategy. The 3D continuously interconnected porous structure, derived from the NaCl template, provides 3D conductive network, and the reduced SnSb nanoparticles were tightly trapped by graphene nanosheets; such microstructure ensures good mechanical stability and electronic conductivity [80].

Hydrothermal synthesis is a simple method that can generate high pressure under high temperature and fixed volume to prepare graphene/inorganic composites, and it is also a high-efficiency and low-cost method for industry application. Perera et al. prepared TiO_2/graphene composite materials by an alkaline hydrothermal method and found that the average diameter of TiO_2 nanotubes was about 9 nm, which were evenly dispersed on the graphene nanosheets [96]. By using one-pot hydrothermal method, Ye et al. prepared graphene/metal sulfides ($SnCoS_4$, CoS_2, and SnS_2) composites. The cobalt sulfides nanoparticles nucleation and growth on graphene surface make the nanoparticles dispersion and size more uniform. The structure of $SnCoS_4$/graphene composites, the high intrinsic conductivity of graphene, and large specific surface area give it superior electrochemical properties when used as lithium-ion battery anode materials [97].

Electrodeposition of inorganic nanomaterials on graphene matrix is a green and efficient method for preparing graphene/inorganic composites. Kuang et al. prepared Ni/graphene composites by electrochemical deposition method and found that when the content of graphene was 0.2%, the thermal conductivity of the composites was 15% higher than that of pure Ni [98]. Gao et al. synthesized graphene-wrapped mesoporous $MnCO_3$ composite through dynamic floating electrodeposition method and achieved the reduction of GO, and deposition and embedding of $MnCO_3$ in one-step electrodeposition [99]. Li et al. deposited active amorphous manganese oxide (a-MnO_x) on the surface of 3D graphene–CNT composites, resulting in good conductivity and porosity for the high-performance storage application [100].

Sol–gel method is also commonly used to prepare graphene/inorganic composite materials, which mainly carries out a series of hydrolysis and condensation polymerization reactions of the metal alkoxide or metal chloride precursor. Giampiccolo et al. reported a TiO_2/graphene composite prepared by simple sol–gel method. They employed a promising catalyst for NO_2 sensor [101]. Similarly, Li et al. used tetrabutyl titanate (TBOT) as the Ti precursor, ammonia solution as hydrolysis accelerator, and GO sheets as substrate; and synthesized sandwiched TiO_2/graphene composite by sol–gel and annealing route [102]. Besides these, the sol–gel method usually can be used to prepare other graphene/inorganic composites, such as $Li_4Ti_5O_{12}$/graphene [103], SiO_2/graphene [104], and ZnO/graphene [105].

1.8 Applications of Graphene

Since the discovery and successful preparation, graphene with excellent physical and chemical properties, such as high Young's modulus (~1100 GPa), high conductivity (200 S m^{-1}), high coefficient of thermal conductivity (~5000 W m^{-1} K^{-1}), high specific surface area (2630 m^2 g^{-1}), high carrier mobility (200 000 cm^2 V^{-1} s^{-1}), and high light transmittance, has been thought to trigger the next industrial revolution [26, 28, 33]. In all applied fields, especially in new-energy vehicles, aerospace, intelligent electronic equipment, and other high-tech fields, graphene is considered a key material to break through existing technical barriers.

Chen et al. reported good electrochemical lithium storage properties (at current density of 0.1 C, 1 C = 744 mA g^{-1}) of graphene with high degree of crystallization [106]. Zhao et al. systematically studied the capacitance characteristics of rGO with a specific capacitance of 260.5 F g^{-1} [107]. Graphene also has unique advantages in solar cell applications. Graphene has good light transmittance and electrical conductivity, and is considered as a potential substitute material for expensive and poisonous indium tin oxide (ITO). Lewis et al. used graphene as the electrode and obtained highly efficient organic solar cells (1.18%), very close to that of ITO (1.21%) [108]. In recent years, with the realization of graphene-controllable preparation and the deepening of application research, the efficiency of graphene-based solar cells has been improved continuously. 3D graphene materials have been proved as a promising counter electrode for dye-sensitized solar cells, which showed a desirable PCE of 7.63% and the possibility of replacing Pt [109]. Hwang et al. prepared graphene-reinforced copper matrix composites by means of molecular mixing. Compared with pure Cu, the elastic modulus and yield strength of the composites were increased by 30% and 80%, as the addition of GO is 2.5 vol% [110]. Li et al. prepared Al/3D graphene foam composite by variable speed-ball milling and adding pore forming. The graphene-reinforced aluminum matrix composites exhibited intelligent compression performance and energy absorption capacity [111].

Although many companies and enterprises claim to have prepared the new generation of graphene batteries, cars, supercapacitors, and so on, most of them are more of a gimmick than reality. The age of graphene is approaching, but not yet. There is still a lot of work to be done in the research of graphene, which needs to be further studied in depth. It still needs the continuous efforts and exploration of the majority of scientific research workers to promote it from the laboratory to the industry and market and benefit human society.

References

1 Li, Y., Xu, L., Liu, H. et al. (2014). Graphdiyne and graphyne: from theoretical predictions to practical construction. *Chemical Society Reviews* 43 (8): 2572–2586.

2 Raccichini, R., Varzi, A., Passerini, S. et al. (2015). The role of graphene for electrochemical energy storage. *Nature Materials* 14 (3): 271–279.

3 Shehzad, K., Xu, Y., Gao, C. et al. (2016). Three-dimensional macro-structures of two-dimensional nanomaterials. *Chemical Society Reviews* 45 (20): 5541–5588.

4 Rajib, P., Feng, D., Liming, D. et al. (2019). 3D heteroatom-doped carbon nanomaterials as multifunctional metal-free catalysts for integrated energy devices. *Advanced Materials* 31: 1805598.

5 Li, G., Li, Y., Liu, H. et al. (2010). Architecture of graphdiyne nanoscale films. *Chemical Communication* 46 (19): 3256–3258.

6 Kroto, H.W., Heath, J.R., O'Brien, S.C. et al. (1985). C60: Buckminsterfullerene. *Nature* 318 (6042): 162–163.

7 Krätschmer, W., Lamb, L.D., Fostiropoulos, K. et al. (1990). Solid C60: a new form of carbon. *Nature* 347 (6291): 354–358.

8 Iijima, S. (1991). Helical microtubules of graphitic carbon. *Nature* 354 (6348): 56–58.

9 Iijima, S. and Ichihashi, T. (1993). Single-shell carbon nanotubes of 1-nm diameter. *Nature* 363 (6430): 603–605.

10 Chen, X., Paul, R., and Dai, L. (2017). Carbon-based supercapacitors for efficient energy storage. *National Science Review* 4 (3): 453–489.

11 Novoselov, K.S., Geim, A.K., Morozov, S.V. et al. (2004). Electric field effect in atomically thin carbon films. *Science* 306 (5696): 666–669.

12 Wallace, P.R. (1947). The band theory of graphite. *Physical Review* 71 (9): 622–634.

13 Boehm, H.P., Setton, R., and Stumpp, E. (1994). Nomenclature and terminology of graphite intercalation compounds. *Pure and Applied Chemistry. Chimie pure et appliqueé* 66 (9): 1893.

14 Dreyer, D.R., Ruoff, R.S., and Bielawski, C.W. (2010). From conception to realization: an historial account of graphene and some perspectives for its future. *Angewandte Chemie* 49 (49): 9336–9344.

15 Geim, A.K. and Novoselov, K.S. (2007). The rise of graphene. *Nature Materials* 6 (3): 183–191.

16 Bianco, A., Cheng, H.-M., Enoki, T. et al. (2013). All in the graphene family – a recommended nomenclature for two-dimensional carbon materials. *Carbon* 65: 1–6.

17 Chen, Z., Ren, W., Gao, L. et al. (2011). Three-dimensional flexible and conductive interconnected graphene networks grown by chemical vapour deposition. *Nature Materials* 10 (6): 424–428.

18 Li, X., Cai, W., An, J. et al. (2009). Large-area synthesis of high-quality and uniform graphene films on copper foils. *Science* 324 (5932): 1312–1314.

19 Wang, Z., Gao, H., Zhang, Q. et al. (2019). Recent advances in 3D graphene architectures and their composites for energy storage applications. *Small* 15 (3): e1803858.

20 Lv, W., Li, Z., Deng, Y. et al. (2016). Graphene-based materials for electrochemical energy storage devices: Opportunities and challenges. *Energy Storage Materials* 2: 107–138.

21 Randviir, E.P., Brownson, D.A.C., and Banks, C.E. (2014). A decade of graphene research: production, applications and outlook. *Materials Today* 17 (9): 426–432.

22 Ji, L., Meduri, P., Agubra, V. et al. (2016). Graphene-based nanocomposites for energy storage. *Advanced Energy Materials* 6 (16): 1502159.

23 Novoselov, K.S., Fal'ko, V.I., Colombo, L. et al. (2012). A roadmap for graphene. *Nature* 490 (7419): 192–200.

24 Han, S., Wu, D., Li, S. et al. (2014). Porous graphene materials for advanced electrochemical energy storage and conversion devices. *Advanced Materials* 26 (6): 849–864.

25 Rao, C.N., Sood, A.K., Subrahmanyam, K.S. et al. (2009). Graphene: the new two-dimensional nanomaterial. *Angewandte Chemie* 48 (42): 7752–7777.

26 Soldano, C., Mahmood, A., and Dujardin, E. (2010). Production, properties and potential of graphene. *Carbon* 48 (8): 2127–2150.

27 Geim, A.K. and MacDonald, A.H. (2007). Graphene: Exploring carbon flatland. *Physics Today* 60 (8): 35–41.

28 Balandin, A.A., Ghosh, S., Bao, W. et al. (2008). Superior thermal conductivity of single-layer graphene. *Nano Letters* 8 (3): 902–907.

29 Lee, C., Wei, X., Kysar, J.W. et al. (2008). Measurement of the elastic properties and intrinsic strength of monolayer graphene. *Science* 321 (5887): 385–388.

30 Zhang, Y., Tan, Y.W., Stormer, H.L. et al. (2005). Experimental observation of the quantum Hall effect and Berry's phase in graphene. *Nature* 438 (7065): 201–204.

31 Novoselov, K.S., Geim, A.K., Morozov, S.V. et al. (2005). Two-dimensional gas of massless Dirac fermions in graphene. *Nature* 438 (7065): 197–200.

32 Morozov, S.V., Novoselov, K.S., Katsnelson, M.I. et al. (2008). Giant intrinsic carrier mobilities in graphene and its bilayer. *Physical Review Letters* 100 (1): 016602.

33 Bolotin, K.I., Sikes, K.J., Jiang, Z. et al. (2008). Ultrahigh electron mobility in suspended graphene. *Solid State Communications* 146 (9–10): 351–355.

34 Schedin, F., Geim, A.K., Morozov, S.V. et al. (2007). Detection of individual gas molecules adsorbed on graphene. *Nature Materials* 6 (9): 652–655.

35 Du, X., Skachko, I., Barker, A. et al. (2008). Approaching ballistic transport in suspended graphene. *Nature Nanotechnology* 3 (8): 491–495.

36 Novoselov, K.S., Jiang, Z., Zhang, Y. et al. (2007). Room-temperature quantum Hall effect in graphene. *Science* 315 (5817): 1379.

37 Yu, C., Shi, L., Yao, Z. et al. (2005). Thermal conductance and thermopower of an individual single-wall carbon nanotube. *Nano Letters* 5 (9): 1842–1846.

38 Berber, S., Kwon, Y.K., and Tomanek, D. (2000). Unusually high thermal conductivity of carbon nanotubes. *Physical Review Letters* 84 (20): 4613–4616.

39 Zhu, Y., Murali, S., Cai, W. et al. (2010). Graphene and graphene oxide: synthesis, properties, and applications. *Advanced Materials* 22 (35): 3906–3924.

40 Dikin, D.A., Stankovich, S., Zimney, E.J. et al. (2007). Preparation and characterization of graphene oxide paper. *Nature* 448 (7152): 457–460.

41 Gomez-Navarro, C., Burghard, M., and Kern, K. (2008). Elastic properties of chemically derived single graphene sheets. *Nano Letters* 8 (7): 2045–2049.

42 Chen, H., Müller, M.B., Gilmore, K.J. et al. (2008). Mechanically strong, electrically conductive, and biocompatible graphene paper. *Advanced Materials* 20 (18): 3557–3561.

43 Nair, R.R., Blake, P., Grigorenko, A.N. et al. (2008). Fine structure constant defines visual transparency of graphene. *Science* 320 (5881): 1308.

44 Mak, K.F., Sfeir, M.Y., Wu, Y. et al. (2008). Measurement of the optical conductivity of graphene. *Physical Review Letters* 101 (19): 196405.

45 Ugeda, M.M., Brihuega, I., Hiebel, F. et al. (2012). Electronic and structural characterization of divacancies in irradiated graphene. *Physical Review B* 85 (12): 121402.

46 Nourbakhsh, A., Cantoro, M., Vosch, T. et al. (2010). Bandgap opening in oxygen plasma-treated graphene. *Nanotechnology* 21 (43): 435203.

47 Zhang, C., Fu, L., Liu, N. et al. (2011). Synthesis of nitrogen-doped graphene using embedded carbon and nitrogen sources. *Advanced Materials* 23 (8): 1020–1024.

48 Panchakarla, L.S., Subrahmanyam, K.S., Saha, S.K. et al. (2009). Synthesis, structure, and properties of boron- and nitrogen-doped graphene. *Advanced Materials* 21: 4726–4730.

49 Wang, S., Wang, R., Wang, X. et al. (2012). Nanoscale charge distribution and energy band modification in defect-patterned graphene. *Nanoscale* 4 (8): 2651–2657.

50 Meyer, J.C., Kisielowski, C., Erni, R. et al. (2008). Direct imaging of lattice atoms and topological defects in graphene membranes. *Nano Letters* 8 (11): 3582–3586.

51 Terrones, M., Botello-Méndez, A.R., Campos-Delgado, J. et al. (2010). Graphene and graphite nanoribbons: morphology, properties, synthesis, defects and applications. *Nano Today* 5 (4): 351–372.

52 Ma, J., Alfe, D., Michaelides, A. et al. (2009). Stone-Wales defects in graphene and other planar sp(2)-bonded materials. *Physical Review B* 80 (3): 1132–1136.

53 Yazyev, O.V. and Louie, S.G. (2010). Electronic transport in polycrystalline graphene. *Nature Materials* 9 (10): 806–809.
54 Banhart, F., Kotakoski, J., and Krasheninnikov, A.V. (2011). Structural defects in graphene. *ACS Nano* 5 (1): 26–41.
55 Krasheninnikov, A.V., Lehtinen, P.O., Foster, A.S. et al. (2006). Bending the rules: contrasting vacancy energetics and migration in graphite and carbon nanotubes. *Chemical Physics Letters* 418 (1–3): 132–136.
56 Wintterlin, J. and Bocquet, M.L. (2009). Graphene on metal surfaces. *Surface Science* 603 (10–12): 1841–1852.
57 Lusk, M.T. and Carr, L.D. (2008). Nanoengineering defect structures on graphene. *Physical Review Letters* 100 (17): 175503.
58 Mkhoyan, K.A., Contryman, A.W., Silcox, J. et al. (2009). Atomic and electronic structure of graphene-oxide. *Nano Letters* 9 (3): 1058–1063.
59 Yanjie, G., Litao, S., and Florian, B. (2010). One- and two-dimensional diffusion of metal atoms in graphene. *Small* 4 (5): 587–591.
60 Vinayan, B.P., Nagar, R., Rajalakshmi, N. et al. (2012). Novel platinum-cobalt alloy nanoparticles dispersed on nitrogen-doped graphene as a cathode electrocatalyst for PEMFC applications. *Advanced Materials* 22 (16): 3519–3526.
61 Zhang, X., Yan, F., Zhang, S. et al. (2018). Hollow N-doped carbon polyhedron containing CoNi alloy nanoparticles embedded within few-layer N-doped graphene as high-performance electromagnetic wave absorbing material. *ACS Applied Materials & Interfaces* 10 (29): 24920–24929.
62 Muhich, C.L., Westcott, J.Y., Morris, T.C. et al. (2013). The effect of N and B doping on graphene and the adsorption and migration behavior of Pt atoms. *Journal of Physical Chemistry C* 117 (20): 10523–10535.
63 Bhaviripudi, S., Jia, X., Dresselhaus, M.S. et al. (2010). Role of kinetic factors in chemical vapor deposition synthesis of uniform large area graphene using copper catalyst. *Nano Letters* 10 (10): 4128–4133.
64 Niu, T., Zhou, M., Zhang, J. et al. (2013). Growth intermediates for CVD graphene on Cu(111): carbon clusters and defective graphene. *Journal of the American Chemical Society* 135 (22): 8409–8414.
65 Dutta, S. and Pati, S.K. (2010). Novel properties of graphene nanoribbons: a review. *Journal of Materials Chemistry* 20 (38): 8207.
66 Nakada, K., Fujita, M., Dresselhaus, G. et al. (1996). Edge state in graphene ribbons: Nanometer size effect and edge shape dependence. *Physical Review B* 54 (24): 17954.
67 Son, Y.W., Cohen, M.L., and Louie, S.G. (2006). Half-metallic graphene nanoribbons. *Nature* 444 (7117): 347–349.
68 Xiaolin, L., Xinran, W., Li, Z. et al. (2008). Chemically derived, ultrasmooth graphene nanoribbon semiconductors. *Science* 319 (5867): 1229–1232.
69 Liu, G., Wu, Y., Lin, Y.-M. et al. (2012). Epitaxial graphene nanoribbon array fabrication using BCP-assisted nanolithography. *ACS Nano* 6 (8): 6786–6792.
70 Sokolov, A.N., Yap, F.L., Liu, N. et al. (2013). Direct growth of aligned graphitic nanoribbons from a DNA template by chemical vapour deposition. *Nature Communications* 4: 2402.

71 Fan, X., Peng, Z., Yang, Y. et al. (2015). Atomic H-induced cutting and unzipping of single-walled carbon nanotube carpets with teepee structure and their enhanced supercapacitor performance. 3 (18): 10077–10084.

72 Jinming, C., Pascal, R., Rached, J. et al. (2010). Atomically precise bottom-up fabrication of graphene nanoribbons. *Nature* 466 (7305): 470–473.

73 Qin, K., Kang, J., Li, J. et al. (2016). Continuously hierarchical nanoporous graphene film for flexible solid-state supercapacitors with excellent performance. *Nano Energy* 24: 158–164.

74 Qin, K., Kang, J., Li, J. et al. (2016). Free-standing 3D nanoporous duct-Like and hierarchical nanoporous graphene films for micron-level flexible solid-state asymmetric supercapacitors. *Advanced Energy Materials* 6 (18), 1600755.

75 Zhang, J., Li, C., Peng, Z. et al. (2017). 3D free-standing nitrogen-doped reduced graphene oxide aerogel as anode material for sodium ion batteries with enhanced sodium storage. *Scientific Reports* 7 (1): 4886.

76 Zhu, S., Li, J., Deng, X. et al. (2017). Ultrathin-nanosheet-induced synthesis of 3D transition metal oxides networks for lithium ion battery anodes. *Advanced Functional Materials* 27 (9), 1605017.

77 Qin, J., Zhang, X., Zhao, N. et al. (2015). In situ preparation of interconnected networks constructed by using flexible graphene/Sn sandwich nanosheets for high-performance lithium-ion battery anodes. *Journal of Materials Chemistry A* 3 (46): 23170–23179.

78 Zhu, S., Taberna, P.-L., Zhao, N. et al. (2018). Salt-template synthesis of mesoporous carbon monolith for ionogel-based supercapacitors. *Electrochemistry Communications* 96: 6–10.

79 Zhu, S., Li, J., Li, Q. et al. (2016). Space-confined synthesis of three-dimensional boron/nitrogen-doped carbon nanotubes/carbon nanosheets line-in-wall hybrids and their electrochemical energy storage applications. *Electrochimica Acta* 212: 621–629.

80 Qin, J., Wang, T., Liu, D. et al. (2018). A top-down strategy toward SnSb in-Plane nanoconfined 3D N-doped porous graphene composite microspheres for high performance Na-ion battery anode. *Advanced Materials* 30 (9), 1704670.

81 Wang, Y., Zhang, Y., Liu, J. et al. (2020). Boosting areal energy density of 3D printed all-solid-state flexible microsupercapacitors via tailoring graphene composition. *Energy Storage Materials* 30: 412–419.

82 Jiang, Y., Xu, Z., Huang, T. et al. (2018). Direct 3D printing of ultralight graphene oxide aerogel microlattices. *Advanced Functional Materials* 28 (16), 1707024.

83 Sha, J., Li, Y., Villegas Salvatierra, R. et al. (2017). Three-dimensional printed graphene foams. *ACS Nano* 11 (7): 6860–6867.

84 Wang, H., Wu, L., Xu, L. et al. (2021). Synthetic multiscale graphene aerogel polymer composites with high-conductive performances for hyperthermia equipment. *Advanced Engineering Materials* 23 (7): 2001429.

85 Guo, Y., Ruan, K., Shi, X. et al. (2020). Factors affecting thermal conductivities of the polymers and polymer composites: a review. *Composites Science and Technology* 193: 108134.

86 Fan, X., Nie, W., Tsai, H. et al. (2019). PEDOT:PSS for flexible and stretchable electronics: modifications, strategies, and applications. *Advanced Science* 6 (19): 1900813.

87 Lv, P. (2018). Highly compressible graphene/polypyrrole aerogel for superelastic pseudocapacitors. *Fullerenes, Nanotubes, and Carbon Nanostructures* 26 (1): 23–29.

88 Zhang, Y., Bakenov, Z., Tan, T. et al. (2018). Three-dimensional hierarchical porous structure of PPy/porous-graphene to encapsulate polysulfides for lithium/sulfur batteries. *Nanomaterial* 8 (8): 606.

89 Hu, N., Zhang, L., Yang, C. et al. (2016). Three-dimensional skeleton networks of graphene wrapped polyaniline nanofibers: an excellent structure for high-performance flexible solid-state supercapacitors. *Scitific Reports* 6: 19777.

90 Geng, H., Peng, Y., Qu, L. et al. (2020). Structure design and composition engineering of carbon-based nanomaterials for lithium energy storage. *Advanced Energy Materials* 10 (10).

91 Yu, P., Cao, G., Yi, S. et al. (2018). Binder-free 2D titanium carbide (MXene)/carbon nanotube composites for high-performance lithium-ion capacitors. *Nanoscale* 10 (13): 5906–5913.

92 Meng, T., Li, B., Wang, Q. et al. (2020). Large-scale electric-field confined silicon with optimized charge-transfer kinetics and structural stability for high-rate lithium-ion batteries. *ACS Nano* 14 (6): 7066–7076.

93 Chen, B., Wang, T., Zhao, S. et al. (2021). Efficient reversible conversion between MoS_2 and Mo/Na_2S enabled by graphene-supported single atom catalysts. *Advanced Materials* 33 (12): 2007090.

94 Liu, Y., Zhang, A., Shen, C. et al. (2017). Red phosphorus nanodots on reduced graphene oxide as a flexible and ultra-fast anode for sodium-ion batteries. *ACS Nano* 11 (6): 5530–5537.

95 Lv, Y., Fang, Y., Wu, Z. et al. (2015). In-situ confined growth of monodisperse pt nanoparticle@graphene nanobox composites as electrocatalytic nanoreactors. *Small* 11 (8): 1003–1010.

96 Perera, S.D., Mariano, R.G., Vu, K. et al. (2012). Hydrothermal synthesis of graphene-TiO_2 nanotube composites with enhanced photocatalytic activity. *ACS Catalysis* 2 (6): 949–956.

97 Perera, S.D., Mariano, R.G., Vu, K. et al. (2016). Facile hydrothermal synthesis of $SnCoS_4$/graphene composites with excellent electrochemical performance for reversible lithium ion storage. *Journal of Materials Chemistry A* 4 (34): 13194–13202.

98 Kuang, D.A., Liye, X.U., Lei, L. et al. (2013). Graphene-nickel composites. *Applied Surface Science* 273 (6): 484–490.

99 Gao, M., Cui, X., Wang, R. et al. (2015). Graphene-wrapped mesoporous $MnCO_3$ single crystals synthesized by a dynamic floating electrodeposition method for high performance lithium-ion storage. *Journal of Materials Chemistry A* 3 (27): 14126–14133.

100 Li, S.-M., Wang, Y.S., Yang, S.Y. et al. (2013). Electrochemical deposition of nanostructured manganese oxide on hierarchically porous graphene-carbon nanotube structure for ultrahigh-performance electrochemical capacitors. *Journal of Power Sources* 225: 347–355.

101 Giampiccolo, A., Tobaldi, D.M., Leonardi, S.G. et al. (2019). Sol-gel graphene/ TiO_2 nanoparticles for the photocatalytic-assisted sensing and abatement of NO_2. *Applied Catalysis B: Environmental* 243: 183–194.

102 Li, W., Wang, F., Liu, Y. et al. (2015). General strategy to synthesize uniform mesoporous TiO_2/graphene/mesoporous TiO_2 sandwich-like nanosheets for highly reversible lithium storage. *Nano Letters* 15 (3): 2186–2193.

103 Zhang, F., Yi, F., Meng, T. et al. (2019). In situ supramolecular self-assembly assisted synthesis of $Li_4T_{i5}O_{12}$-carbon-reduced graphene oxide microspheres for lithium-ion batteries. *ACS Sustainable Chemistry & Engineering* 7 (1): 916–924.

104 Wu, H., Tang, Q., Fan, H. et al. (2017). Dual-confined and hierarchical-porous graphene/C/SiO_2 hollow microspheres through spray drying approach for lithium-sulfur batteries. *Electrochimica Acta* 255: 179–186.

105 Sun, Z., Guo, Y., Li, B. et al. (2019). ZnO/carbon nanotube/reduced graphene oxide composite film as an effective interlayer for lithium/sulfur batteries. *Solid State Sciences* 95: 105924.

106 Chen, S., Bao, P., Xiao, L. et al. (2013). Large-scale and low cost synthesis of graphene as high capacity anode materials for lithium-ion batteries. *Carbon* 64 (11): 158–169.

107 Zhao, B., Liu, P., Jiang, Y. et al. (2012). Supercapacitor performances of thermally reduced graphene oxide. *Journal of Power Sources* 198: 423–427.

108 Lewis, G.D.A., Yi, Z., Schlenker, C.W. et al. (2010). Continuous, highly flexible, and transparent graphene films by chemical vapor deposition for organic photovoltaics. *ACS Nano* 4 (5): 2865–2873.

109 Nechiyil, D., Vinayan, B.P., and Ramaprabhu, S. (2017). Tri-iodide reduction activity of ultra-small size PtFe nanoparticles supported nitrogen-doped graphene as counter electrode for dye-sensitized solar cell. *Journal of Colloid and Interface Science* 488: 309–316.

110 Hwang, J., Yoon, T., and Jin, S.H. (2013). Enhanced mechanical properties of graphene/copper nanocomposites using a molecular-level mixing process. *Advanced Materials* 25 (46): 6724–6729.

111 Li, W., Yang, X., He, C. et al. (2019). Compressive responses and strengthening mechanisms of aluminum composite foams reinforced by graphene nanosheets. *Carbon* 153: 396–406.

2

Graphene Synthesis: An Overview of Current Status

Simi Sui

Tianjin University, School of Materials Science and Engineering and Tianjin Key Laboratory of Composites and Functional Materials, 135 Yaguan Rd., Jinnan District, Tianjin 300350, P.R. China

Since the synthesis process dominates the morphologies and properties of graphene, the proper method is very essential for graphene when applied to particular applications. As it is well-known that a simple try of mirco-cleaving from highly oriented pyrolytic graphite (HOPG) led to the discovery of graphene, rebutting the presumption that planar graphene cannot exist in the free state. After that, numerous methods have been developed to meet the diverse needs, boosting the graphene industry. On a general observation, the synthesis method of graphene can be divided into two categories – top-down and bottom-up approaches. As the name implies, top-down approaches refer to obtaining graphene layers from cleavage or exfoliation, which is by weakening the van der Waals force between the existing graphite layers to obtain dispersed single/few-layer graphene. On the contrary, bottom-up approaches refer to synthesizing graphene through the rearrangement of carbon atoms, which focus more on the "growth" process. The classification of methods is listed in Figure 2.1 and the corresponding introduction is described in this chapter.

2.1 Top-Down Approaches

2.1.1 Mechanical Cleavage

Graphite is made of graphene sheets in which each carbon atom is in a layer connected by strong covalent bonds, while different layers are held by weak van der Waals forces. If the interaction between layers can be broken, it is possible to obtain graphene from purified graphite theoretically. As mentioned above, the discovery of graphene was made by mechanical cleavage. A.K. Geim and K.S. Novoselov used HOPG sheets as raw materials and pretreated them to stick to a photoresist. Then, peeled off layers from the graphite sheet with scotch tape. After that, they transferred the residual flakes to a Si substrate and the resulting thin carbon films were graphene. Through such a simple and kind of "violent" method, high-quality graphene was produced and lots of tests were able to be done, making graphene a

Templated Fabrication of Graphene-Based Materials for Energy Applications, First Edition.
Edited by Chunnian He, Naiqin Zhao, and Junwei Sha.

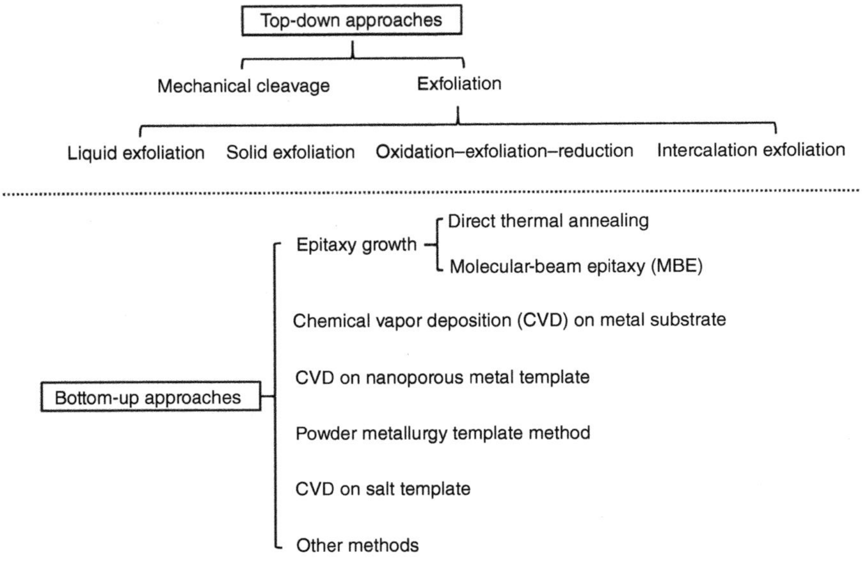

Figure 2.1 Classification of graphene preparation methods.

superstar famous for its unique electric field effect. Mechanical cleavage approaches (not limited to taping from HOPG) were found to be highly reliable and repeatable, thus researchers all over the world were able to conduct this study further [1] and promote the global investigation of graphene.

2.1.2 Exfoliation

Although mechanical cleavage is an easy way to produce graphene, the relatively low productivity and controllability hamper its widespread use. Therefore, methods by chemical means to separate out individual graphene sheets from graphite are used. Based on the reaction system, it can be divided into four categories, including liquid exfoliation, solid exfoliation, oxidation–reduction exfoliation, and intercalation exfoliation [2] (Figure 2.2).

2.1.2.1 Liquid Exfoliation

Based on exposing graphite to a solvent that favors an increase in the total area of graphite crystallites, liquid exfoliation is rather a convenient way to produce graphene flakes in the form of liquid suspensions, inks, or dispersions that cater for some important applications, such as printed electronics, conductive coatings, and composite fillers [3, 4]. The solvent used here could be aqueous, organic [3, 5, 6], or even ionic liquid [7]. Either with or without surfactant, the principle is to ensure the surface energy matches that of graphene thus the energy required to exfoliate graphene could be balanced by the solvent–graphene interaction. Since the raw material is graphite, it is necessary to provide the relatively stable liquid system with external energy to induce the exfoliation reaction. Ultrasonication and high-shear

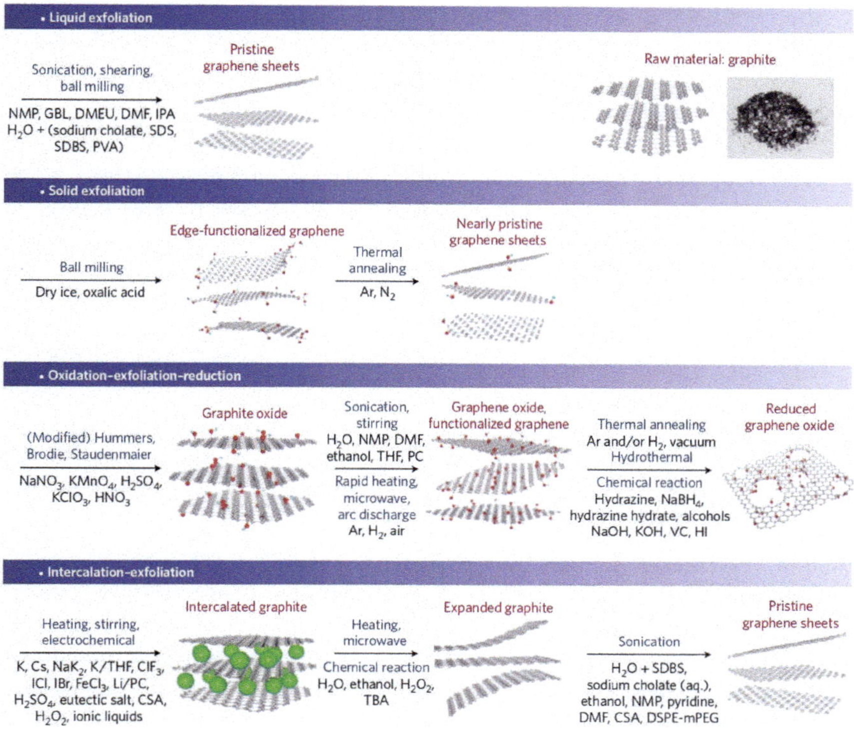

Figure 2.2 Four typical methods for graphene by exfoliation of bulk graphite. Source: Ren and Cheng [2]. Reproduced with permission of Springer Nature.

mixing are typical routes. A considerable number of few-layered graphene and part number of monolayer graphene are found in the products.

2.1.2.2 Solid Exfoliation

As mentioned above, the key point to isolating graphene sheets from graphite is to destruct the interactions among the layers. Edge functionalization is one of the workable ways. By ball milling of graphite with ammonia borane (NH_3BH_3), high-quality single/few-layered (no more than five layers) graphene sheets can be produced, resulting from the strong interaction of NH_3BH_3 and sp^2 carbon lattice network of graphite that breaks the van der Waals forces between graphite layers [8]. This kind of solid ball-milling process can be applied to the high-yield production of graphene on a large scale. Besides, dry ice as an eco-friendly chemical can also be used in ball-milling process to produce high-quality graphene at an unprecedented low cost [9]. The resultant edge-carboxylated graphite can be dispersed not only in protic solvents but also in polar aprotic solvents, facilitating its further applications.

2.1.2.3 Oxidation–Exfoliation–Reduction

Indirect ways can also be used. Researchers find it efficient to produce functionalized graphene, especially graphene oxides (GOs), by exfoliation of pretreated graphite, since the functional groups can weaken the van der Waals interaction between stacked layers thus facilitating their delamination. After reduction, graphene flakes can be obtained. This process is called oxidation–exfoliation–reduction and is regarded as the most efficient method used in industrial production.

The oxidation step has been studied for over the last 150 years [10] in graphite oxides research, among which Hummers method (or its modified methods) is one of the prominent methods that found favor from the past to now. By mixing graphite with concentrated sulfuric acid, sodium nitrate, and potassium permanganate, Hummers accomplished the oxidation of graphite to graphitic oxide. Recently, considering the reliability, cost, and environmental issues, other options such as electrochemical processes, have been developed and widely used [11], the as-obtained GO shows similar chemical composition, structure, and properties to those achieved by traditional Hummers method.

The following exfoliation step is rather simple because of the existence of oxygen-containing functional groups in graphite oxides, which make the GO layers hydrophilic and water molecules can readily intercalate into the interlayer galleries [12]. Therefore, either a mechanical or chemical treatment can exfoliate or expand GO flakes [13]. It has been proved that sonication and stirring in water or organic solvent (NMP, DMF, THF, etc.) are capable of separating GO layers. Chemical treatment, rapid heating, microwave, and arc discharge are used to provide external energy to trigger the pyrolysis of oxygen-containing functional groups. Consequently, GO flakes are expanded and delaminated caused by the evolution of decomposition products.

Efforts have been directed toward the reduction of GO and are successful. There are four main methods available, namely thermal reduction [14, 15], chemical reduction [16–19], photocatalytic reduction [20, 21], and electrochemical reduction [22–24]. In all these four methods, the same reaction is completed by the removal of the oxygen-containing functional groups. Thermal reduction is simple yet found to be either sufficient because of inadequate pressure buildup during the thermal heat treatment. Reduction by chemical agents such as hydrazine hydrate is effective, while photocatalytic and electrochemical ways are environmental-friendly. Besides, facile reduction methods have been studied in recent years, one to two seconds of microwave pulses are reported that can reduce GO into pristine graphene [25].

2.1.2.4 Intercalation Exfoliation

Defects and functional group residuals that exist in the reduced GO are beneficial to applications, such as catalysis, reinforcements in composite materials, and electrodes in electrochemical energy-storage devices, but unsuitable for the applications requiring superior electrical and thermal conductivity. Therefore, productive strategies of high-quality pristine graphene without the use of super-strong acid nor strong oxidation processes are required. The method of intercalation exfoliation emerges. Different from direct liquid exfoliation, intercalation exfoliation is defined as exfoliation from expanded graphite in which the interplanar spacing is broadened by intercalators. Various substances, such as molten metal or salt [26], sulfuric acid [27], and ammonia [28], are successively intercalated by electrochemical means or microwave radiation. Finally, using ultrasonication, single and few-layered graphene peels off from the expanded graphite and disperses in the solvent [27]. Without intense reaction, the high quality of graphene products obtained by intercalation exfoliation is maintained, such method is also an efficient choice for industrial production with less pollution and lower consumption than the formers.

2.2 Bottom-up Approaches

2.2.1 Epitaxy Growth

2.2.1.1 Direct Thermal Annealing

Silicon carbide (SiC) is a common material used for high-power electronics [4]. It has been demonstrated that the surfaces of SiC annealed at high temperature under vacuum tend to graphitize [29, 30]. This kind of graphitization consists of simultaneous sublimation of Si atoms and reorganization of the remaining C atoms. Since graphene on SiC can be used for electronic transport studies and related applications directly, it skips the transferring process during which the damage of graphene is almost inevitable. But it is limited by a lack of control over graphene growth [31]. Local scattering caused by charge buildup at the SiC substrate step edges can degrade electron transport properties [32] and step edges give rise to the growth of additional one or two monolayers that deteriorate the graphene thickness uniformity [33, 34]. At the same time, the high cost of single-crystal SiC substrate and severe experimental conditions lead to hesitation of investors.

2.2.1.2 Molecular-Beam Epitaxy (MBE)

Molecular-beam epitaxy (MBE) is a term used to denote the epitaxial growth of compound semiconductor films by a process involving the reaction of one or more thermal molecular beams with a crystalline surface under ultrahigh vacuum conditions [35]. It is now broadly used in the production of wafer-scale epitaxial layers [31, 36] and works well in producing graphene with high quality, controllability, and efficiency. To provide carbon molecular beam, C_{60}, as well as pyrolytic graphite filament, are usually used as carbon sources. Graphene growth occurs on the SiC substrate that is exposed to carbon flux in the MBE system. Different solid carbon sources result in different growth mechanisms. The main advantage of MBE is to reserve the initial SiC surface morphology and to allow precise control of the thickness, especially at the beginning of growth [37].

2.2.2 Chemical Vapor Deposition on Metal Substrate

Chemical vapor deposition (CVD) is a process in which the substrate is exposed to one or more volatile precursors, which react and/or decompose on the surface of substrate to produce the desired thin-film deposit [38]. The first attempt of growing graphene by CVD was reported in 2006 [39] using nickel as a substrate. Since then, the use of transition metals in the CVD production of graphene was extremely extended. With very good layer homogeneity and electrical properties comparable to those of mechanically exfoliated samples, CVD process is regarded as a clean and efficient way to obtain graphene on large scale and with high quality under precise control.

There are three main steps involved in CVD process, namely the activation of gaseous reactants, the subsequent chemical reaction, and the formation of a stable solid deposit over a suitable substrate [40]. Parameters settings are crucial for CVD, thus it can be classified into – e.g. atmospheric-pressure chemical vapor deposition (APCVD), low-pressure chemical vapor deposition (LPCVD), and ultrahigh-vacuum chemical vapor deposition (UHVCVD) based on different chamber pressure; hot-wall CVD, cold-wall CVD, and rapid thermal CVD based on substrate heating mode; plasma-enhanced chemical vapor deposition (PECVD) and remote plasma-enhanced chemical vapor deposition (RPECVD) based on plasma condition [41].

Substrate is another factor that determines the quality of graphene. Products obtained on ruthenium (Ru), iridium (Ir), nickel (Ni), and copper (Cu) surfaces follow different reaction mechanisms. For instance, it appears that graphene interacts strongly with step edges on Ru while on Ir, the interaction with steps is much smaller [42]. Cu and Ni foils are the most widely used substrates in the CVD growth of graphene. Based on the C-metal binary phase diagram, on which it can be read that the solubility of C in Cu is much lower than that in Ni, only a small amount of carbon can be dissolved in Cu. As illustrated in Figure 2.3, the gaseous C decomposes and then migrates on the Cu surface. This confined-to-surface process facilitated monolayer graphene growth, with negligible dissolution and precipitation of C from the substrate. In contrast, bulk dissolution along with surface migration was observed on Ni for it can dissolve many more carbon atoms than Cu. The graphene growth occurs mainly due to the precipitation during the cool-down process and in most cases, the product formed is multilayer graphene.

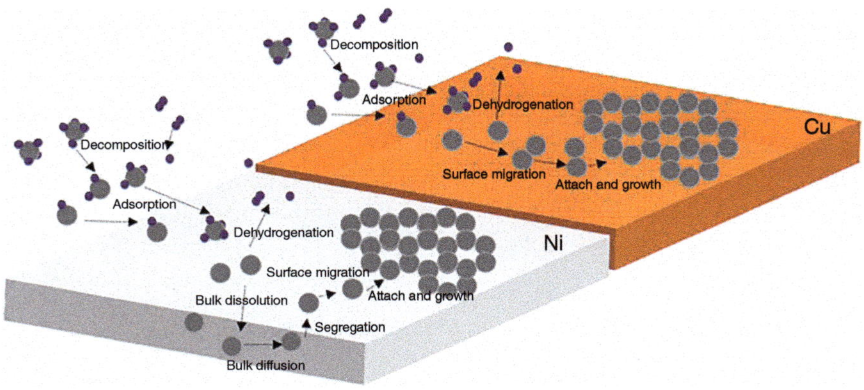

Figure 2.3 Growth kinetics in CVD-produced graphene on various catalysts: case of CH_4 on Ni and Cu. Source: Munoz and Gomez-Aleixandre [40]. Reproduced with permission of John Wiley and Sons.

2.2.3 CVD on Nanoporous Metal Template

Similar to the former section, CVD can also be performed on the metal template with nanoporous structures. There are two main ways to obtain nanoporous metal. One is by fabrication, which means preparing a nanoporous template first and then treating it with various pretreatments such as gold deposition, ion treatment in solutions, and inductively-coupled plasma [43]. The other is by dealloying, which means dissolving the soluble part of an infinitely intersoluble alloy. Binary alloys, such as Zn-Co [44], Cu-Pt [45], Cu-Ni [46], and Ni-Mn [47], have been used in graphene growth. Recently, ternary alloys have also attracted lots of attention [48]. Systems based on Ag-Au-Pt [49], Ni-Cu-Mn [50], Ni-Fe-Mn [51], and Mg-Cu-Al [52] are developed and show good application prospects in energy-storage devices.

Apart from the difference of nanostructure, graphene prepared on nanoporous metal template follows the same principles of graphene grown on metal foil. Maintaining the template morphology, as-prepared graphene shows well-developed porosity and high specific surface area. For example, Qin et al. prepared nanoporous Cu (NPC) by dealloying $Cu_{40}Mn_{60}$ alloy (shown in Figure 2.4) [53]. The graphene was grown on NPC by a short-time CVD method (at 900 °C for two minutes). As mentioned above, graphene with nanopores would be prepared on the Cu surface. After removing the Cu template, graphene with continuous duct-like structure (3D-DG) was obtained, showing high electrical conductivity.

2.2.4 Powder Metallurgy Template Method

Another way to obtain graphene with 3D structure is by combining traditional powder metallurgy and CVD, which is known as powder metallurgy template method. In this method, the catalysts used are metal powders, which have a

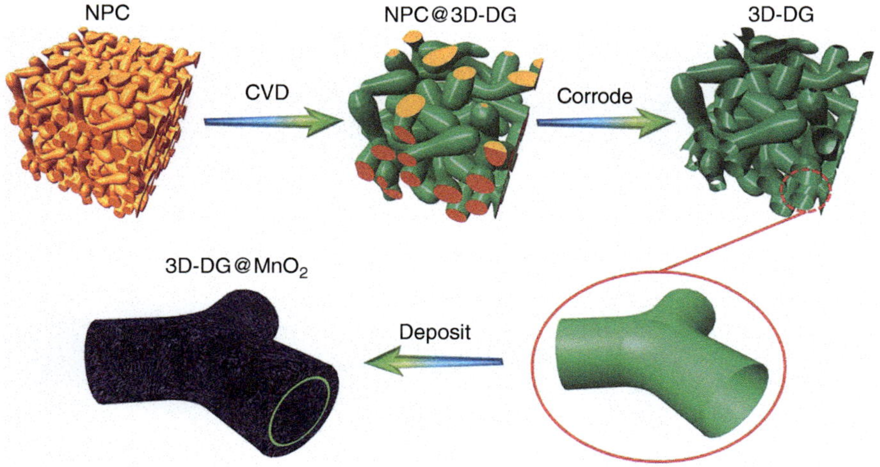

Figure 2.4 Schematic of the fabrication process of 3D-DG@MnO_2 film. Source: Qin et al. [53]. Reproduced with permission of John Wiley and Sons.

catalysis mechanism similar to CVD on metal substrate. At the same time, the metal powders are sintered by heat treatment and become a 3D skeleton for graphene growth. After removing the metal template, 3D structural graphene can be prepared. Sha et al. first demonstrate this powder metallurgy template method by using Ni powders and sucrose [54]. The procedure to obtain 3D graphene can be concluded as "mixing-cold pressing-heating-etching," which is simple and scalable. During the synthesis, sucrose wrapped in Ni powers was decomposed and carbonized into graphene. By removing the Ni template, a free-standing product (PMT-GF) was obtained, consisting of carbon shells and graphene layers. The procedure can be adjusted by changing the type of metal powders or designing the structure of the die to prepare various types of 3D carbon materials. Furthermore, the powder metallurgy template method is promising to combine with 3D-printing technology, catering to the need of manufacturing. Since the raw materials are both metal powders and sucrose, the key difference is the change from heating to laser induction. Sha et al. reported a metal powder-templated 3D-printing technology by using a commercial CO_2 laser to convert the Ni powders/sucrose mixture into 3D graphene foams [55]. Absorbed the laser energy, Ni powders can be sintered and catalyze the growth of graphene (Figure 2.5).

2.2.5 Soluble-Salt-Template Methods

The process can be further simplified as the discard of metal substrates. As shown in Figure 2.6, researchers use sodium chloride (NaCl) and stannic chloride ($SnCl_2$), both are soluble salts, as templates and catalysts to promote the fabrication of graphene by pyrolysis of glucose ($C_6H_8O_7$) [56]. All precursors are dissolved in water and then freeze-dried during which NaCl particles are self-assembled into a 3D structure coated with a complex film. The pyrolysis process is conducted

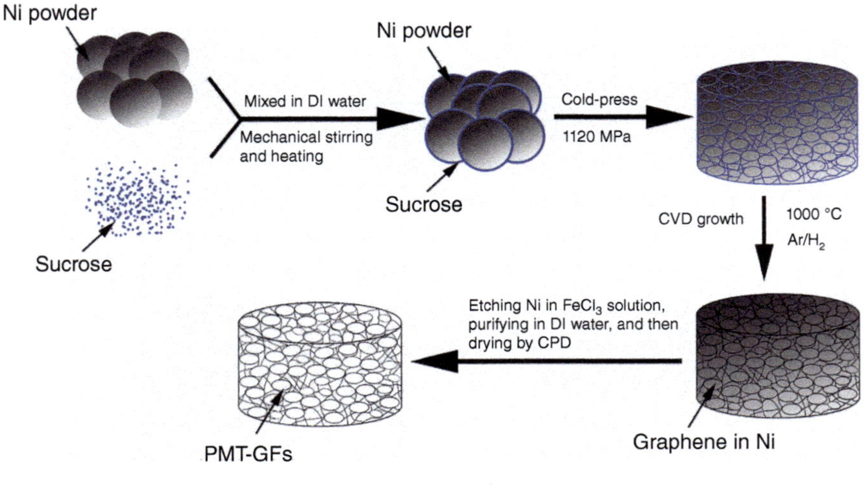

Figure 2.5 Schematics of powder metallurgy. Source: Sha et al. [54]. Reproduced with permission of American Chemical Society.

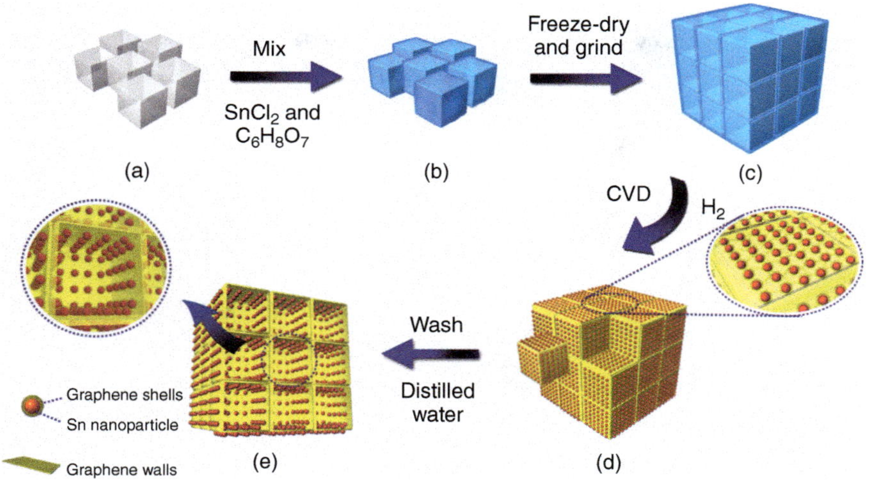

Figure 2.6 Schematic illustration of the in situ CVD process for the one-step synthesis of 3D Sn@G-PGNWs using 3D NaCl self-assembly as a template. (a) NaCl particles. (b) $SnCl_2-C_6H_8O_7$-coated NaCl particles. (c) $SnCl_2-C_6H_8O_7$-coated NaCl self-assembly. (d) Sn@G-GNW-coated NaCl self-assembly. (e) 3D Sn@G-PGNWs. Source: Qin et al. [56]. Reproduced with permission of American Chemical Society.

under hydrogen. The final products are obtained by removing the NaCl particles through washing with deionized water. Carbon precursors such as sucrose and biomass, inorganic salts such as Na_2CO_3 and $CaCO_3$ [57, 58], and even more ordinary templates such as sand, seawater can also be used in this way [59], thus the template-assisted sugar pyrolysis can be extended to scalable production [60].

2.2.6 Other Methods

2.2.6.1 CNTs Unzipping

After the discovery of graphene, carbon nanotube is typically described as a cylinder that is rolled by graphene sheets. It was until 2009 when Kosynkin [61] fabricated graphene nanoribbons by unzipping CNTs treated with potassium permanganate in a concentrated sulphuric acid solution that people realized CNTs can be starting materials of graphene production. With the help of polymers' burial layer, graphene nanoribbons can also be obtained from selective etching of highly crystalline MWCNTs under argon plasma [62]. Metal atoms such as lithium [63] and thermally activated nickel nanoparticles [64] successfully opened CNTs longitudinally by intercalation and exfoliation. Works above were finished in almost the same time, while further research is presuming on carefully controlling the widths and edge patterns of nanoribbons (Figure 2.7).

2.2.6.2 Molecular Self-Assembly

Ultrathin two-dimensional carbon materials are synthesized by cross-linking of self-assembled monolayers on a variety of solid surfaces [66, 67]. Followed

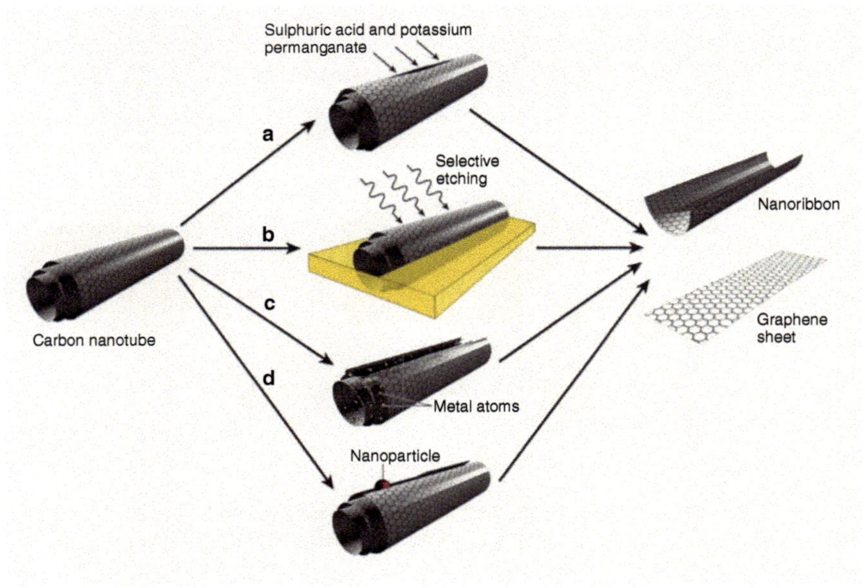

Figure 2.7 Methods for unzipping carbon nanotubes. (a) Treated with sulphuric acid and potassium permanganate. (b) Partially embedded in a polymer film and etched by argon plasma. (c) Inserting alkali-metal atoms between the concentric cylinders of CNTs. (d) Using catalytic metal nanoparticles to cut along the length of a CNT. Source: Terrones [65]. Reproduced with permission of Springer Nature.

by electron irradiation and pyrolysis, although the products obtained were not defined as graphene, the thickness of 1 nm and monolayer nature disclosed its "quasi-graphene" character besides the low crystallinity. This approach provides a powerful method to explore other 2D monolayer films as well as 3D topologies.

2.2.6.3 Laser Ablation

Neither severe process conditions nor multiple complex chemical reactions are required, laser ablation emerges as a facile way to fabricate porous graphene films. The starting material can be commercial polymer films [68], woods [69, 70], leaves [71] clothes, and foods [72]. By pulsed irradiation of CO_2 infrared laser, the sp^3-carbon atoms are photothermally converted to sp^2-carbon atoms, forming the framework of graphene. Researchers in James M. Tour's group affirmed that any carbon precursor that can be converted into amorphous carbon can be converted into graphene by laser ablation, which may be a generally applicable technique for forming graphene on diverse substrates, such as flexible or even biodegradable and edible electronics (Figure 2.8).

2.2.6.4 Pyrolysis of Solid Carbon Sources

Since the gaseous raw materials were making CVD difficult to apply to a wider variety of potential feedstocks, researchers attempted to grow graphene from solid carbon sources and finally succeed. Pyrolysis of carbon-containing materials, such

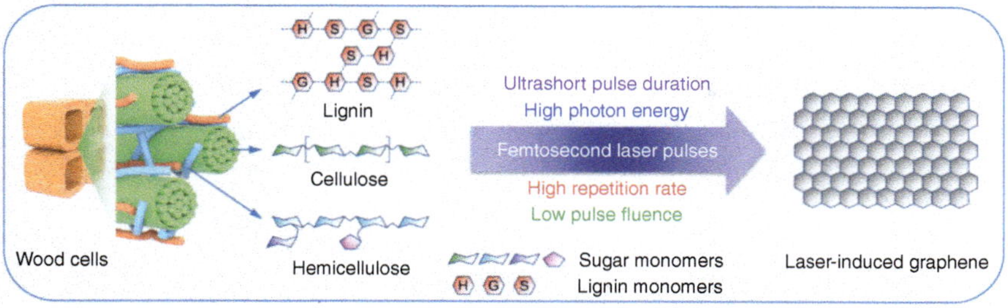

Figure 2.8 Schematic illustration of femtosecond laser direct writing for one-step laser-induced graphene synthesis on wood. All the chemical compositions (cellulose, hemicellulose, and lignin) in wood can be transformed into graphene by the irradiation of UV femtosecond pulses in ambient conditions. Source: Le et al. [71]. Reproduced with permission of John Wiley and Sons.

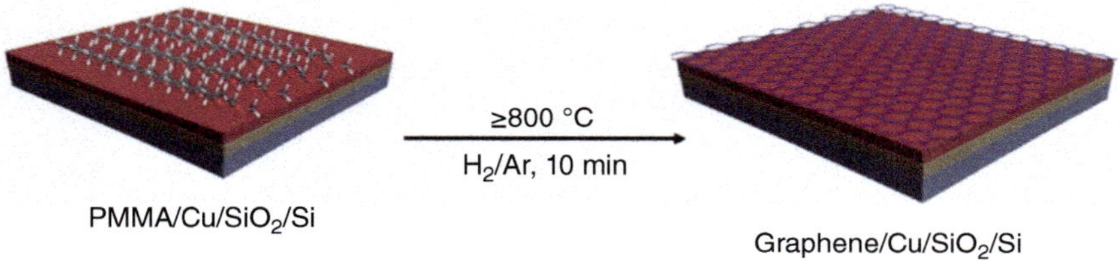

Figure 2.9 Monolayer graphene is derived from solid PMMA films on Cu substrates by heating in an H_2/Ar atmosphere at 800 °C or higher (up to 1000 °C). Source: Sun et al. [73]. Reproduced with permission of Springer Nature.

as commercial polymers [73], foods [74], industrial waste [75], and biomass [76], on copper or nickel substrate under either argon or ambient-air environment can produce quality-controlled graphene films. Similar to CVD process, the metal functioned as catalysts that induce graphene growth. This simple but effective method breaks new ground in graphene production that low-valued or negative-valued solid sources can be successfully transformed into high-valued graphene, which brings new solutions for recycling carbon from impure sources [74] (Figure 2.9).

References

1 Castro Neto, A.H., Guinea, F., Peres, N.M.R. et al. (2009). The electronic properties of graphene. *Reviews of Modern Physics* 81 (1): 109–162.

2 Ren, W. and Cheng, H.-M. (2014). The global growth of graphene. *Nature Nanotechnology* 9 (10): 726–730.

3 Paton, K.R., Varrla, E., Backes, C. et al. (2014). Scalable production of large quantities of defect-free few-layer graphene by shear exfoliation in liquids. *Nature Materials* 13 (6): 624–630.

4 Novoselov, K.S., Fal'ko, V.I., Colombo, L. et al. (2012). A roadmap for graphene. *Nature* 490 (7419): 192–200.

5 Lotya, M., Hernandez, Y., King, P.J. et al. (2009). Liquid phase production of graphene by exfoliation of graphite in surfactant/water solutions. *Journal of the American Chemical Society* 131 (10): 3611–3620.

6 Hernandez, Y., Nicolosi, V., Lotya, M. et al. (2008). High-yield production of graphene by liquid-phase exfoliation of graphite. *Nature Nanotechnology* 3 (9): 563–568.

7 Lu, J., Yang, J.-X., Wang, J. et al. (2009). One-pot synthesis of fluorescent carbon nanoribbons, nanoparticles, and graphene by the exfoliation of graphite in ionic liquids. *ACS Nano* 3 (8): 2367–2375.

8 Liu, L., Xiong, Z., Hu, D. et al. (2013). Production of high quality single- or few-layered graphene by solid exfoliation of graphite in the presence of ammonia borane. *Chemical Communications* 49 (72): 7890–7892.

9 Jeon, I.-Y., Shin, Y.-R., Sohn, G.-J. et al. (2012). Edge-carboxylated graphene nanosheets via ball milling. *Proceedings of the National Academy of Sciences of the United States of America* 109 (15): 5588–5593.

10 Feicht, P., Biskupek, J., Gorelik, T.E. et al. (2019). Brodie's or Hummers' method: oxidation conditions determine the structure of graphene oxide. *Chemistry – A European Journal* 25 (38): 8955–8959.

11 Pei, S., Wei, Q., Huang, K. et al. (2018). Green synthesis of graphene oxide by seconds timescale water electrolytic oxidation. *Nature Communications* 9 (1): 145.

12 Stankovich, S., Dikin, D.A., Piner, R.D. et al. (2007). Synthesis of graphene-based nanosheets via chemical reduction of exfoliated graphite oxide. *Carbon* 45 (7): 1558–1565.

13 Schniepp, H.C., Li, J.L., McAllister, M.J. et al. (2006). Functionalized single graphene sheets derived from splitting graphite oxide. *Journal of Physical Chemistry B* 110 (17): 8535–8539.

14 Zhu, Y., Stoller, M.D., Cai, W. et al. (2010). Exfoliation of graphite oxide in propylene carbonate and thermal reduction of the resulting graphene oxide platelets. *ACS Nano* 4 (2): 1227–1233.

15 Dubin, S., Gilje, S., Wang, K. et al. (2010). A one-step, solvothermal reduction method for producing reduced graphene oxide dispersions in organic solvents. *ACS Nano* 4 (7): 3845–3852.

16 Gao, X., Jang, J., and Nagase, S. (2010). Hydrazine and thermal reduction of graphene oxide: reaction mechanisms, product structures, and reaction design. *Journal of Physical Chemistry C* 114 (2): 832–842.

17 Dreyer, D.R., Murali, S., Zhu, Y. et al. (2011). Reduction of graphite oxide using alcohols. *Journal of Materials Chemistry* 21 (10): 3443–3447.

18 Zhou, T., Chen, F., Liu, K. et al. (2011). A simple and efficient method to prepare graphene by reduction of graphite oxide with sodium hydrosulfite. *Nanotechnology* 22 (4).

19 Fernández-Merino, M.J., Guardia, L., Paredes, J.I. et al. (2010). Vitamin C is an ideal substitute for hydrazine in the reduction of graphene oxide suspensions. *Journal of Physical Chemistry C* 114 (14): 6426–6432.

20 Cote, L.J., Cruz-Silva, R., and Huang, J. (2009). Flash reduction and patterning of graphite oxide and its polymer composite. *Journal of the American Chemical Society* 131 (31): 11027–11032.

21 Williams, G., Seger, B., and Kamt, P.V. (2008). TiO_2-graphene nanocomposites. UV-assisted photocatalytic reduction of graphene oxide. *ACS Nano* 2 (7): 1487–1491.

22 Shao, Y., Wang, J., Engelhard, M. et al. (2010). Facile and controllable electrochemical reduction of graphene oxide and its applications. *Journal of Materials Chemistry* 20 (4): 743–748.

23 An, S.J., Zhu, Y., Lee, S.H. et al. (2010). Thin film fabrication and simultaneous anodic reduction of deposited graphene oxide platelets by electrophoretic deposition. *Journal of Physical Chemistry Letters* 1 (8): 1259–1263.

24 Guo, H.L., Wang, X.F., Qian, Q.Y. et al. (2009). A green approach to the synthesis of graphene nanosheets. *ACS Nano* 3 (9): 2653–2659.

25 Voiry, D., Yang, J., Kupferberg, J. et al. (2016). High-quality graphene via microwave reduction of solution-exfoliated graphene oxide. *Science* 353 (6306): 1413–1416.

26 Wang, S., Wang, C., and Ji, X. (2017). Towards understanding the salt-intercalation exfoliation of graphite into graphene. *RSC Advances* 7 (82): 52252–52260.

27 Wu, L., Li, W., Li, P. et al. (2014). Powder, paper and foam of few-layer Graphene prepared in high yield by electrochemical intercalation exfoliation of expanded graphite. *Small* 10 (7): 1421–1429.

28 Jiang, F., Yu, Y., Wang, Y. et al. (2017). A novel synthesis route of graphene via microwave assisted intercalation-exfoliation of graphite. *Materials Letters* 200: 39–42.

29 Forbeaux, I., Themlin, J.M., and Debever, J.M. (1999). High-temperature graphitization of the 6H-SiC (0001) face. *Surface Science* 442 (1): 9–18.

30 Hass, J., De Heer, W.A., and Conrad, E.H. (2008). The growth and morphology of epitaxial multilayer graphene. *Journal of Physics: Condensed Matter* 20 (32).

31 Park, J., Mitchel, W.C., Grazulis, L. et al. (2010). Epitaxial graphene growth by carbon molecular beam epitaxy (CMBE). *Advanced Materials* 22 (37): 4140–4145.

32 Yakes, M.K., Gunlycke, D., Tedesco, J.L. et al. (2010). Conductance anisotropy in epitaxial graphene sheets generated by substrate interactions. *Nano Letters* 10 (5): 1559–1562.

33 Emtsev, K.V., Bostwick, A., Horn, K. et al. (2009). Towards wafer-size graphene layers by atmospheric pressure graphitization of silicon carbide. *Nature Materials* 8 (3): 203–207.

34 Virojanadara, C., Watcharinyanon, S., Zakharov, A.A. et al. (2010). Epitaxial graphene on 6H-SiC and Li intercalation. *Physical Review B - Condensed Matter and Materials Physics* 82 (20): 205402.

35 Cho, A.Y. and Arthur, J.R. (1975). Molecular beam epitaxy. *Progress in Solid State Chemistry* 10: 157–191.

36 Cho, A.Y. (1999). How molecular beam epitaxy (MBE) began and its projection into the future. *Journal of Crystal Growth* 201: 1–7.

37 Moreau, E., Godey, S., Ferrer, F.J. et al. (2010). Graphene growth by molecular beam epitaxy on the carbon-face of SiC. *Applied Physics Letters* 97 (24).

38 Choy, K.L. (2003). Chemical vapour deposition of coatings. *Progress in Materials Science* 48 (2): 57–170.

39 Somani, P.R., Somani, S.P., and Umeno, M. (2006). Planer nano-graphenes from camphor by CVD. *Chemical Physics Letters* 430 (1–3): 56–59.

40 Munoz, R. and Gomez-Aleixandre, C. (2013). Review of CVD synthesis of graphene. *Chemical Vapor Deposition* 19 (10–12): 297–322.

41 Deng, B., Liu, Z., and Peng, H. (2019). Toward mass production of CVD Graphene films. *Advanced Materials (Deerfield Beach, Fla.)* 31 (9): e1800996.

42 Batzill, M. (2012). The surface science of graphene: metal interfaces, CVD synthesis, nanoribbons, chemical modifications, and defects. *Surface Science Reports* 67 (3–4): 83–115.

43 Fang, J., Levchenko, I., Kumar, S. et al. (2014). Vertically-aligned graphene flakes on nanoporous templates: morphology, thickness, and defect level control by pre-treatment. *Science and Technology of Advanced Materials* 15 (5): 055009.

44 Lu, Z., Li, C., Han, J. et al. (2018). Three-dimensional bicontinuous nanoporous materials by vapor phase dealloying. *Nature Communications* 9 (1).

45 Chen, Q., Ding, Y., and Chen, M. (2018). Nanoporous metal by dealloying for electrochemical energy conversion and storage. *MRS Bulletin* 43 (1): 43–48.

46 Lührs, L. and Weissmüller, J. (2018). Nanoporous copper-nickel – macroscopic bodies of a strong and deformable nanoporous base metal by dealloying. *Scripta Materialia* 155: 119–123.

47 Li, Z., Bian, H., Lee, C.M. et al. (2018). Nickel nanotube array via electroplating and dealloying. *Thin Solid Films* 658: 1–6.

48 Zhu, S., Zhao, N., Li, J. et al. (2019). Hard-template synthesis of three-dimensional interconnected carbon networks: rational design, hybridization and energy-related applications. *Nano Today* 29: 100796.

49 Nguyen, N.T., Ozkan, S., Tomanec, O. et al. (2018). Nanoporous AuPt and AuPtAg alloy co-catalysts formed by dewetting-dealloying on an ordered TiO_2 nanotube surface lead to significantly enhanced photocatalytic H_2 generation. *Journal of Materials Chemistry A* 6 (28): 13599–13606.

50 Hakamada, M. and Mabuchi, M. (2009). Preparation of nanoporous Ni and Ni-Cu by dealloying of rolled Ni-Mn and Ni-Cu-Mn alloys. *Journal of Alloys and Compounds* 485 (1–2): 583–587.

51 Detsi, E., Cook, J.B., Lesel, B.K. et al. (2016). Mesoporous $Ni_{60}Fe_{30}Mn_{10}$-alloy based metal/metal oxide composite thick films as highly active and robust oxygen evolution catalysts. *Energy and Environmental Science* 9 (2): 540–549.

52 Aburada, T., Fitz-Gerald, J.M., and Scully, J.R. (2011). Synthesis of nanoporous copper by dealloying of Al-Cu-Mg amorphous alloys in acidic solution: the effect of nickel. *Corrosion Science* 53 (5): 1627–1632.

53 Qin, K., Liu, E., Li, J. et al. (2016). Free-standing 3d nanoporous duct-like and hierarchical nanoporous graphene films for micron-level flexible solid-state asymmetric supercapacitors. *Advanced Energy Materials* 6 (18).

54 Sha, J., Gao, C., Lee, S.K. et al. (2016). Preparation of three-dimensional graphene foams using powder metallurgy templates. *ACS Nano* 10 (1): 1411–1416.

55 Sha, J., Li, Y., Villegas Salvatierra, R. et al. (2017). Three-dimensional printed graphene foams. *ACS Nano* 11 (7): 6860–6867.

56 Qin, J., He, C., Zhao, N. et al. (2014). Graphene networks anchored with Sn@graphene as lithium ion battery anode. *ACS Nano* 8 (2): 1728–1738.

57 Zhu, S., Li, J., Ma, L. et al. (2016). Three-dimensional network of N-doped carbon ultrathin nanosheets with closely packed mesopores: controllable synthesis and application in electrochemical energy storage. *ACS Applied Materials and Interfaces* 8 (18): 11720–11728.

58 Qiu, D., Cao, T., Zhang, J. et al. (2019). Precise carbon structure control by salt template for high performance sodium-ion storage. *Journal of Energy Chemistry* 101–106.

59 Gupta, S.S., Sreeprasad, T.S., Maliyekkal, S.M. et al. (2012). Graphene from sugar and its application in water purification. *ACS Applied Materials and Interfaces* 4 (8): 4156–4163.

60 Zhu, S., Li, J., He, C. et al. (2015). Soluble salt self-assembly-assisted synthesis of three-dimensional hierarchical porous carbon networks for supercapacitors. *Journal of Materials Chemistry A* 3 (44): 22266–22273.

61 Kosynkin, D.V., Higginbotham, A.L., Sinitskii, A. et al. (2009). Longitudinal unzipping of carbon nanotubes to form graphene nanoribbons. *Nature* 458 (7240): 872–876.

62 Jiao, L., Zhang, L., Wang, X. et al. (2009). Narrow graphene nanoribbons from carbon nanotubes. *Nature* 458 (7240): 877–880.

63 Cano-Márquez, A.G., Rodríguez-Macías, F.J., Campos-Delgado, J. et al. (2009). Ex-MWNTs: graphene sheets and ribbons produced by lithium intercalation and exfoliation of carbon nanotubes. *Nano Letters* 9 (4): 1527–1533.

64 Campos, L.C., Manfrinato, V.R., Sanchez-Yamagishi, J.D. et al. (2009). Anisotropic etching and nanoribbon formation in single-layer graphene. *Nano Letters* 9 (7): 2600–2604.

65 Terrones, M. (2009). Nanotubes unzipped. *Nature* 458 (7240): 845–846.

66 Schultz, M.J., Zhang, X., Unarunotai, S. et al. (2008). Synthesis of linked carbon monolayers: films, balloons, tubes, and pleated sheets. *Proceedings of the National Academy of Sciences of the United States of America* 105 (21): 7353–7358.

67 Turchanin, A., Beyer, A., Nottbohm, C.T. et al. (2009). One nanometer thin carbon nanosheets with tunable conductivity and stiffness. *Advanced Materials* 21 (12): 1233–1237.

68 Lin, J., Peng, Z., Liu, Y. et al. (2014). Laser-induced porous graphene films from commercial polymers. *Nature Communications* 5.

69 Ye, R., Chyan, Y., Zhang, J. et al. (2017). Laser-induced graphene formation on wood. *Advanced Materials* 29 (37), 1702211.

70 Trusovas, R., Ratautas, K., Račiukaitis, G. et al. (2019). Graphene layer formation in pinewood by nanosecond and picosecond laser irradiation. *Applied Surface Science* 471: 154–161.

71 Le, T.S.D., Park, S., An, J. et al. (2019). Ultrafast laser pulses enable one-step graphene patterning on woods and leaves for green electronics. *Advanced Functional Materials*, 1902771.

72 Chyan, Y., Ye, R., Li, Y. et al. (2018). Laser-induced graphene by multiple lasing: toward electronics on cloth, paper, and food. *ACS Nano* 12 (3): 2176–2183.

73 Sun, Z., Yan, Z., Yao, J. et al. (2010). Growth of graphene from solid carbon sources. *Nature* 468 (7323): 549–552.

74 Ruan, G., Sun, Z., Peng, Z. et al. (2011). Growth of graphene from food, insects, and waste. *ACS Nano* 5 (9): 7601–7607.

75 Akhavan, O., Bijanzad, K., and Mirsepah, A. (2014). Synthesis of graphene from natural and industrial carbonaceous wastes. *RSC Advances* 4 (39): 20441–20448.

76 Seo, D.H., Pineda, S., Fang, J. et al. (2017). Single-step ambient-air synthesis of graphene from renewable precursors as electrochemical genosensor. *Nature Communications* 8: 14217.

3

Nanoporous Metal Template Methods

Kaiqiang Qin

George Mason University, Department of Chemistry and Biochemistry, Planetary Hall, 4400 University Drive, Fairfax, VA, 22030, USA

3.1 Introduction

It is very important to choose porous templates for the preparation of 3D graphene with suitable porous structure. Porous materials, which have large amount of interconnected porosity, contain porous metallic materials and porous nonmetallic materials. It has been developed rapidly since its discovery in the twentieth century. As we all know, graphene with high quality can be obtained by chemical vapor deposition (CVD) approach by using metals as the catalysts, among them, Cu and Ni are the most commonly used substrates. Hence, porous metals are the ideal templates for the fabrication of high-quality 3D porous graphene materials. The pore size of nanoporous metals obtained through dealloying approach is distributed in the range of nano to micrometer scales, which are the ideal candidates for the synthesis of high-quality 3D porous graphene. In this chapter, the fabrication of nanoporous metal foils and their application for the synthesis of high-quality 3D porous graphene are discussed in detail.

3.2 Dealloying Method for the Preparation of Nanoporous Metal Foil

There are many kinds of methods for the fabrication of nanoporous metals. Among them, dealloying method shows the best controllability and connectivity of pore structure. A dealloying method is an approach of dissolving active components and leaving stable components in bulk structures. Firstly, alloy ingots content of active and stable metal elements are prepared by melting approach. After composition homogenization annealing, the ingots are rolled to a thin foil through the combination of hot- and cold-rolling processes. Finally, the nanoporous metal foil is obtained after dealloying treatment.

The first paper about dealloying method was published in *Nature* in 2001 and is widely concerned by researchers [1]. In this paper, the $Au_{10\%}Ag_{90\%}$ (atom%) thin

Templated Fabrication of Graphene-Based Materials for Energy Applications, First Edition.
Edited by Chunnian He, Naiqin Zhao, and Junwei Sha.

foil was used as an example to explain the evolution of nanoporosity in dealloying. In the beginning after Au–Ag thin foil is immersed into the corrosive liquid, the Ag dissolves into the solution and the Au clusters are exposed, which have high surface energy. As more and more Ag dissolves, the Au clusters gradually aggregate to bigger sizes and finally propagate into the bulk to form nanoporous structure.

Until now, nanoporous copper [2, 3], nanoporous gold [4–7], nanoporous silver [8], nanoporous palladium [9, 10], nanoporous platinum [11, 12], and nanoporous nickel [13] were successfully obtained by dealloying method. Among them, nanoporous copper and nanoporous nickel can be utilized for the synthesis of 3D nanoporous graphene due to their excellent catalytic activity for the growth of graphene and other advantages, such as low cost, rich raw materials, simple preparation process, and wide range of application.

3.3 Nanoporous Ni as the Substrate for the Growth of 3D Nanoporous Graphene

3.3.1 3D Nanoporous Graphene

3D nanoporous graphene was firstly synthesized using nanoporous Ni as the substrate by the Cheng group in 2014 [14]. For preparing nanoporous Ni substrates, $Ni_{30}Mn_{70}$ ingots were obtained by melting pure Ni and Mn using an Ar-protected arc-melting furnace. After annealing at 900 °C for 24 hours for homogenization, the ingots were rolled to foil with a thickness of ~50 μm at room temperature. The nanoporous Ni was prepared by chemical dealloying in 1 M $(NH_4)_2SO_4$ aqueous solution at 50 °C and then rinsed thoroughly with water and ethanol and dried in vacuum. The obtained nanoporous Ni shows uniformly porous structure with an average pore size of 10 nm.

For the growth of 3D nanoporous graphene, the prepared nanoporous Ni foils were loaded in a quartz tube and inserted into the center of a quartz tube furnace and annealed at 900 °C under 2500 sccm Ar and 100 sccm H_2. After the reduction pretreatment, benzene was introduced with the gas flow of Ar (2500 sccm) and H_2 (100 sccm) at 900 °C for graphene growth. The furnace was immediately opened and the inner quartz tube was cooled with a fan to room temperature. The pore size of the nanoporous Ni@graphene was increased to 0.1–2 μm. After that, the nanoporous Ni substrates were dissolved by 3.0 M HCl, and the free-standing 3D nanoporous graphene films were repeatedly washed in water before drying for structural characterization and property measurements. The Raman spectroscopy shows the sharp 2D bands, and the high ratio of I_{2D}/I_G indicates the high quality of 3D nanoporous graphene. The structure of 3D nanoporous graphene was investigated by scanning transmission electron microscopy (STEM), which proves the complex 3D morphology with concave and convex curvatures and nanopores. The high-resolution transmission electron microscopy (HR-TEM) and bright field scanning transmission electron microscopy (BF-STEM) indicate that the majority of the 3D nanoporous graphene shows a perfect hexagonal structure. Meanwhile,

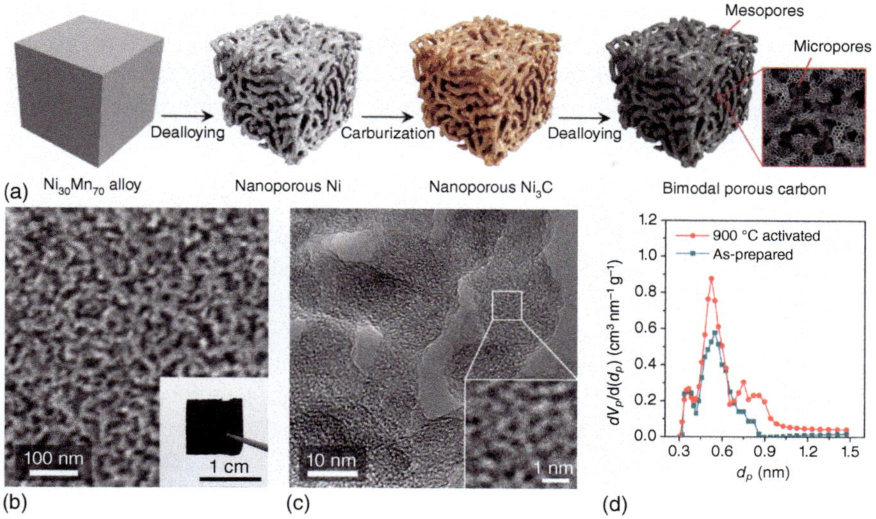

Figure 3.1 (a) Schematic illustration of the synthetic procedure of 3D bimodal porous carbon. (b) SEM image of the bimodal porous carbon. Inset shows an optical image of a free-standing bimodal porous carbon sheet with a size of ~1 cm × 1 cm. (c) High-magnification TEM image of bimodal porous carbon showing the bicontinuous micropores in the ligaments of primary mesopores. The inset is a zoom-in image of the white box area. (d) Micropore size distribution of the bimodal porous carbon from CO_2 gas sorption using the DFT method. Source: Han et al. [15]. Reproduced with permission of American Chemical Society.

pronounced topological defects can be obviously observed on the curved surface. This work paves a new way for the preparation of high-quality 3D nanoporous graphene with preserved intrinsic 2D electronic characteristics.

Although the 3D nanoporous graphene synthesized above delivers excellent physics and chemical performance, its pore size is still limited at mesoporous scale. The preparation of 3D porous carbon materials with microporous even atomic-scale control is of great significance but remains hugely challenging. Currently, Han et al. developed a sequential metastable-carbide-mediated dealloying method to obtain 3D bimodal porous carbon material with self-similar meso- and microporosities, in which micropores can be regulated at angstrom scale [15]. As shown in Figure 3.1a, nanoporous Ni foil with a thickness of 50 μm was obtained by dealloying process using $Ni_{30}Mn_{70}$ alloy foil as the precursor. Afterward, the prepared nanoporous Ni was added into triethylene glycol and NaOH aqueous solution to synthesize nanoporous Ni_3C. The bimodal porous carbon was obtained after dropping nanoporous Ni_3C into $FeCl_3$ solution for overnight and rinse/dry process. The low magnification TEM image shows a free-standing uniform 3D primary continuous porous film with an average pore size of 20–40 nm, which was obtained after removing Ni from Ni_3C precursor (Figure 3.1b). The HRTEM image clearly verifies that the ligaments of the primary porous structure consist of micropores with sub-nm feature size. The microporous structure is proved by CO_2 gas sorption measurement, which shows that the obtained porous carbon has micropores of ~0.55 nm and specific surface area of ~785 $m^2 g^{-1}$ (Figure 3.1c,d). The prepared 3D bimodal porous carbon delivers rapid ion diffusion and excellent electrochemical performance for Na-ion batteries, such as rate capacities and long cycling stability.

3.3.2 Heteroatoms-Doped 3D Nanoporous Graphene

3.3.2.1 N-Doped 3D Nanoporous Graphene

Heteroatoms-doped graphene has been widely investigated in the energy storage and conversion fields due to its unique physics and electrochemical properties. Based on the process to synthesize 3D nanoporous graphene mentioned above, Ito et al. developed a CVD method using nanoporous Ni as the substrate to prepare free-standing bi-continuous 3D nanoporous nitrogen-doped graphene [16]. In detail, nanoporous Ni foils with an average pore size of 10–20 nm were obtained according to the method mentioned above. Afterward, nanoporous Ni foils were inserted into the center of a quartz tube furnace and annealed at 900 °C under 2500 sccm Ar and 100 sccm H_2. After the reduction pretreatment, pyridine (0.2 mbar, 99.8%, anhydrous) was introduced with the gas flow of Ar (2500 sccm) and H_2 (100 sccm) at 900 °C for graphene growth. The furnace was immediately opened and the inner quartz tube was cooled with a fan to room temperature. The nanoporous Ni foils with graphene were dissolved by 3.0 M HCl and the samples were repeatedly washed in water and dried. The pore size of 3D N-doped graphene increases to 1–1.5 μm after removing the Ni substrate. The Brunauer–Emmett–Teller (BET) method indicates that the specific surface area of 3D N-doped graphene is about 1000 m^2g^{-1}. Raman spectra did not show obvious difference in the D, G, and 2D bands before and after removing Ni substrates, indicating the high quality of the obtained 3D N-doped nanoporous graphene films. However, the intensity of D band is decreased along with the increase of annealing temperature, suggesting that N atoms are released at higher temperature. Furthermore, the existence of D band is mainly contributed by N atoms doping according to the result of X-ray photoelectron spectroscopy (XPS) measurements.

The BF-TEM image and corresponding selected area diffraction (SAED) pattern of the prepared 3D N-doped nanoporous graphene show that the graphene sheet presents smoothly connected bicontinuous duct-like structure, which can facilitate the mass and electronic transport through the nanoporous structure and fully utilize the large specific surface area. The HRTEM and corresponding Fourier-transform pattern indicate the high crystallinity and the existence of defects. Meanwhile, the pyridinic N positions can be identified. In addition, the N atoms are homogenously dispersed in the graphene skeleton according to the chemical mappings of electron energy-loss spectroscopy (EELS). The XPS shows that the concentration of pyridinic N decreases from 800 °C (1.47 at.%) to 1000 °C (0.1 at.%), so does the graphitic N (2.23 at.% in 800 °C, 1.49 at.% in 1000 °C). Afterward, the prepared 3D N-doped nanoporous graphene films were used as the electrode for oxygen reduction reaction (ORR) testing. The result indicates that the higher the nitrogen concentration and smaller pore size, the higher ORR activity. Moreover, these free-standing 3D N-doped nanoporous graphene films can be directly used as the electrodes for fuel cells with conductive additives and binders.

To maximize the catalytic activity of N-doped nanoporous graphene of ORR, it is necessary to increase the concentration of N dopants and the surface area of nanoporous graphene. However, the pore size of 3D nanoporous graphene by common CVD approach is larger than 100 nm due to the aggregation of nanoporous metal at high temperature, which limits its specific surface area. To effectively decrease the pore size of nanoporous graphene and maintain the concentration of N dopants, Han et al. reported a 3D N-doped mesoporous graphene with pore size of 25 nm by a low-temperature carbide-mediated graphene-growth process [17]. Detailly, the prepared nanoporous Ni substrates according to the process mentioned above were added into the mixture solution of triethylene glycol and NaOH, then kept in an oil bath at 185 °C for 14 hours to synthesize nanoporous Ni_3C. The prepared nanoporous Ni_3C samples were annealed at 500 °C for one hour under Ar to form the nanoporous Ni@graphene composite films. After removing the Ni substrates, the prepared nanoporous graphene was fluorinated using XeF_2 as the precursor, then the N dopants were introduced under NH_3 at 600–900 °C for one hour. The prepared nanoporous graphene after removing Ni substrate at 500 °C show a smaller pore size of 20–30 nm with a high crystalline compared with that using common CVD approach (200–300 nm) due to the low growth temperature and provides a large specific surface area of 1051 $m^2\ g^{-1}$. After the introduction of N dopants at high temperature using NH_3 as the nitrogen source, the N element is uniformly dispersed in the nanoporous graphene backbone. XPS spectra indicated that the N-doped nanoporous graphene synthesized at 600 °C (N-MG-600) contains 5.83 at.% N element, including 0.44 at.% oxidized N, 1.50 at.% pyrrolic N, and 3.86 at.% pyridinic N. The pyridinic N plays the main contribution for ORR, which toward N-MG-600 delivers excellent ORR performance and can be used as an ideal electrode for Zn–air batteries.

To effectively suppress the aggregation of nanoporous Ni at high temperature and reduce the pore size of nanoporous graphene, Wang et al. coated a thin layer of hydrogenated graphite (HG) on the surface of nanoporous Ni by CVD approach at low temperature of 230 °C under C_2H_2, Ar, and H_2. Then, the samples were annealed at 700 °C under Ar and NH_3 to prepare N-doped nanoporous graphene. After removing the Ni substrate and further treatment in concentrated nitric acid, the N, O co-doped nanoporous graphene (NO-3DNG) was obtained (Figure 3.2a) [18]. The SEM and TEM images indicate that the obtained NO-3DNG shows pore size less than 20 nm, and the STEM mapping reveal that N and O elements are homogenously dispersed in graphene backbone (Figure 3.2b–d). This NO-3DNG shows a high specific surface area of 737.3 $m^2\ g^{-1}$ with pore volume of 1.793 $cc\ g^{-1}$ and consists of C, N, O, Ni, and Mn elements with 90.6 at.%, 1.74 at.%, 7.5 at.%, 0.11 at.%, and 0.05 at.%, respectively. The small pore size and heteroatoms doping toward the NO-3DNG exhibit excellent electrochemical performance as the electrode for supercapacitors.

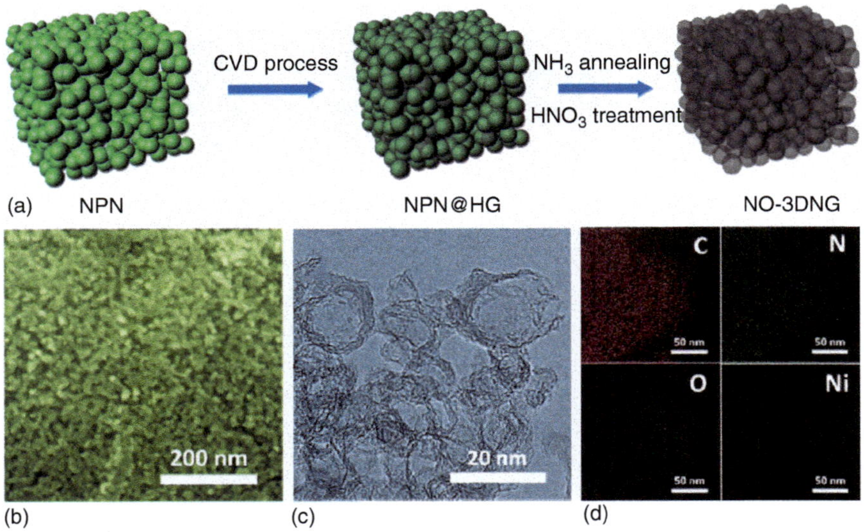

Figure 3.2 (a) Schematic diagram of the fabrication of NO-3DNG film. (b) SEM image of NO-3DNG, demonstrating the continuous sub-20-nm structure. (c) High-resolution TEM images of nanoporous structure of NO-3DNG. (d) STEM image and corresponding element mapping of C, N, O, and Ni. Source: Wang et al. [18]. Reproduced with permission of Elsevier B.V.

3.3.2.2 N, S Co-Doped 3D Nanoporous Graphene

Although the 3D N-doped nanoporous graphene is an efficient catalyst for the ORR, its hydrogen evolution reaction (HER) is rather poor. To further improve the HER performance of 3D heteroatoms-doped nanoporous graphene, Ito et al. reported nitrogen and sulfur co-doped 3D nanoporous graphene films using nanoporous Ni as the substrate, which provides a high catalytic activity in HER and low operating potential [19]. Similarly, 5 kg $Ni_{30}Mn_{70}$ ingots were prepared firstly. Then, the NiMn alloy thin foil with a thickness of ~50 μm was cold-rolled at room temperature. After dealloying in a 1 M $(NH_4)_2SO_4$ aqueous solution at 50 °C, the nanoporous Ni substrates were obtained. For the preparation of N, S co-doped nanoporous graphene, nanoporous Ni foils were inserted into the center of a quartz tube furnace and annealed at 800 °C under Ar and H_2 for three minutes, respectively. After that, pyridine and thiophene were introduced with the gas flow of Ar and H_2 at 800 °C for two minutes of graphene growth. The furnace was immediately opened and the samples were cooled with a fan to room temperature. The 3D N, S co-doped nanoporous graphene, was obtained after removing nanoporous Ni substrates in HCl solution. The obtained 3D nanoporous graphene shows high density of defective structure due to the heteroatoms doping. The BET indicates that the specific surface area of N, S co-doped nanoporous graphene is 762 m^2 g^{-1} and average pore size is 230 nm for NS-800. Raman spectra prove that an intense D band was obtained for NS-800 sample, which suggests that the heteroatoms doping introduces structure defects into the 3D nanoporous graphene lattices.

The TEM micrograph indicates that the NS-800 presents a duct-like graphene structure with nanoporosity. Moreover, sharp diffraction spots and crystal lattices were shown in the HRTEM and SAED, verifying the high crystallinity of N, S co-doped graphene. Meanwhile, the EELS mappings prove that nitrogen atoms are coupled with sulfur atoms and homogenously disperse in the 3D nanoporous graphene. The N 1s XPS spectra indicate that the nitrogen in nanoporous graphene consists of graphitic, pyridinic, and oxide nitrogen structures, and the S 2p XPS spectra demonstrate the binding states of sulfur are —C—S—C—, —C=S—, —C—SO—, and —C—SO_2—. The density functional theory (DFT) calculations verify that the coupling of nitrogen and sulfur atoms with geometric defects in 3D nanoporous graphene generates a synergistic effect to change the Gibbs free energy of H* absorption, toward N, S co-doped nanoporous graphene an excellent HER catalysis.

3.3.2.3 N, S, P Tri-Doped 3D Nanoporous Graphene

To further improve the HER performance of heteroatoms-doped 3D nanoporous graphene to a comparable level with Pt or other highly transition metal dichalcogenides, Ito et al. developed an N, S, P tri-doped nanoporous graphene structure [20]. In detail, the NiO nanoparticles with an average size of 7–8 nm were obtained onto the surface of Cu foil via supercritical hydrothermal synthesis and annealed at 350 °C for the preparation of nanoporous Ni substrate. Sequentially, the nanoporous Ni was inserted into a quartz tube furnace and annealed at 750 °C for two minutes to clean its surface under Ar and H_2. Then, heteroatoms-doped graphene was grown at 750 °C for 30 seconds under a mixed H_2, Ar, benzene, pyridine, thiophene, and triphenylphosphine. The furnace was immediately opened and the inner quartz tube was cooled with a fan to room temperature. The N, S, P tri-doped 3D nanoporous graphene films were obtained after removing the nanoporous Ni substrates. Smaller pore size and higher curvature accommodate the 3D nanoporous morphology, which is also suitable for the host of the chemical dopants. The N, S, P tri-doped 3D nanoporous graphene obtained at 750 °C (abbreviated as NSP750) shows smoothly interconnected nanoporosity with pore size of 25–50 nm, three to six atomic layers in thickness, and a specific surface area of 790 m^2 g^{-1}. The HRTEM image shows that the topological 5–7 defect can be obviously detected to accommodate the high curvature and curvature gradient. In addition, the chemical dopants and topological defects indicate the successful introduction of heteroatoms doping. The EELS mapping proves the uniform distribution of the N, S, and P dopants elements in the 3D nanoporous graphene. The XPS spectra indicate that the NSP750 has high contents of 1.93 at.% N, 3.59 at.% S, and 0.84 at.% P. The high density of topological defects, highly curved graphene, and synergistic effect of N, S, P tri-dopants effectively tune the electronic structures and dramatically enhance the H* adsorption Gibbs free energy of nanoporous graphene close to Pt catalysts.

3.3.2.4 N and Ni Single Atoms Co-Doped 3D Nanoporous Graphene

Bifunctional electrocatalysts with high electrochemical performance for both ORR and oxygen evolution reaction (OER) show high desirability for practical application of metal–air batteries. Qiu et al. prepared a nanoporous graphene material doped with N and Ni single atoms [21]. Similarly, 3D N-doped nanoporous graphene films with pore size of 200–500 nm were fabricated by CVD approach using nanoporous Ni as substrates at 850 °C under Ar, H_2, and pyridine. After partially removing the Ni substrates in 2 M HCl aqueous solution at room temperature, the N, Ni co-doped nanoporous graphene films were obtained. The dark-field HAADF (high-angle annular dark field)-STEM image and Ni 2p XPS spectrum indicate the doping of Ni single atoms and clusters in the nanoporous graphene. The existence of N dopants and defects effectively enhance the binding energy with Ni atoms, which facilitates the Ni atoms doping in the nanoporous graphene backbone. Moreover, the synergistic effect of N and Ni significantly enhances the catalytic activities of nanoporous graphene for both OER and ORR, which is an ideal electrode for Zn–air battery.

3.3.2.5 Li Metal Anode Application of 3D Nanoporous Graphene

In addition, the 3D nanoporous graphene prepared using nanoporous Ni as the substrate also can be used as the 3D current collectors for Li metal anode due to their bicontinuous structure and rich porosity for the accommodation of metallic Li. Zhang et al. and Huang et al. reported the preparation of pure 3D nanoporous graphene and N-doped 3D nanoporous graphene for dendrite-free Li metal anode using nanoporous Ni as the substrate, respectively [22, 23]. Both the nanoporous graphene and N-doped nanoporous graphene have a pore size range from 100 to 900 nm and show a bicontinuous structure. The high surface area, high conductivity, and large porosity toward them effectively lower the local current density, suppress the growth of Li dendrite, and accommodate the large volume fluctuation of metallic Li during plating and stripping process.

3.3.3 3D Nanoporous rGO

Although 3D nanoporous graphene delivers excellent electrochemical performance in energy storage and conversion fields, its application in photonics and optoelectronics fields is unsatisfactory due to its limited bandgap. Reduced graphene oxide (rGO) is an ideal candidate for these areas. Hence, Ito et al. developed a bicontinuous nanoporous rGO film using 3D nanoporous graphene as the precursor, which provides high optoelectronic performance [24]. Detailly, the 3D nanoporous graphene film with a thickness of 30–35 μm was prepared by CVD approach using nanoporous Ni as the substrate, which is similar to the process mentioned above. A 3D nanoporous graphene oxide was obtained after treating the 3D nanoporous graphene films via a modified Hummer method. Afterward, the 3D nanoporous graphene oxide films were transferred onto a Si substrate and heated at 150 °C to synthesize 3D nanoporous rGO films after reduction treatment. The 3D nanoporous rGO shows a high specific surface area of $1000\,m^2\,g^{-1}$ with a pore size of 200 nm and a ratio of oxygen of 19.6 at.% after reduction for 260 minutes. Consequently, the prepared 3D nanoporous rGO films deliver ultrahigh photoresponse of $3.10 \times 10^4\,A\,W^{-1}$ at room temperature.

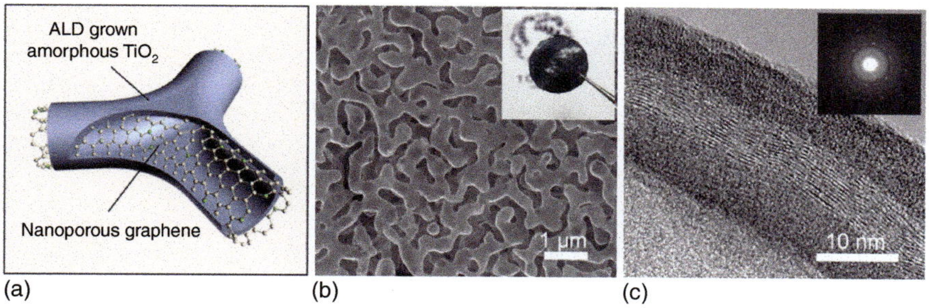

Figure 3.3 Microstructural and chemical characterization of amorphous and crystalline TiO_2/graphene composites. (a) Schematic illustration of the structure of a-TiO_2/G. (b) SEM image of a-TiO_2/G. Inset shows the photo of the free-standing a-TiO_2/G electrode with a diameter of 13 mm and a thickness of 30 μm. (c) Cross-sectional HRTEM image of the a-TiO_2/graphene/a-TiO_2 sandwich structure. Inset: SAED pattern shows the amorphous feature of TiO_2. The diffraction spots of hex-fold symmetry are from the graphene substrate. Source: Han et al. [25]. Reproduced with permission of Elsevier.

3.3.4 3D Nanoporous Graphene-Based Composite Materials

Besides modifying the structure of 3D nanoporous graphene, the 3D nanoporous graphene can be used as the substrate to composite with other active materials. Han et al. grew amorphous TiO_2 onto the surface of the wall of 3D nanoporous graphene (Figure 3.3a) [25]. 3D nanoporous graphene frameworks with an average pore size of 300 nm were prepared by CVD approach using nanoporous Ni as the substrate and catalyst, which has been mentioned above. Afterward, amorphous TiO_2 was coated on the surface of 3D nanoporous graphene using atomic layer deposition (ALD) system controlled by LabVIEW software. After ALD process, the TiO_2@3D nanoporous graphene composite films still show bicontinuous microstructure and the thickness of TiO_2 is ~7.39 nm with 60 ALD cycles (Figure 3.3b,c). The composite electrode can be directly used as the electrode for supercapacitors, which delivers excellent electrochemical performance even at high rates due to the synergistic effect of pseudocapacitive and intercalation/de-intercalation mechanisms.

3.4 Nanoporous Cu as the Substrate for the Growth of 3D Nanoporous Graphene

3.4.1 Continuously Hierarchical Nanoporous Graphene

Nanoporous Cu substrate shows a different mechanism for the growth of graphene compare with that of nanoporous Ni, which catalyzes the pyrolysis of carbon sources and the rearrangement of carbon atoms to form the graphene structure. Hence, the morphology of graphene can be controlled using nanoporous Cu as the substrate. Qin et al. prepared continuous hierarchical nanoporous graphene (hnp-G) films by a combination of low-temperature CVD and pyrolysis at high temperature [26]. Detailly, $Cu_{40}Mn_{60}$ ingots were prepared by arc melting and

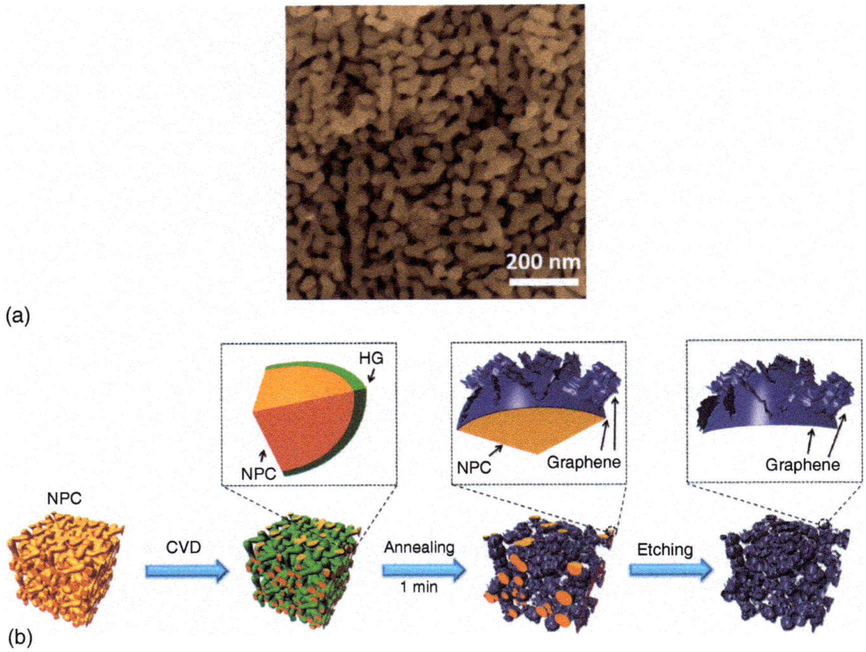

Figure 3.4 (a) Schematic diagram of the fabrication process of the hnp-G. (b) SEM image of nanoporous Cu. Source: Qin et al. [26]. Reproduced with permission of Elsevier.

homogenization. The CuMn alloy foils with a thickness of 100 μm were obtained after cold-rolling process. Afterward, the nanoporous Cu foils with a pore size of 40 nm and thickness of 100 μm were prepared via chemical dealloying in HCl aqueous solution at room temperature (Figure 3.4a). For the growth of hnp-G films, the as-prepared nanoporous Cu foils were inserted in the center of tube furnace and annealed at 200 °C under Ar, H_2, and C_2H_2 for 10 minutes to coat a thin layer of HG on the surface of nanoporous Cu. Then, the samples were pulled out of the hot zone, but the furnace was heated to 800 °C. Once the hot zone of the furnace arrived at 800 °C, the samples were pushed into the center of the furnace again to synthesize the graphene and then pulled out from the hot zone and cooled down to room temperature. Finally, the hnp-G films were obtained after removing the nanoporous Cu substrates (Figure 3.4b).

After the low-temperature CVD process, the nanoporous Cu still shows a similar pore size. The TEM image of nanoporous Cu@HG indicates the successful coating of HG on the surface of nanoporous Cu (Figure 3.5a). In addition, the HRTEM image of HG shows that the lattice space of HG is 0.4 nm, which is much larger than that of graphene due to the insertion of hydrogen elements (Figure 3.5b). The morphologies of as-prepared hnp-G after pyrolysis at high temperature and removing the nanoporous Cu substrate are shown in Figure 3.5c–f. Obviously, the hnp-G shows an interconnected structure of spherical graphene with pore size of 100–150 nm and a thickness of less than 10 atoms layers, moreover, abundant graphene nanosheets erect on the surface of porous graphene backbone. The BET reveals that the hnp-G

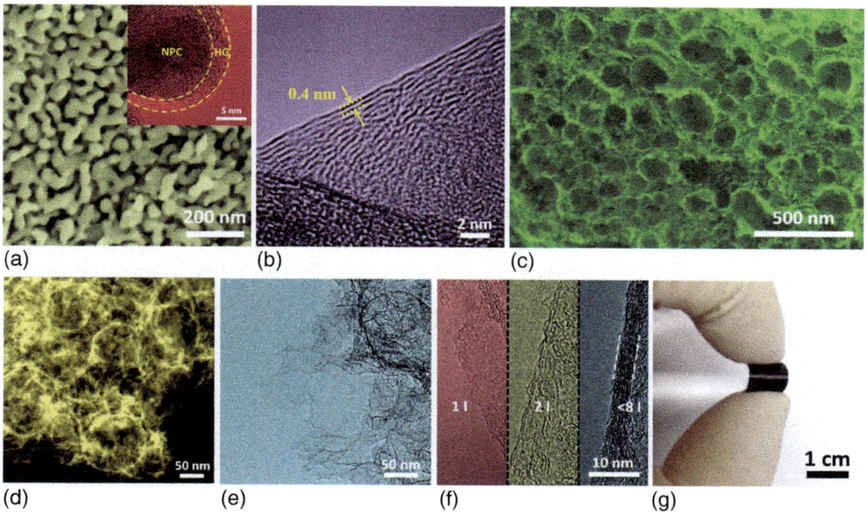

Figure 3.5 (a) SEM image of the nanoporous copper (NPC) coated with HG, the insert is TEM image. (b) HRTEM image of the HG after removing the NPC. (c), (d) SEM and HAADF-STEM images of the hnp-G after being treated by HNO_3. (e) High-magnification TEM image of ultra-thin graphite nanosheets located onto the surface of hnp-G. (f) High-magnification TEM images of graphene sheets with different number of layer in hnp-G. (g) Photograph of hnp-G film, demonstrating its integrity and flexibility. Source: Qin et al. [26]. Reproduced with permission of Elsevier.

films deliver a high specific surface area of 1160 $m^2\,g^{-1}$, which is mainly contributed by the mesoporous structure. Moreover, the obtained hnp-G films show high flexibility (Figure 3.5g). The advantages of the as-prepared hnp-G films, such as high flexibility, high specific surface area, small pore size, toward it directly used as the electrodes for solid-state supercapacitors and provides excellent electrochemical performance.

3.4.2 Heteroatoms-Doped 3D Nanoporous Graphene

Similar to the nanoporous Ni substrate, the heteroatoms-doped 3D nanoporous graphene also can be synthesized using nanoporous Cu as the catalyst and template. For example, Zhang et al. prepared N-doped nanoporous graphene using nanoporous Cu as the substrate by CVD approach [27]. The nanoporous Cu foils with the pore size of 30–50 nm and a thickness of 100 μm were prepared using $Cu_{30}Mn_{70}$ alloy foil as the precursor. After pre-annealing at 900 °C for three minutes, a mixed gas of C_2H_2, NH_3, Ar, and H_2 was introduced into the tube furnace to grow the N-doped graphene on the surface of nanoporous Cu by the CVD approach (Figure 3.6a). The nanoporous Cu@N-doped graphene shows a large pore size of 1–2 μm due to the aggregation of nanoporous Cu during the pre-annealing process (Figure 3.6b). The HRTEM image of N-doped nanoporous graphene indicates the high quality of the graphene with a thickness of less than 1 atoms layers (Figure 3.6c). The bicontinuous structure and high conductivity of

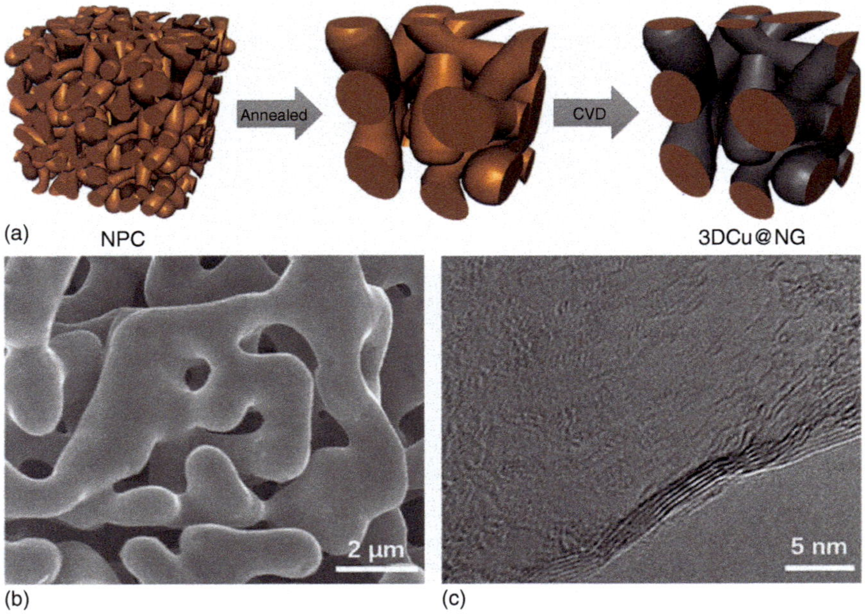

Figure 3.6 (a) Schematic illustrations of the fabrication of 3DCu@NG. (b) The SEM images of the cross-section of 3DCu@NG. (c) The TEM images of N-doped graphene obtained by dissolving Cu. Source: Zhang et al. [27]. Reproduced with permission of WILEY-VCH Verlag GmbH & Co. KGaA.

such framework can effectively lower the local current density and accommodate the volume change during Li plating and stripping process. Moreover, the N-doped graphene shows high lithiophilicity, which can facilitate the uniform plating of Li metal. The mentioned advantages toward the nanoporous Cu@N-doped graphene deliver excellent electrochemical performances used as the 3D current collector for Li metal anode.

Besides the N-doped nanoporous graphene, the heteroatoms co-doped nanoporous graphene shows unique electrochemical performance in energy storage fields. Qin et al. reported a flexible N, O co-doped bicontinuous hierarchical nanoporous graphene that consists of duct-like graphene and carbon nanocage by CVD approach using nanoporous Cu as the substrate [28]. Detailly, the as-prepared nanoporous Cu foils by the process mentioned above were added in the cold zone of tube furnace, and the furnace was heated to 900 °C under the mixed gas of C_2H_2, NH_3, Ar, and H_2. After the hot zone of the furnace get to 900 °C, the samples were pushed into the center of the furnace. After reacting for three minutes, the samples were pulled out and rapidly cooled to room temperature. The N, O co-doped 3D nanoporous graphene@carbon nanocages films were obtained after removing the Cu substrate and further treatment with concentrated nitric acid (Figure 3.7).

After the growth of N-doped graphene on the surface of nanoporous Cu, the size of the ligaments is enlarged to 300 nm, and abundant of small particles with a size of 20 nm are located on the surface of Cu ligaments (Figure 3.8a). The N-doped

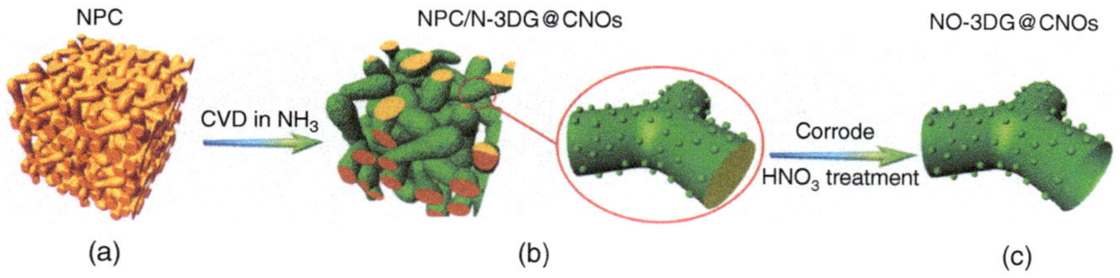

Figure 3.7 Schematic of the fabrication process of the NO-3DG@CNC films. (a) The as-prepared NPC substrate. (b) The NPC/N-3DG@CNCs structure obtained at 900 °C via CVD process. (c) NO-3DG@CNCs hybrids obtained after the removal of copper substrate and further modified with HNO_3. Source: Qin et al. [28]. Reproduced with permission of The Royal Society of Chemistry.

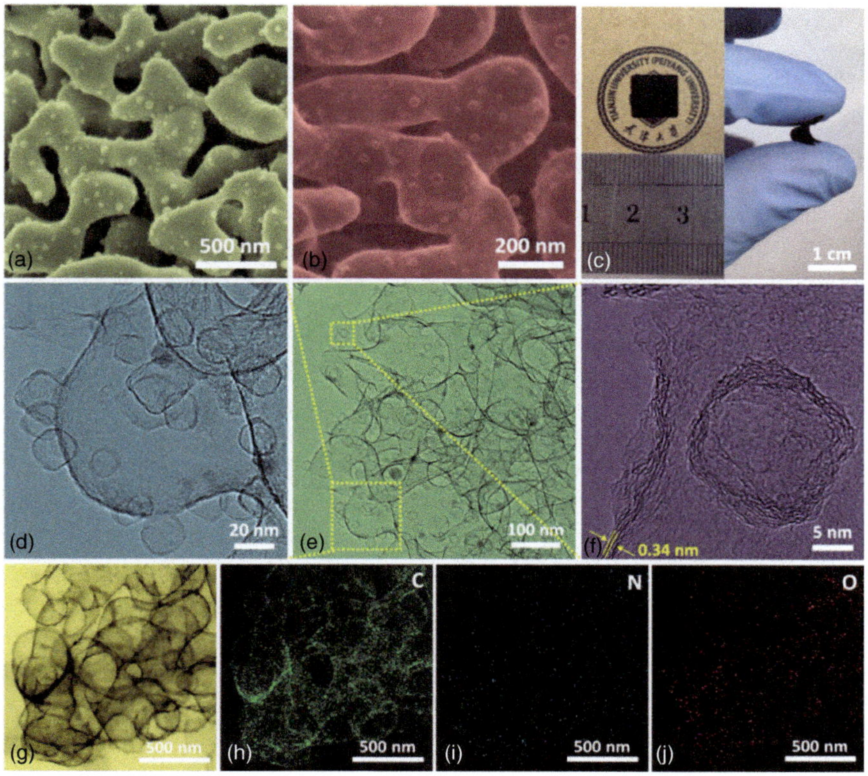

Figure 3.8 (a) SEM image of the cross-section of the NPC@N-3DG/CNCs grown at 900 °C for three minutes under an Ar, H_2, C_2H_2, and NH_3 atmosphere. (b) SEM image of the NO-3DG@CNC hybrid film after the removal of the NPC substrate and further HNO_3 modification. (c) Photograph of the free-standing flexible NO-3DG@CNC film. (e) TEM image of the NO-3DG@CNCs. (d, f) High-resolution TEM images of the NO- 3DG@CNCs. (g–j) STEM image and the corresponding elemental mapping images of C, N, and O. Source: Qin et al. [28]. Reproduced with permission of Royal Society of Chemistry.

nanoporous graphene maintains the interconnected structure after removing the Cu substrate and shows a bicontinuous structure. Moreover, many spheres are still grown on the surface of graphene (Figure 3.8b). The digital image of the obtained graphene indicates its high flexibility and free-standing features (Figure 3.8c). The TEM images with different magnifications verify the high quality of such N, O co-doped nanoporous graphene and the structure of the carbon nanocages grown onto the out wall of graphene. Both the nanoporous graphene and carbon nanocages show thickness of less than 10 atoms layers (Figure 3.8d–f). The STEM mapping was introduced to prove the successful doping of N and O elements, which is shown in Figure 3.8g–j. The bicontinuous N, O co-doped nanoporous graphene structure shows high conductivity and abundant defects. The existence of carbon nanocages effectively increase the specific surface area of the composite sample and introduce more defect. Such a superior structure toward the free-standing N, O co-doped nanoporous graphene@carbon nanocages composite films excellent electrochemical performance used as the electrodes of multi-style supercapacitors, such as aqueous supercapacitors, ionic supercapacitors, and lithium-ion capacitors.

3.4.3 3D Nanoporous Graphene-Based Composites

Although 3D nanoporous graphenes have unique architecture and deliver excellent conductivity, their specific capacitance and energy density for supercapacitors are still limited by their intrinsic attributes. Transition metal oxides and sulfides provide high specific capacitance and energy density according to the Faradic reaction. However, their electrochemical performance, such as rate capability, cycling stability, and power density, are not satisfied due to the poor conductivity. Using the 3D nanoporous graphene as the conductive substrate and composite with transition metal inorganics is an effective route to enhance their energy density but not affect the power density. To stabilize the electrochemical performance of MnO_2 as the electrode of supercapacitors, Qin et al. fabricated a free-standing 3D nanoporous graphene film and deposited MnO_2 onto the inner and outer wall of graphene using electrodeposition approach [29]. Firstly, the nanoporous Cu substrates with a thickness of 100 μm and pore size of 30–50 nm were prepared by chemical dealloying method using $Cu_{40}Mn_{60}$ alloy foils. After the CVD approach at 900 °C for two minutes under the mixed C_2H_2, Ar, and H_2, a thin layer of graphene was grown onto the surface of nanoporous Cu substrates. Afterward, the 3D nanoporous graphene films were obtained after removing the Cu substrates. Finally, the as-prepared 3D nanoporous graphene films were used as the bicontinuous hosts for the deposition of MnO_2 (Figure 3.9).

As shown in Figure 3.10a,b, the pore size of the as-prepared 3D nanoporous graphene films is 500–800 nm and shows bicontinuous structure. In addition, this high-quality 3D nanoporous graphene has a high specific surface area of 586.3 $m^2\ g^{-1}$ and extremely low sheet electrical resistance of 23.3 Ω per square. The optical image of 3D nanoporous graphene films proves their free-standing architecture and high flexibility (Figure 3.10c). After electrochemical deposition approach, flower-like MnO_2 nanosheets with a thickness of 4 nm are uniformly deposited

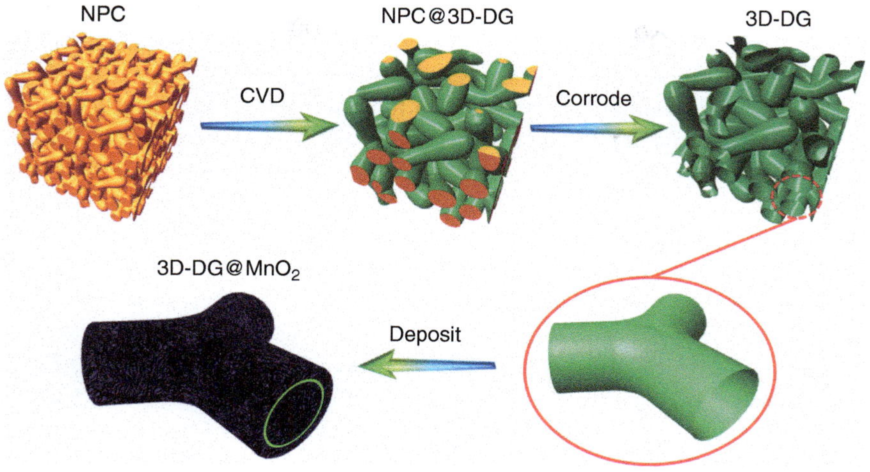

Figure 3.9 Schematic of the fabrication process of 3D nanoporous graphene@MnO_2 films. Source: Qin et al. [29]. Reproduced with permission WILEY-VCH Verlag GmbH & Co. KGaA.

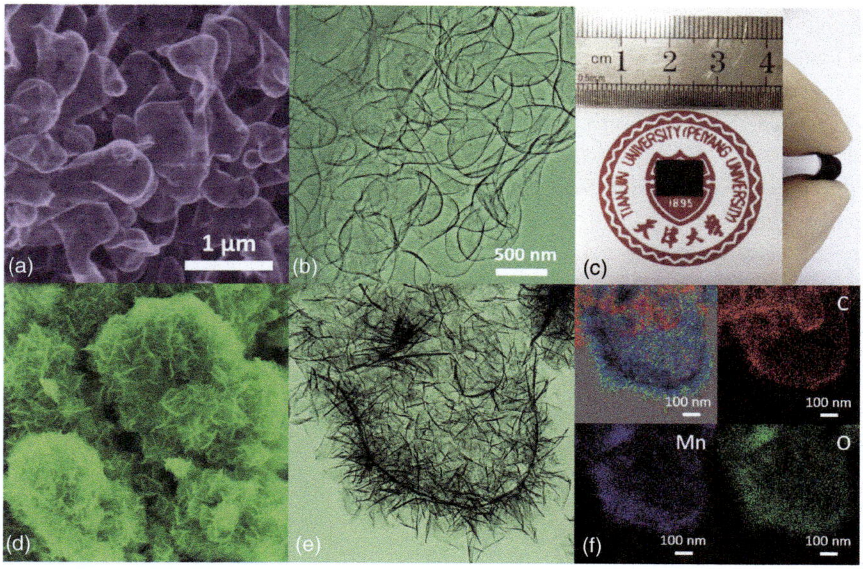

Figure 3.10 (a, b) SEM and TEM images of 3D nanoporous graphene after removing the Cu substrate. (c) Photographs of 3D nanoporous graphene film, demonstrate its integrity and flexibility. (d) SEM image of 3D-DG@MnO_2 electrode with the MnO_2 plating time of 60 minutes. (e) TEM images of the 3D nanoporous graphene@MnO_2. (f) STEM image and corresponding elemental mapping images of C, Mn, and O. Source: Qin et al. [29]. Reproduced with permission WILEY-VCH Verlag GmbH & Co. KGaA.

onto the inner and outer wall of the 3D nanoporous graphene (Figure 3.10d–f). The high conductivity and bicontinuous structure of 3D nanoporous graphene toward fast ion and electron transfer, and the flower-like MnO_2 nanosheets show large specific surfaces that can contact with the electrolyte. Consequently, the 3D nanoporous@MnO_2 composite electrodes based solid-state supercapacitors deliver excellent electrochemical performance, such as high volumetric energy density and power density.

To achieve high electrochemical performance in both energy storage and convention fields. Qin et al. prepared a heteroatom-doped edge-enriched 3D rivet graphene (H-3DRG) film and used it as the conductive substrate for the deposition of NiCo-LDH/NiCo-S [30]. Firstly, nanoporous Cu foils were added in the cold zone of the tube furnace and the furnace was directly heated to 900 °C under a mixed gas of Ar and H_2. Once the furnace gets to 900 °C, the samples were pushed into the hot zone for one minute. Partial of the nanoporous Cu ligaments were aggregated to the larger size but there still have small size of Cu nanoparticles on the surface of larger Cu ligaments after annealing for one minute. Afterward, C_2H_2 and NH_3 were introduced into the furnace for three minutes. The larger Cu ligament and Cu nanoparticles catalyze the formation of carbon materials at the same time. With the enlarged time of annealing, the small Cu nanoparticles were gradually aggregated together with the larger Cu ligaments and finally catalyze the formation of carbon nanoparticles with a gradient structure located on the surface of 3D nanoporous graphene (Figure 3.11a and 3.12a,b). For the preparation of H-3DRG@NiCo-LDH, the aqueous solution contents of $Ni(NO_3)_2$, $Co(NO_3)_2$, and urea were transferred into a Teflon-lined stainless steel autoclave, and the as-prepared H-3DRG films were immersed into the solution and reacted at 90 °C (Figure 3.11b). Abundant of needle-like NiCo-LDH with a diameter of 10–20 nm and length of 1 μm was deposited on the graphene (Figure 3.12d,e). Afterward, the H-3DRG@NiCo-LDH samples were immersed into a Na2S aqueous solution and reacted for 20 hours at 90 °C, the H-3DRG@$NiCo_2S_4$ was obtained by anion exchange process (Figure 3.11c). Obviously, the needle-like structure of NiCo-LDH changed to a fluffy nanorod but still preserve the general array morphologies

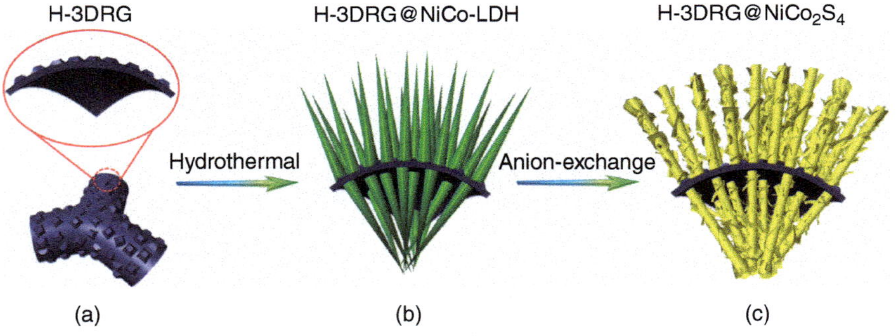

Figure 3.11 Schematic illustration of the fabrication of H-3DRG@NiCo-LDH and H-3DRG@$NiCo_2S_4$. Source: Qin et al. [30]. Reproduced with permission of The Royal Society of Chemistry.

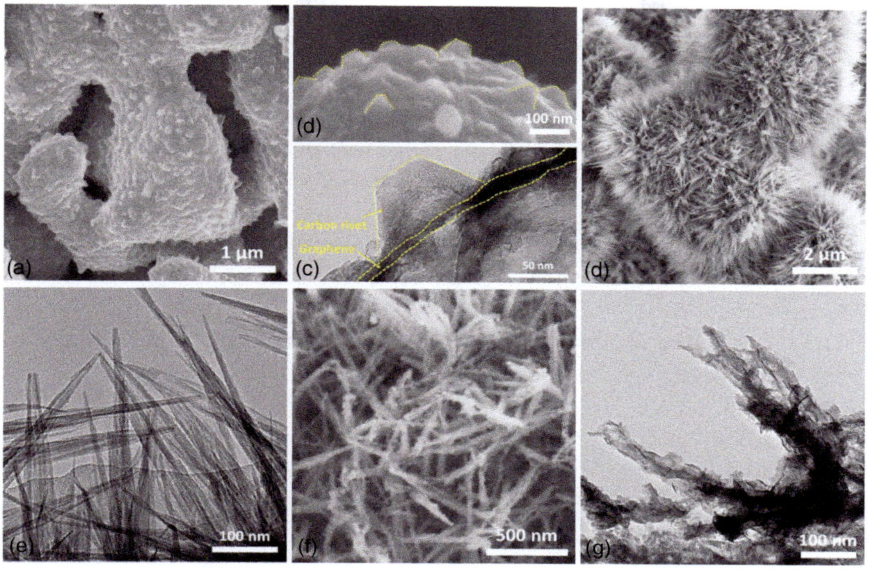

Figure 3.12 (a, b) SEM images of the cross section of H-3DRG film with different magnifications. (c) TEM image of H-3DRG. (d) SEM image of the cross section of H-3DRG@NiCo-LDH hybrid film. (e) TEM image of NiCoLDH nanoneedles. (f) High-resolution SEM image of $NiCo_2S_4$ hierarchical nanorods grown onto the surface of H-3DRG films. (g) TEM image of $NiCo_2S_4$ hierarchical nanorods. Source: Qin et al. [30]. Reproduced with permission of Royal Society of Chemistry.

(Figure 3.12f,g). Such a unique structure accelerates the reaction kinetics and toward the composite as ideal electrodes for supercapacitors and electrocatalysis.

References

1 Erlebacher, J., Aziz, M., Karama, A. et al. (2001). Evolution of nanoporosity in dealloying. *Nature* 410: 450–453.

2 Hayes, H., Hodge, A., Biener, J. et al. (2006). Monolothic nanoporous copper by dealloying Mn-Cu. *Journal of Materials Research* 21 (10): 2611–2616.

3 Lu, H., Li, Y., and Wang, F. (2007). Synthesis of porous copper from nanocrystalline two-phase Cu-Zr film by dealloying. *Script Materials* 56 (2): 165–168.

4 Yoo, S. and Park, S. (2007). Platinum-coated, nanoporous gold nanorod arrays: synthesis and characterization. *Advanced Materials* 19 (12): 1612–1615.

5 Lu, X., Bischoff, E., Spolenak, R. et al. (2007). Investigation of dealloying in Au-Ag thin films by quantitative electron probe microanalysis. *Script Materials* 56 (7): 557–560.

6 Hodge, A., Hayes, J., Caro, J. et al. (2006). Characterization and mechanical behavior of nanoporous gold. *Advanced Engineering Materials* 8 (9): 853–857.

7 Hodge, A., Biener, J., Hayes, J. et al. (2007). Scaling equation for yield strength of nanoporous open-cell foams. *Acta Materials* 55 (4): 1343–1349.

8 Zhang, Z., Wang, Y., Qi, Z. et al. (2009). Generalized fabrication of nanoporous metals (Au, Pd, Pt, Ag, and Cu) through chemical dealloying. *The Journal of Physical Chemistry C* 113 (29): 12629–12636.

9 Meyerheim, H., Soyka, E., and Kirschner, H. (2006). Alloying and dealloying in pulsed laser deposited Pd films on Cu (100). *Physics Review B* 74: 085405.

10 Schofield, E. (2005). Anodic routes to nanoporous materials. *Transactions of the Institute of Metal Finishing* 83: 35–42.

11 Pugh, D., Dursun, A., and Corcoran, S. (2003). Formation of nanoporous platinum by selective dissolution of Cu from $Cu_{0.75}Pt_{0.25}$. *Journal of Materials Research* 18 (1): 216–221.

12 Thorp, J., Sieradzki, K., and Tang, L. (2006). Formation of nanoporous noble metal thin by electrochemical dealloying of Pt_xSi_{1-x}. *Applied Physics Letters* 88: 033110.

13 Qiu, H., Kang, J., Liu, P. et al. (2014). Fabrication of large-scale nanoporous nickel with tunable pore size for energy storage. *Journal of Power Sources* 247: 896–905.

14 Ito, Y., Tanabe, Y., Qiu, H. et al. (2014). High-quality three-dimensional nanoporous graphene. *Angewandte Chemie* 126 (19): 4922–4926.

15 Han, J., Li, H., Lu, Z. et al. (2021). 3D bimodal porous amorphous carbon with self-similar porosity by low-temperature sequential chemical dealloying. *Chemical Materials* 33 (3): 1013–1021.

16 Ito, Y., Qiu, H., Fujita, T. et al. (2014). Bicontinuous nanoporous N-doped graphene for the oxygen reduction reaction. *Advanced Materials* 26 (24): 201400570.

17 Han, J., Huang, G., Wang, Z. et al. (2018). Low-temperature carbide-mediated growth of bicontinuous nitrogen-doped mesoporous graphene as an efficient oxygen reduction electrocatalyst. *Advanced Materials* 30 (38): 1803588.

18 Wang, L., Qin, K., Li, J. et al. (2018). Doping and controllable pore size enhanced electrochemical performance of free-standing 3D graphene films. *Applied Surface Science* 427 (Part B): 598–604.

19 Ito, Y., Cong, W., Fujita, T. et al. (2015). High catalytic activity of nitrogen and sulfur co-doped nanoporous graphene in the hydrogen evolution reaction. *Angewandte Chemie* 127 (7): 2159–2164.

20 Ito, Y., Shen, Y., Hojo, D. et al. (2016). Correlation between chemical dopants and topological defects in catalytically active nanoporous graphene. *Advanced Materials* 28 (48): 10644–10651.

21 Qiu, H., Du, P., Hu, K. et al. (2019). Metal and nonmetal codoped 3D nanoporous graphene for efficient bifunctional electrocatalysis and rechargeable Zn-air batteries. *Advanced Materials* 31 (19): 1900843.

22 Zhang, R., Wang, N., Shi, C. et al. (2020). Spatially uniform Li deposition realized by 3D continuous duct-like graphene host for high energy density Li metal anode. *Carbon* 161: 198–205.

23 Huang, G., Han, J., Zhang, F. et al. (2019). Lithiophilic 3D nanoporous nitrogen-doped graphene for dendrite-free and ultrahigh-rate lithium-metal anodes. *Advanced Materials* 31 (2): 1805334.

24 Ito, Y., Zhang, W., Li, J. et al. (2016). 3D bicontinuous nanoporous reduced graphene oxide for highly sensitive photodetectors. *Advanced Functional Materials* 26 (8): 1271–1277.

25 Han, J., Hirate, A., Du, J. et al. (2018). Intercalation pseudocapacitance of amorphous titanium dioxide@ nanoporous graphene for high-rate and large-capacity energy storage. *Nano Energy* 49: 354–362.

26 Qin, K., Kang, J., Li, J. et al. (2016). Continuously hierarchical nanoporous graphene film for flexible solid-state supercapacitors with excellent performance. *Nano Energy* 24: 158–164.

27 Zhang, R., Wen, S., Wang, N. et al. (2018). N-doped graphene modified 3D porous Cu current collector toward microscale homogeneous Li deposition for Li metal anodes. *Advanced Energy Materials* 8 (23): 1800914.

28 Qin, K., Wang, L., Wang, N. et al. (2017). Nitrogen and oxygen co-doped 3D nanoporous duct-like graphene@carbon nano-cage hybrid films for high-performance multi-style supercapacitors. *Journal of Materials Chemistry A* 5: 18535–18541.

29 Qin, K., Liu, E., Li, J. et al. (2016). Free-standing 3D nanoporous duct-like and hierarchical nanoporous graphene films for micron-level flexible solid-state asymmetric supercapacitors. *Advanced Energy Materials* 6 (18): 1600755.

30 Qin, K., Wang, L., Wen, S. et al. (2018). Designed synthesis of NiCo-LDH and derived sulfide on heteroatom-doped edge-enriched 3D rivet graphene films for high-performance asymmetric supercapacitor and efficient OER. *Journal of Materials Chemistry A* 6: 8109–8119.

4

Soluble-Salt-Template Methods

Ming Liang[1] *and Chunnian He*[1,2]

[1] *Tianjin University, School of Materials Science and Engineering, Tianjin Key Laboratory of Composite and Functional Materials, No.135 Yaguan Road, Haihe Education Park, Tianjin, 300350, P.R. China*
[2] *Joint School of National University of Singapore and Tianjin University, International Campus of Tianjin University, Jinbin Rd, Binhai New City, Fuzhou, 350207, P.R. China*

Graphene, which possesses large surface area, excellent chemical stability, high electrical conductivity, and thermal conductivity in combination with lightweight and relatively low cost, has inspired tremendous research interest in view of its extensive applications in medicine, biology, catalysis, energy storage and conversion, etc. [1–4]. To date, a number of ways have been developed to fabricate graphene. One of the most effective protocols for precisely modulating the graphene structures is template carbonization approach with nanostructured and/or self-assembled materials as sacrificed templates [5–8]. Among them, hard templates (such as anodic aluminum oxide, zeolites, silica, colloidal crystals, metals, and metal oxides) and soft templates (such as ionic surfactants and nonionic amphiphilic block copolymers) are the most widely adopted templates for preparing graphene [5–8]. However, the major hurdle associated with the hard template methods requires the synthesis of porous inorganic materials or special nanoparticles as hard templates in advance, which involves time-consuming multiple steps and high expense. Moreover, the final removal of the hard template by either firing or etching is rather technically complicated and costly, and frequently accompanied by the release of environmental hazards leading to safety risks. As for the soft template approaches, they are usually carried out on the basis of self-assembly of surfactants and block copolymers, which are highly expensive and nonrenewable and thus very unsuitable for scalable industrial production of graphene [5–8]. Consequently, the search for a green, low-cost, and easy-to-removal template method is quite challenging but necessary.

4.1 Salt-Template Methods

Recently, a truly facile and good method, i.e. salt-template strategy, has been developed for the large-scale and low-cost fabrication of graphene and graphene

Templated Fabrication of Graphene-Based Materials for Energy Applications, First Edition.
Edited by Chunnian He, Naiqin Zhao, and Junwei Sha.

composites for various applications [9–14]. In a typical process, carbon precursors (such as biomass and ionic liquids), non-carbonizable inorganic salts (such as NaCl, $ZnCl_2$), and even metal (or metal oxide, etc.) precursors, which can be fully miscible, are firstly mixed together to form homogeneous solution. During the subsequent carbonization course, the carbon precursors are condensed and scaffolded in the presence of the salts, while miscibility between the salts and the carbonizing materials is maintained over a main part of the reaction pathway. After the removal of the salt templates with simple water washing, pure graphene or graphene composites are produced, which often demonstrate a high surface area and porosity with the pore size corresponding to the salt clusters and their percolation structures [9–14]. Compared with conventional hard and soft template techniques, the salt-template methods have several obvious advantages. First, salt-template route can be regarded as quite a green process because most of the commercial salts are environmentally benign, very low cost besides being not volatile or harmful. Furthermore, the salts usually can be easily recycled. Industrialization of this process therefore seems very attractive and feasible. Second, the salt phases can be readily removed by simple water washing, while the graphene is not etched as such, thus avoiding any damage to the graphene or graphene composites in this process and fulfilling optimal structural retention. Third, a considerable number of miscellaneous combinations of carbon precursors, template salts, and even metal (or metal oxide, etc.) precursors can be well selected and mediated, enabling a great potential for the design and fabrication of a variety of unexpected graphene-based materials with tailor-made morphologies and dimensions, including (doped) graphene as well as metal (or metal oxide, etc.)-containing graphene composites, which have not yet been achieved by conventional methods [9–14]. Motivated by the prospects described above, building graphene-based materials via salt-template strategy has seen unprecedented achievements over the past few years. However, there is still lack of an overview regarding the advances in this booming field, so we considered it worthwhile to summarize recent progress gained in salt-templated graphene-based materials and to highlight the existing challenges to be addressed in the near future. In this review, we will focus on the typical salt-template methods for constructing 2D graphene, 3D porous graphene, and their composites toward energy storage (e.g. batteries, supercapacitors) and conversion (e.g. fuel cells). Furthermore, the existing challenges and possible future trends in this burgeoning field will be discussed on the basis of the current progress.

4.1.1 The Effects of Different Kinds of Salts

Because one of the advantages of the salt-template method is that the salt phases can be readily removed by simple water washing, the salt used as the template should be soluble salts. Meanwhile, the different properties of soluble salts always bring some unique effects on the graphene-based materials obtained from the salt-template strategy. The properties of soluble salts generally include the melting point, size, and crystal structure.

The state of the salt at different temperatures was determined by the melting point. Salts remain in crystalline state when the temperature is below the melting point of the salt, or will turn to liquid state at the temperature above the melting point. If the temperature further increases, the salt will reach the gas state or even decompose. Different salts have different melting points. For example, the melting point of NaCl (801 °C) is higher than that of KCl (773 °C). In addition, mixing two or more salts together in certain proportions can change the melting point. Therefore, according to the melting point of salts, the salt-template method can be split into crystal salt-template pathway and molten-salt pathway.

For crystal salt-template pathway, the salt determines the maximum temperature of the calcination process, and the high temperature is beneficial to the fabrication of graphene-based materials with high crystallinity. Generally, the salt with high melting point is used as the template to prepare graphene-based materials.

The effect of the salts' size manifests in two aspects. The first is the surface size of the salt crystal, which can be used to grow 2D graphene-based materials. Recently, Liu and colleagues showed NaCl crystal planes can be applied as substrate for the growth of free-standing few-layer graphene via a facile chemical vapor deposition (CVD) method (Figure 4.1a) [15]. The synthesis process was ethylene thermal cracked in the high-temperature region, which then flowed to the low-temperature region and was deposited on the surface NaCl crystal to form graphene. The authors found that the synthesis process of graphene in this study was significantly affected by the NaCl crystal size. The graphene preferred to grow on the surface of NaCl with crystal size in the range of 1–30 μm (Figure 4.1b–d), and there was almost no graphene on the commercial NaCl salt (100–500 μm). This might be attributed to the fact that the bigger-size NaCl crystals have a higher melting point than the smaller-size ones; under the same CVD temperature, the surface of the smaller-size NaCl crystals can be melted and the carbon species can nucleate on them to form graphene.

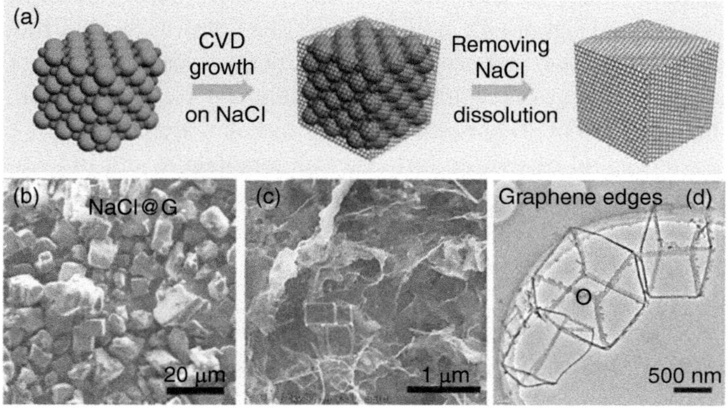

Figure 4.1 (a) Schematic view of graphene growth on a NaCl crystal. (b, c) scanning electron microscope (SEM) images of NaCl@graphene crystals and graphene powder, respectively. (d) TEM images of graphene facets and polycrystalline graphene edges. Source: Shi et al. [15]. Reproduced with permission of Wiley-VCH.

Another aspect is the particle size of the salt crystal, which can be used as the hard template to fabricate 3D porous graphene-based materials. Our group developed the freeze-drying salt-template solid calcination strategy to prepare 3D porous graphene-based materials [13, 14]. In this process, we found that the morphology of the 3D carbon network, especially the distribution of pores, was controllable by selecting different salts as the template [16]. With the same experiment parameter, NaCl templates can produce the micrometer-scale pores, Na_2CO_3 templates can generate the pores with the size of 50–100 nm, the smaller nanopores with the size of 5–10 nm can be got by Na_2SiO_3 templates (Figure 4.2a–i). When mixing these three kinds of salts in a reasonable ratio as template, the obtained carbon product showed a 3D hierarchical porous network structure, and the pore size distribution is controllable by changing the ratio of different salts (Figure 4.2j–m).

The crystal structure of salts can also influence the final graphene-based materials and we can use the unique crystal structure of some salts to prepare some 2D nanomaterials, such as $FeCl_3{\cdot}6H_2O$, the crystal structure of which is layered structure. Li et al. employed $FeCl_3{\cdot}6H_2O$ as an active salt template, dopamine hydrochloride (DA), catechol (Cat), or aniline (An) as the carbon precursor to prepare the heteroatom-doped 2D porous carbon nanosheets (CNs) [17]. During the preparation process, the organic carbon precursor is inserted into the interlayer space of $FeCl_3{\cdot}6H_2O$ and interacts with Fe^{3+} to form organic–inorganic hybrid material with a layered structure. After the carbonization process, the layered organic–inorganic hybrid transformed into a layered carbon–inorganic hybrid. Finally, the heteroatom-doped 2D porous carbon nanosheets (CNs) were collected after removing the inorganic template (Figure 4.3).

Montmorillonite (MMT) also has a similar characteristic, which is a layered compound and consisted of water-bearing aluminosilicate. In the research of Fan et al., MMT was used as the template and nanoreactor [18]. The bio-source carbon precursor, gelatin, can be intercalated into MMT. After thermal treatment and template removal, 2D carbon nanosheets (CNs), formed by gelatin pyrolysis in the interlayer space of MMT, were obtained (Figure 4.4). In this process, MMT with a layered structure was used as a nano-template to produce CNs. Especially, the layered nano-space located in the interlayer of MMT can confine and modulate the pyrolysis of gelatin, finally leading to the formation of 2D nanosheet-shaped carbon materials.

Some special reactions could be used to form the salt template in situ in addition to the direct utilization of soluble salts as templates, or some special reactions could make the salt act as both the template and the reactant for in situ synthesis of graphene-based materials. Hu et al. sintered the mixture of 1,5-diphenylcarbazide and $MgCl_2{\cdot}6H_2O$ at 800 °C under N_2. After the calcination products were etched with the template by 1 mol l^{-1} HCl solution, 2D nitrogen-doped carbon nanosheets could be prepared [19]. In this research, 1,5-diphenylcarbazide was regarded as the carbon and nitrogen source, the real template was not $MgCl_2{\cdot}6H_2O$, but MgO. At 800 °C, $MgCl_2{\cdot}6H_2O$ transformed to MgO, which acted as the template and catalyst to prepare 2D porous N-doped carbon nanosheets (Figure 4.5a).

In the application of Li–S batteries, carbon–sulfur composite is an ideal cathode material. However, the conventional method for the preparation of carbon–sulfur

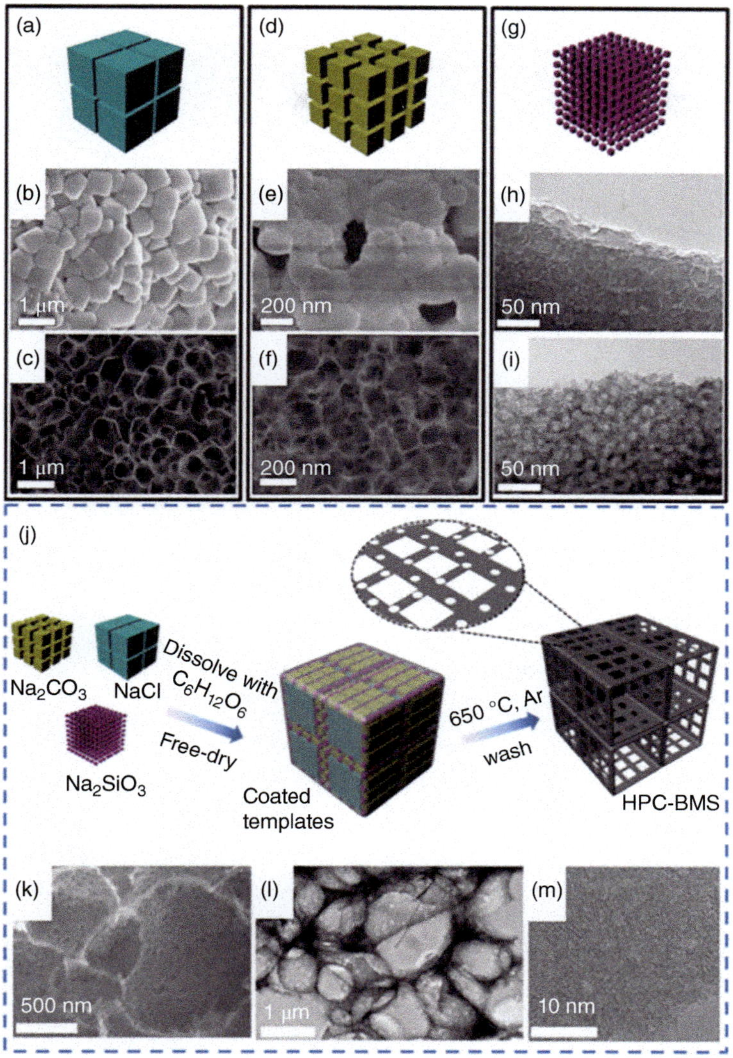

Figure 4.2 (a) Schematic of NaCl template, (b) SEM image of carbon-coated micro-sized NaCl, and (c) SEM image of the obtained 3DCN. (d) Schematic of Na_2CO_3 template, (e) SEM image of carbon-coated Na_2CO_3, and (f) SEM image the obtained 3DCN after washing step. (g) Schematic of Na_2SiO_3 template, (h) SEM image of carbon-coated Na_2SiO_3, and (i) TEM image of the obtained 3DCN after washing step. (j) Schematics demonstrating the fabrication process of carbon (HPC-BMS) with the assistance of multi-scale soluble salt self-assembly. (k) SEM, (l) TEM, and (m) HRTEM images of HPC-BMS. Source: Zhu et al. [16]. Reproduced with permission of Elsevier.

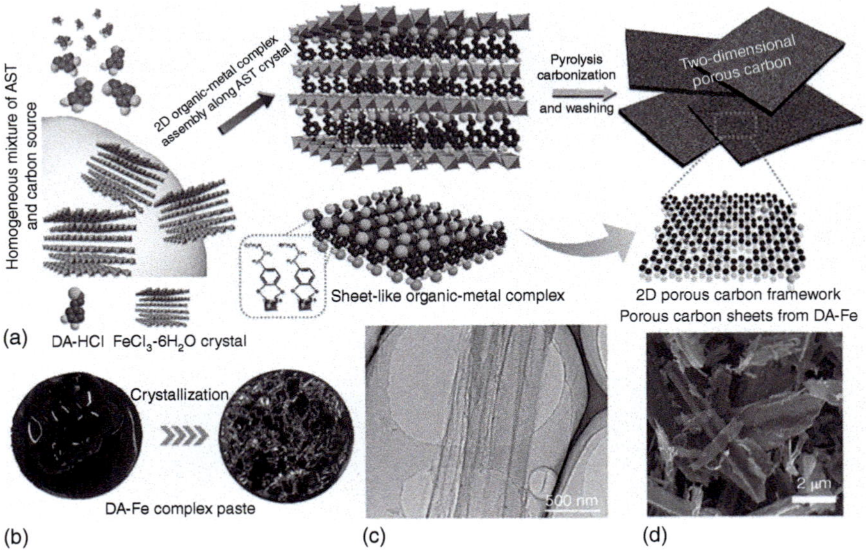

Figure 4.3 (a) Schematic of the preparation procedure of the heteroatom-doped 2D porous carbon nanosheets (CNs). (b) Photographs of the prepared DA–Fe. (c) TEM image and (d) SEM image of the heteroatom-doped CNs obtained from the DA–Fe complex at 700 °C. Source: Li et al. [17]. Reproduced with permission of Wiley-VCH.

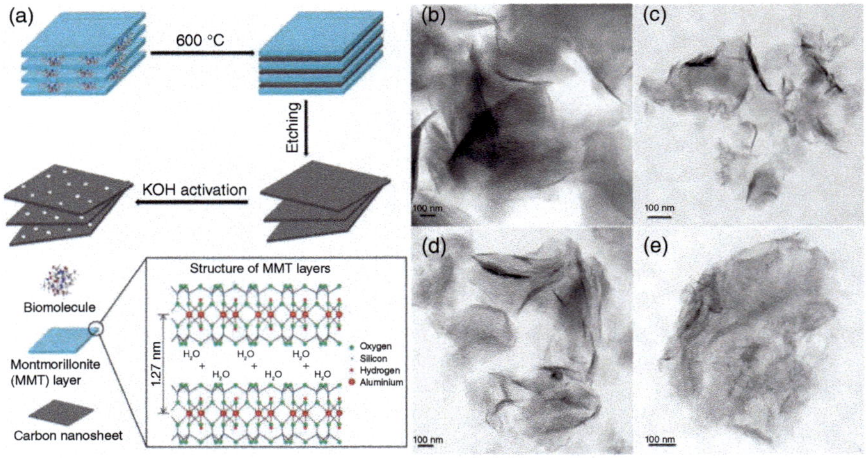

Figure 4.4 (a) Schematic of the synthesis of porous 2D nanosheet-shaped carbon materials from gelatin. TEM images of (b) MMT, (c) G@MMT (gelatin intercalation into MMT), (d) G@MMT-p (after calcination), and (e) CNs. Source: Fan et al. [18]. Reproduced with permission of Wiley-VCH.

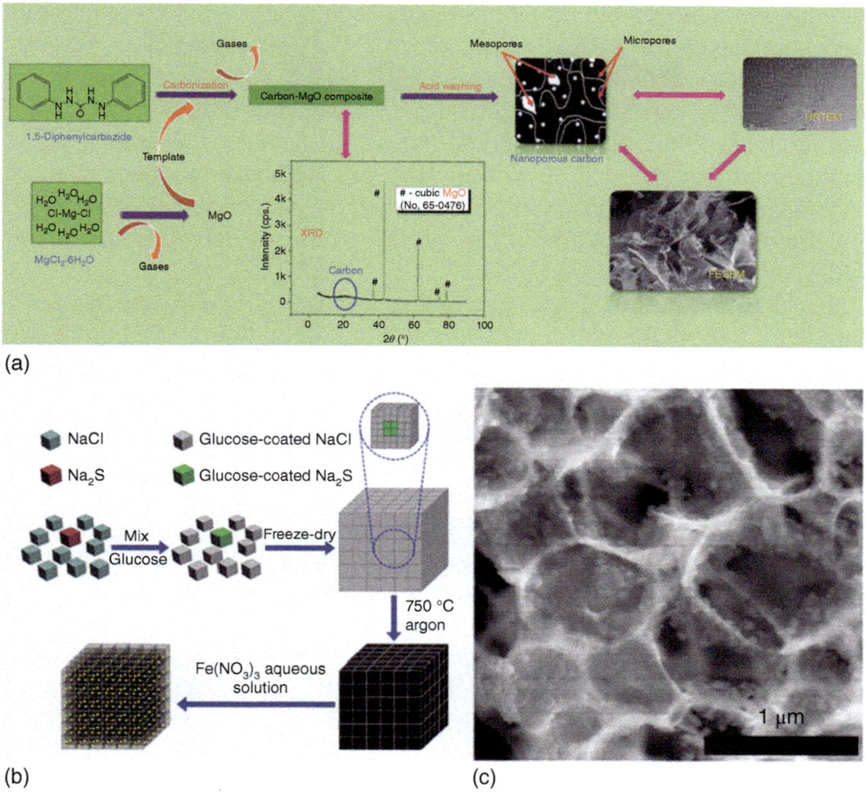

Figure 4.5 (a) Schematic of the production process for porous carbon materials. Source: Hu et al. [19]. Reproduced with permission of American Chemical Society. (b) Schematic of an in situ strategy for the preparation of 3D S@PGC composites. (c) SEM image of 3D S@PGC (90% S) composites. Source: (b, c) Li et al. [20]. Open Access Article/ Springer Nature/CC BY 4.0.

composites includes complexity of multistep operations, in which carbon materials with unique structure were prepared firstly, and then the sulfur was injected by impregnation diffusion method. Li et al. reported a facile and scalable strategy for the in situ synthesis of sulfur nanoparticles in three-dimensional (3D) porous graphitic carbon (PGC) (designated 3D S@PGC) with an adjustable sulfur content [20]. As shown in Figure 4.5b, glucose was used as the carbon source, while the water-soluble NaCl and Na_2S crystals were utilized as hard template for the 3D PGC network. After the heat treatment process in an argon atmosphere, the obtained 3D NaCl–Na_2S@GC was immersed in an aqueous $Fe(NO_3)_3$ solution to remove the NaCl template. At the same time, the Na_2S template could react with $Fe(NO_3)_3$ to in situ form sulfur nanoparticles on the walls of 3D PGC network (Figure 4.5c). Moreover, the sulfur content in the 3D PGC could be easily adjusted by varying the ratio of Na_2S to glucose.

The organic salts can also be used as multifunctional salt templates to fabricate heteroatom-doping hierarchical porous carbon. Wang et al. employed ethylenediaminetetraacetic acid disodium zinc salt ($EDTANa_2Zn$) as a hard template and sucrose as carbon precursor [21]. During the calcination process, $EDTANa_2Zn$ firstly decomposed to the Na_2CO_3 and nano-ZnO particles, then the obtained decomposition can serve as hard templates for mesopore formation and an activation agent for etching some carbon to form the micropores. Moreover, the decomposition of $EDTANa_2Zn$ can introduce nitrogen doping in the carbon matrix.

The molten-salt pathway is also widely used for the synthesis of nanomaterials, which offers a liquid reaction environment and can participate in reactions as precursor. Compared with crystal salt template, the molten-salt method can provide a wide range of reaction temperatures, from the melting point to boiling point or the decomposition temperature [22], and decrease reaction temperature and accelerate reaction dynamics [23].

Liu et al. converted glucose to graphene with a high yield of 75–83% in molten salts LiCl/KCl under inert gases [24]. In the salt melt, the conversion from glucose to graphene gradually proceeded through the elimination of water. It turned out that the carbon precursor/salt ratio and synthesis temperature can obviously impact the structure and porosity of the final carbons (as shown in Figure 4.6a–d). The heteroatomic-doped 2D nanocarbons/graphene can also be synthesized by a similar molten-salt route, such as KNO_3 [26]/$LiNO_3$ [25] and Na_2SO_4/$Na_2S_2O_3$ [25], which could be introduced in a LiCl/KCl molten salt system to lead to the formation of highly porous N (or S or N,S)-doped carbon nanosheets and graphene (as shown in Figure 4.6e–g). With the molten-salt route, the heteroatom doping content and the porosity could be tuned easily via adjusting the glucose/melt salt ratios and synthesis temperature. Generally, high concentration of the melt salts results in a high heteroatom doping content but obviously leads to the reduction of the carbon yield.

4.1.2 The Acquisition Method of Salt Templates

The acquisition of salt templates is another key factor in the preparation of nano-carbon-based materials. The in situ formation of salt is one of the ways to obtain salt template. Zhang et al. reported a method for the preparation of Fe_3C nanoparticles encapsulated in porous graphite (Fe_3C/PG) via a designed in situ salt-template method. When the $FeCl_3$ solution was dropped into sodium oleate solution, NaCl and ferric oleate could be precipitated due to the poor solubility of NaCl and ferric oleate in the mixed solvents of methanol and hexane, which were used as the in situ salt template and precursor, respectively [27] (as shown in Figure 4.7a). Hu et al. made MgO, which was in situ obtained from the decomposition of $MgCl_2 \cdot 6H_2O$ at 800 °C acting as the template for the formation of 2D porous structure [19] (as shown in Figure 4.5a). Liu et al. heated a mixture of sucrose and nitrogen-containing organic salt, i.e. ethylenediaminetetraacetic acid disodium zinc salt ($EDTANa_2Zn$), to synthesize nitrogen-doped, hierarchical porous carbons with high surface area [21]. In this process, $EDTANa_2Zn$ turned into nano-ZnO and

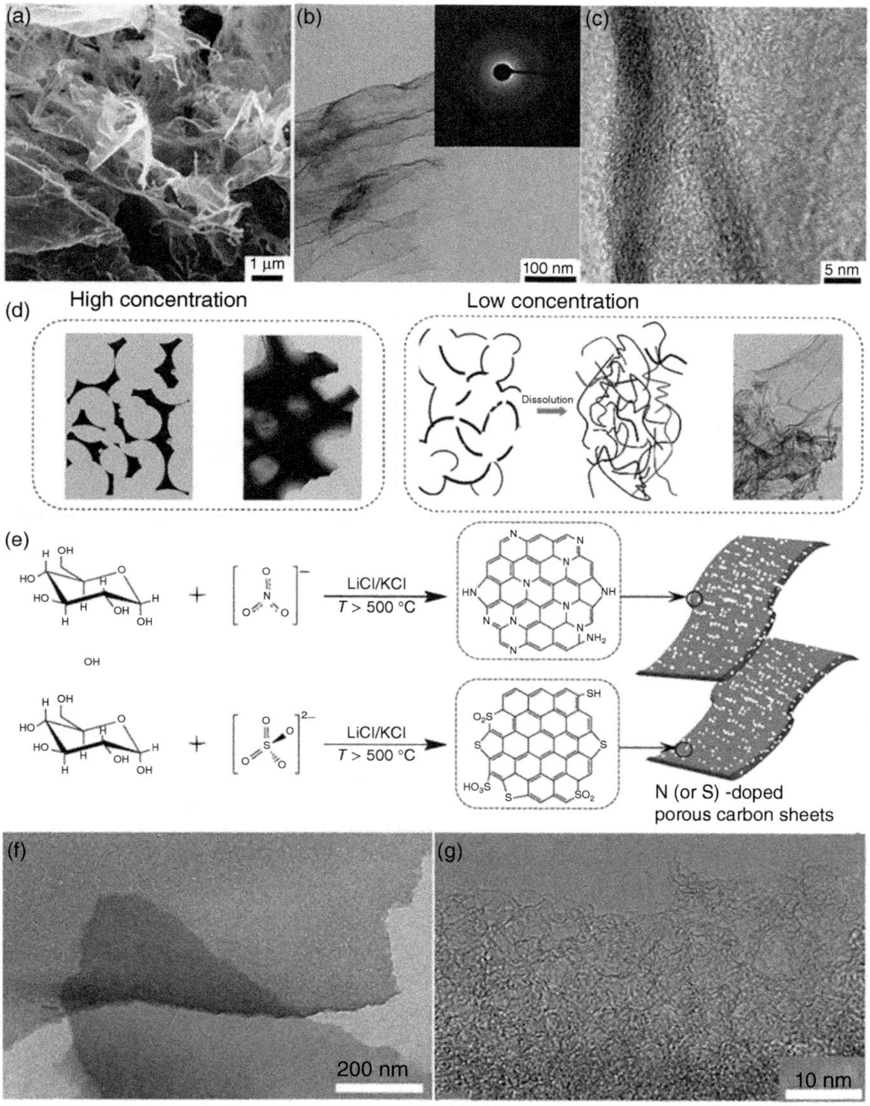

Figure 4.6 (a) SEM, (b) TEM (the inset is the electron diffraction pattern), and (c) HRTEM images of the graphene synthesized from glucose in MS at 800 °C (glucose/salt = 1 : 100). (d) Schematic of the formation of porous carbon in MS at a high glucose concentration and the synthesis of graphene at a low glucose concentration. Source: (a–d) Liu et al. [24]. Reproduced with permission of Wiley-VCH. (e) Schematic for the formation of N- and S-doped porous carbon sheets through the reaction of glucose and nitrate (NO_3^-) or sulfate (SO_4^{2-}) anions in eutectic molten LiCl/KCl. (f) Typical SEM image for CS-800B. (g) Typical HRTEM image taken at the edge of the S-doped sample (CS-800B), indicating clearly graphene-film-like structures. Source: (f–g) Liu and Antonietti [25]. Reproduced with permission of Wiley-VCH.

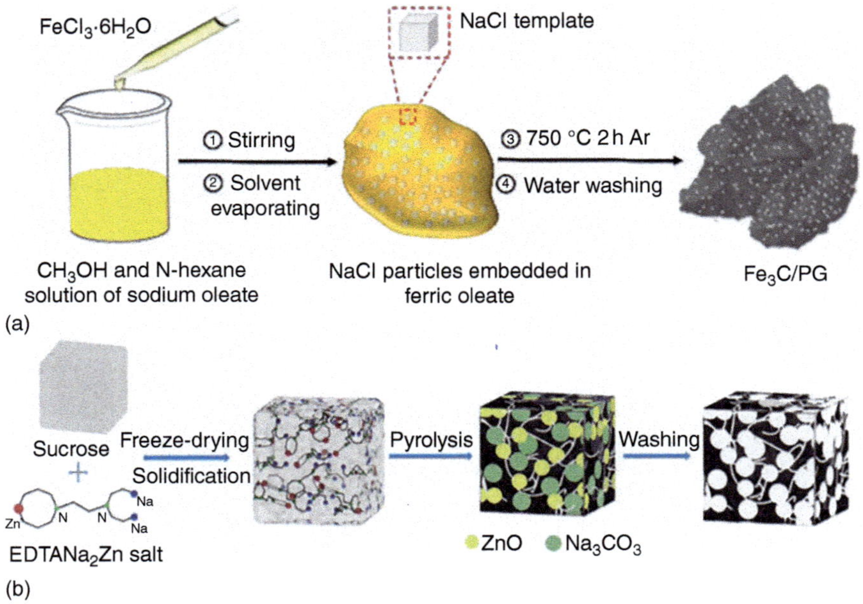

Figure 4.7 (a) Schematic of the preparation process of Fe_3C/PG. Source: Zhang et al. [27]. Reproduced with permission of The Royal Society of Chemistry. (b) Schematic of the preparation of nitrogen-doped hierarchical porous carbons. Source: Wang et al. [21]. Reproduced with permission of Elsevier Inc.

Na_2CO_3 particles and acted as the template for the formation of nitrogen-doped hierarchical porous carbons (as shown in Figure 4.7b).

Generally, the size of original commercial NaCl (100–500 μm) is too big to be employed as a template in some research, or to make the preparation of graphene-based materials more controllable. The original commercial NaCl should be recrystallized. The way of NaCl recrystallization can often affect the micromorphology of the final graphene-based material. Common methods of NaCl recrystallization include heating, freeze drying, spray drying, and alcohol extraction from aqueous solution containing NaCl. When using NaCl as a novel substrate to grow few-layer graphene, Shi et al. found no graphene was grown on the original commercial NaCl (100–500 μm) due to the higher melting point of larger NaCl crystals [15]. To reduce the size of NaCl crystals, ethanol was added to the saturated NaCl solution to produce the recrystallization of NaCl crystals, and the recrystallized NaCl crystals with a size in range of 10–30 μm can be obtained after filtration and drying.

Another way to acquire salt template is by heating the aqueous solution containing NaCl to make dissolve and recrystallize NaCl, which is a common method to produce template for fabrication of 2D graphene-based materials. Our group dried mixed a solution that contained $Fe(NO_3)_3$, glucose, and NaCl in a drying oven [12]. Then the very fine composite powders were annealed to in situ synthesize 2D PGC nanosheets uniformly embedded with carbon-encapsulated Fe_3O_4 nanoparticles on the surface

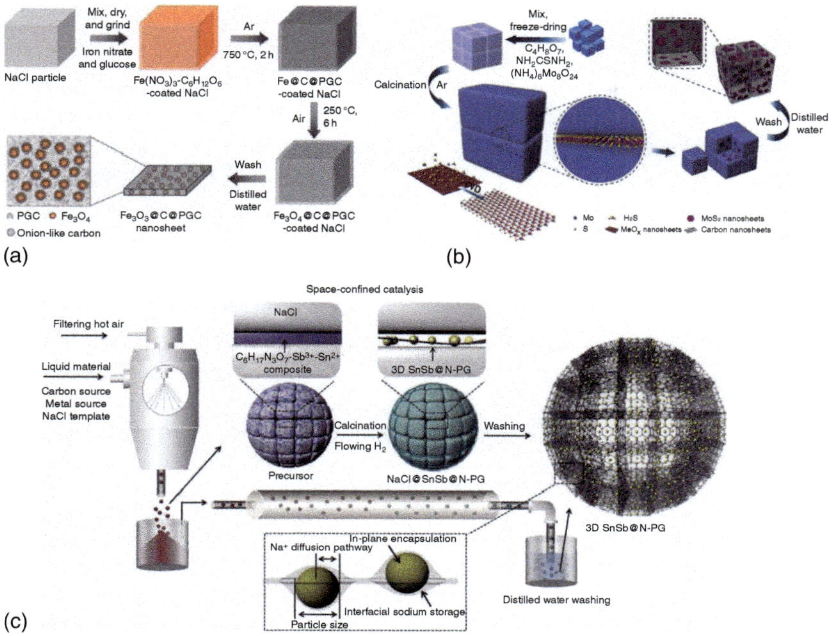

Figure 4.8 (a) Schematic of the in situ technique to fabricate 2D Fe_3O_4@C@PGC nanosheets by using the surface of NaCl particles as the template. Source: He et al. [12]. Reproduced with permission of American Chemical Society. (b) Schematics depicting the fabrication procedure of 3D FL-MoS_2@PCNNs. Source: Zhou et al. [14]. Reproduced with permission of American Chemical Society. (c) Schematic of the top-down in situ synthesis of 3D SnSb@N–PG composite for SIB anodes. Source: Qin et al. [28]. Reproduced with permission of Wiley-VCH.

of NaCl template (as shown in Figure 4.8a). Similar to this, numerous high-quality 2D metal (or metal oxides, metal sulfides, Si, etc.)-containing graphene composites have been successfully achieved by this method [29–50]. More achievements, such as heteroatomic doping (e.g. N, S, and B) [41, 48, 51, 52], have been made through this method.

To construct 3D porous graphene and its composites, our group developed a general and scalable pathway, i.e. 3D salt-templating assembly technique [13, 14, 16, 28, 53–58]. Take the preparation of MoS_2/nanocarbon composites as an example, the precursor solution was frozen by refrigerator or liquid nitrogen, and subsequently was subjected to freeze-drying treatment to remove water and form dry gel. During this process, multi-scale salts uniformly coated with an ultrathin complex film of carbon or carbon–metal (or carbon–metal sulfide, etc.) precursors were self-assembled into a regular 3D structure. After calcination and salt template removal, pure 3D porous nanocarbon-based materials with controllable hierarchical pores, large surface area, and high electron conductivity were generated (as shown in Figure 4.8b) [14]. Different from freeze drying, spray drying is another 3D salt-templating assembly technique, which is a low-cost and up-scaled strategy for preparing 3D porous graphene-based materials. A top-down strategy toward SnSb in-plane nanoconfined three-dimensional N-doped porous graphene composite microspheres (3D SnSb@N-PG) was synthesized by our group using industrialized spray-drying method [28]. As shown in Figure 4.8c, liquid precursor was sprayed as small droplets by a constant output atomizer, and then these small precursor droplets were quickly dried by hot air. During the spray-drying process, the sizes of the NaCl cubes were restricted, and the cubes were uniformly coated with an ultrathin complex film of $C_6H_{17}N_3O_7$-Sb^{3+}-Sn^{2+}, thus yielding micrometer-sized spheres constructed by several tens of NaCl@$C_6H_{17}N_3O_7$-Sb^{3+}-Sn^{2+} cubes. In the following heat treatment process, the 3D interconnected cubic NaCl particles were applied to act as the template to guide the formation of interconnected empty nanoboxes structure, and meanwhile offered continuous 2D-confined space between adjacent NaCl surfaces to achieve the in situ generation of SnSb-nanoconfined ultrathin N-doped graphene walls in the nanoboxes.

4.1.3 The Other Important Influencing Parameters

The temperature is a key parameter in material preparation, which can affect the quality and performance of the prepared material. For graphene-based materials prepared by salt-template methods, generally, the higher temperature will result in a higher quality of carbon products. Wang et al. prepared a high-surface area, nitrogen-doped, and hierarchical porous carbon material from sucrose using $EDTANa_2Zn$ salt as hard template. They investigated the influence of the calcination temperature on the porous carbon material. With the calcination temperature increasing from 700 to 950 °C, the crystalline structure of the carbon rearranges from amorphous state to a more ordered state. Due to the etching effect of nano-ZnO produced from $EDTANa_2Zn$ pyrolysis on carbon, the BET specific surface area increased from 1166 to 2160 $m^2\ g^{-1}$, the micropore volume increased nearly twice when the carbonation temperature increased from 700 to 800 °C, and the small mesopore volume also increased nearly twice from 800 to 950 °C [21]. During the synthesis of doped and highly porous graphene nanoplatelets via molten-salt pathway, Liu et al. found a reduction of sample yield (with respect to carbon) and a steady evolution of local structure toward graphitized carbon with increasing reaction temperature [22, 24, 25]. As for the porosity of the products, it first increased with the carbonization temperature to a maximum value, and then decreased distinctly because of graphitization and enhanced nanosheet stacking [22, 24, 25].

The carbon precursor/salt ratio can also affect the morphology of the final graphene-based material. Li et al. used $FeCl_3{\cdot}6H_2O$ as salt template and dopamine hydrochloride (DA) as carbon source to fabricate 2D porous carbons [17]. Different molar ratios of DA and $FeCl_3{\cdot}6H_2O$ (3 : 1, 1 : 1.5, 1 : 3.5, 1 : 7, and 1 : 14) on the resulting carbon nanostructures were investigated. It was noted that only when the DA–Fe molar ratio decreases to 1 : 3.5 or lower (1 : 7 and 1 : 14), the obtained CNs exhibit a uniform 2D morphology. Our group applied the mixture of NaCl and Na_2SiO_3 as salt template to fabricate 3D network constructed by N-doped carbon ultrathin nanosheets with closely packed mesopores (3D N-MCN) [53]. Given the amount of NaCl, the observed mesopore sizes in N-MCN increase from 0, 5, 8, to 27 nm by elevating the weight ratio of Na_2SiO_3 from 0%, 2%, 4%, to 6%. With the Na_2SiO_3 having greater weight (the weight ratio is above 10%), the formed nanosheets become too thick to be supported for constructing a 3D network, resulting in the architecture collapsing into stacked 2D flat mesopore thick sheets. A similar phenomenon can also be observed in the molten-salt pathway. In the preparation process of graphene through molten-salt route, a high glucose concentration resulted in a carbon network with spherical open voids and thick walls, and a low glucose concentration made the successful synthesis of graphene [24].

The heating rate and sintering duration are two important kinetic parameters during the material synthesis process. Take ferrite/carbon nanosheets prepared on the surface of Na_2SO_4 particles as an example, when the heating rate decreased from 10 to 5 °C min^{-1}, the size of the ferrite nanocrystals increased from ~16 to ~30 nm, while their shape became more spherical as compared to the 16 nm-sized nanocubes [11]. Yolk-shell structured Sb@C nanoconfined nitrogen–sulfur co-doped 3D porous carbon microspheres (Sb@NS-3DPCMSs) were prepared by our group via salt-template method [57]. The formation mechanism of the yolk-shell structure is that Sb nanocrystals would be sublimated during the high-temperature calcination process due to the low melting point of Sb (630 °C). Therefore, with the extension of calcination time, the void space between yolk and shell gradually increased.

4.2 Salt-Template-Directed Graphene-Based Materials

4.2.1 2D Graphene-Based Materials

2D graphene-based materials, including graphene and graphene composites, have drawn vast attention for fundamental research during the last few decades due to their unique physicochemical properties and promising applications [1–4]. Numerous methods have been developed for the synthesis of 2D graphene-based materials with controlled size, thickness, and chemical/electronic properties. Among those, salt-template process seems a very scalable and low-cost way for mass production of 2D graphene-based materials with tunable structures and desirable properties [9–14].

A facile molten-salt pathway for the synthesis of carbon nanosheets and graphene has been reported [24, 25, 59, 60]. In this method, the carbonization of the carbon precursors (such as glucose) in molten metal chloride salts (such as LiCl/KCl) under inert gases proceeded gradually through the dehydration and finally induced the formation of carbon nanosheets or graphene through regulating the experimental conditions. It turned out that the carbon precursor/salt ratio and synthesis temperature exerted a remarkable impact on the structure and porosity of the resultant products. By increasing the carbon precursor/salt ratio and synthesis temperature, thinner and higher graphitized carbon nanosheets could be produced. As for the porosity of the products, it first increased with the carbonization temperature to a maximum value, and then decreased distinctly because of graphitization and enhanced nanosheet stacking. Take the synthesis of graphene as an example, when heating a glucose/salt (LiCl/KCl) mixture (1 : 100 by weight) under N_2 at 800 °C, few-layer graphene with a maximum thickness of 0.65–1.35 nm and a surface area of 595 $m^2\ g^{-1}$ dominated the resultant products [24]. To dope the 2D graphene with *p*-block foreign atoms, which can modify the electronic structure and the Fermi level of the host, a mixture of $LiNO_3$ (and/or K_2SO_4 and $Na_2S_2O_3$) and glucose in the solvent of LiCl/KCl was heat-treated, leading to the formation of highly porous N (or S or N,S)-doped carbon nanosheets and graphene with surface area of up to 3200 $m^2\ g^{-1}$ after washing away the salts (as shown in Figure 4.9). It was noted that the heteroatom doping content could be tuned easily via adjusting the glucose/melt salt ratios and synthesis temperature. Generally, high concentration of the melt salts induces a high heteroatom doping content but obviously leads to the reduction of the carbon yield [25].

Utilizing highly temperature-resistant salt materials as a template during CVD can facilitate the growth of unexpected 2D graphene. One must bear in mind that Liu and colleagues recently proposed a NaCl-template-assisted CVD protocol for the synthesis of free-standing few-layer graphene (as shown in Figure 4.10a) [15]. In this particular case, NaCl powders with crystal size in the range of 1–30 μm, which were firstly obtained through recrystallization of the commercial NaCl salt (~300 μm), were subjected to CVD using ethylene as a carbon source, leading to the generation of graphene over the surfaces of NaCl powder (NaCl@graphene) (Figure 4.10b). After dissolving the cores of NaCl crystals in water, pure large flat free-standing few-layer graphene nanosheets along with some hollow cuboidal graphene were obtained (Figure 4.10c,d). It should be noted that the NaCl crystal size played a role of paramount importance in graphene growth. Smaller-sized NaCl was more helpful for growing graphene, while the larger-sized commercial NaCl powders (~300 μm) caused no graphene synthesis.

Owing to its unique nature, 2D graphene is very attractive since it can integrate with various active organic or inorganic components, which can not only circumvent the restacking of 2D graphene and thus retain the high surface area, but also inherit the full advantages of each of the two components, rendering them favorable for energy storage/conversion applications. The surface of solid salts offers unique benefits for the in situ growth of this kind of material [11, 12]. For instance, self-assembled 2D ferrite/carbon composite nanosheets have been prepared by using sodium sulfate (Na_2SO_4) salt powder as a sacrificial template [11]. In the direct synthesis of these composites, the surface of Na_2SO_4 particles served as the template, while the metal–oleate complex was adopted as the precursor for both ferrite and carbon. The synthesis process is schematically shown in Figure 4.1a. During calcination, in situ-formed metal–oleate complex, evenly coated around the Na_2SO_4 surface, was converted into 2D ferrite/carbon nanosheets with even thickness (Figure 4.1b). As a consequence, pure ferrite/carbon nanosheets were obtained by removing Na_2SO_4 with water. It was revealed that these 2D nanosheets were very uniform and thin (several 10 nm), and their surface profile reflected that the highly crystallized ferrite nanocrystals with uniform size and shape independently concealed within a thin carbon layer were assembled in a highly ordered 2D array, which was evenly

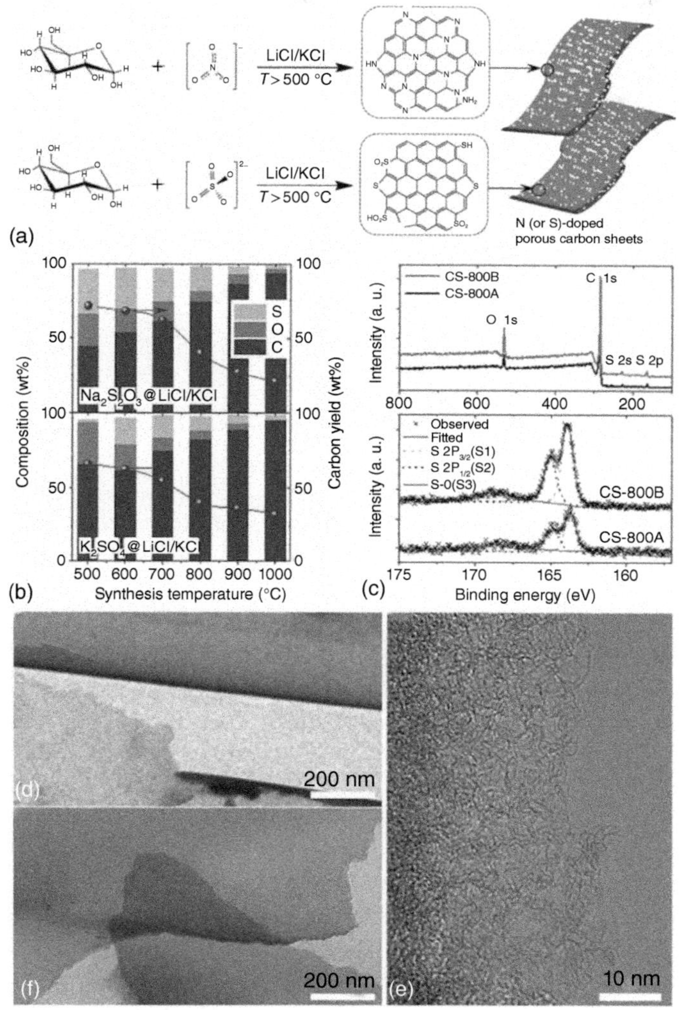

Figure 4.9 (a) Schematic for the formation of N- and S-doped porous carbon sheets through the reaction of glucose and nitrate (NO_3^-) or sulfate (SO_4^{2-}) anions in eutectic molten LiCl/KCl. S-doped samples synthesized in salt system of (thio)sulfate@LiCl/KCl. (b) Temperature dependence of elemental composition and carbon yield. (H% = 100% − C% − S% − O%). (c) Survey scanned XPS spectrum (top) and high-resolution spectrum in S 2p region (bottom) for CS-800A (from K_2SO_4 @LiCl/KCl) and CS-800B (from $Na_2S_2O_3$@LiCl/KCl). Thiophene-S located at 163.9 eV (S1) and 165.1 eV (S2) and oxidized sulfur at 168.5 eV (S3) are considered for spectrum deconvolution. Typical SEM images for (d) CS-800A and (e) CS-800B. (f) Typical HRTEM image taken at the edge of the S-doped sample (CS-800B), indicating clearly graphene-film-like structures. Source: Liu and Antonietti [25]. Reproduced with permission of Wiley-VCH.

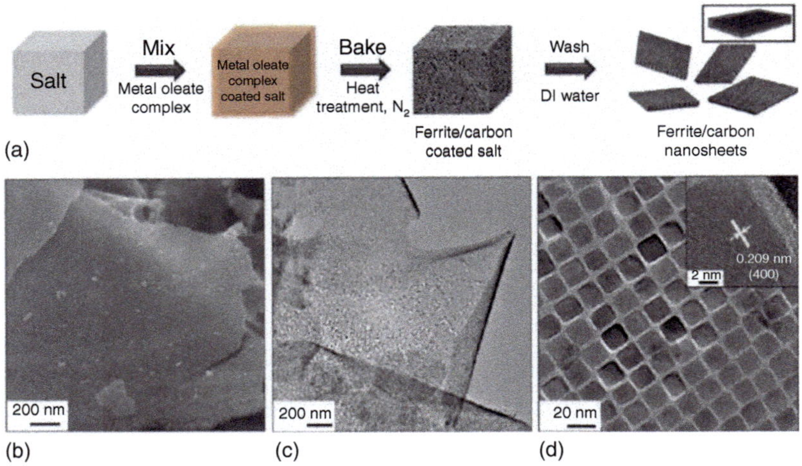

Figure 4.10 (a) Schematic representation of the preparation of ferrite/carbon nanosheets by salt-template process. (b) SEM image and (c, d) TEM images of 16 nm iron-oxide/carbon nanosheets. The inset shows a HRTEM image demonstrating the highly crystalline nature of the nanosheets. Source: Jang et al. [11]. Reproduced with permission of American Chemical Society.

distributed across the carbon matrix (Figure 4.1c,d). Moreover, the size and shape of the ferrite nanocrystals within nanosheets could be controlled through the heating rate and temperature. Take iron oxide as an example, when the heating rate decreased from 10 to 5 °C min^{-1}, the size of the iron oxide nanocrystals increased from ~16 to ~30 nm, while their shape became more spherical as compared to the 16 nm-sized nanocubes.

More recently, to further increase the graphitization degree and porosity of the 2D graphene composites, which are highly beneficial for improving their overall electrical conductivity and rapid diffusion of electrolytes in the electrodes, our group firstly in situ synthesized 2D PGC nanosheets uniformly embedded with carbon-encapsulated Fe nanoparticles (Fe@C@PGC nanosheets) with assistance of the surface of NaCl template by using the $Fe(NO_3)_3$ and $C_6H_{12}O_6$ as the metal and carbon precursor, respectively (as displayed in Figure 4.11a) [10]. In this stage, the continuous decomposition of carbon source and the catalyzation of metallic Fe nanoparticles caused greatly graphitized structure and a large number of micropores and mesopores into the nanosheets. The obtained Fe@C@PGC nanosheets were then annealed under air to accomplish the formation of Fe_3O_4@C@PGC nanosheets, in which Fe_3O_4 nanoparticles (~18.2 nm) coated with conformal and thin onion-like carbon shells were homogeneously embedded in 2D PGC nanosheets with a thickness of less than 30 nm (Figure 4.11b–e). Further study indicated that the resultant Fe_3O_4@C@PGC nanosheets after the NaCl removal had intrinsic dispersibility, large surface area, hierarchical porous structure, high electrical conductivity, and superior structural stability. The surface template effect of solid salts described above is a simple, but efficient and tunable, approach to a large variety of 2D graphene composites.

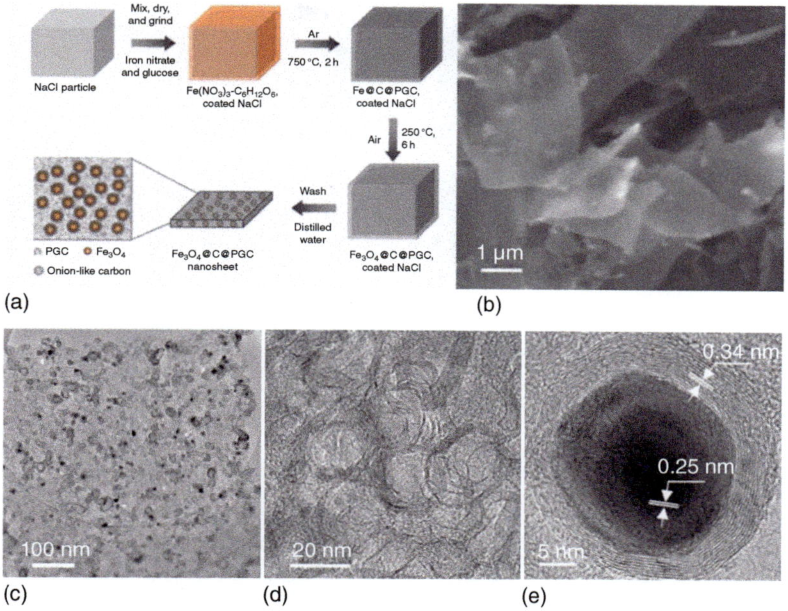

Figure 4.11 (a) Schematic illustration of the in situ technique to fabricate 2D Fe_3O_4@C@PGC nanosheets by using the surface of NaCl particles as the template. (b) SEM image, (c–e) HRTEM images of Fe_3O_4@C@PGC nanosheets. Source: He et al. [12]. Reproduced with permission of American Chemical Society.

In the following, numerous high-quality 2D metal (or metal oxides, metal sulfides, Si, etc.)-containing graphene composites [29–50] have been successfully formed by this method via altering the carbon precursors and metal (or metal oxide, metal sulfides, etc.) precursors. Simultaneously, heteroatomic doping (such as N, S, and B) is always combined with above method to improve the electrochemical activity of graphene composites [41, 48, 51, 52]. For example, by adopting Ni—Fe as metal source and catalyst, CS as carbon source, and NH_3 as nitrogen source, Goodenough and colleagues reported a bifunctional electrocatalyst consisting of Ni_3Fe nanoparticles embedded in porous nitrogen-doped carbon sheets (Ni_3Fe/N-C sheets) with the assistance of NaCl template [41]. It is revealed the NH_3 activation does not merely provide a nitrogen source, it also dramatically improves the porosity of the carbon sheets, leading to more active surface sites in the Ni_3Fe/N-C sheets. In another case, Dai and colleagues found Ni_2SO_4 could serve as nickel and sulfur precursor simultaneously. Briefly, a uniformly formed melamine–nickel sulfate complex was first collected by lyophilization in powder form, which was then mixed well with potassium chloride by ball milling. With the assistance of KCl and KCl-coated Ni template, N, S co-doped graphitic sheets could grow from the Ni and KCl seeds during high-temperature calcination and finally lead to the two-dimensional (2D) N, S co-doped graphitic sheets with a unique hierarchical structure consisting of stereoscopic holes over the graphitic surface (SHG) after removing the nickel and KCl residues, [50] as shown in Figure 4.12. Besides, it is

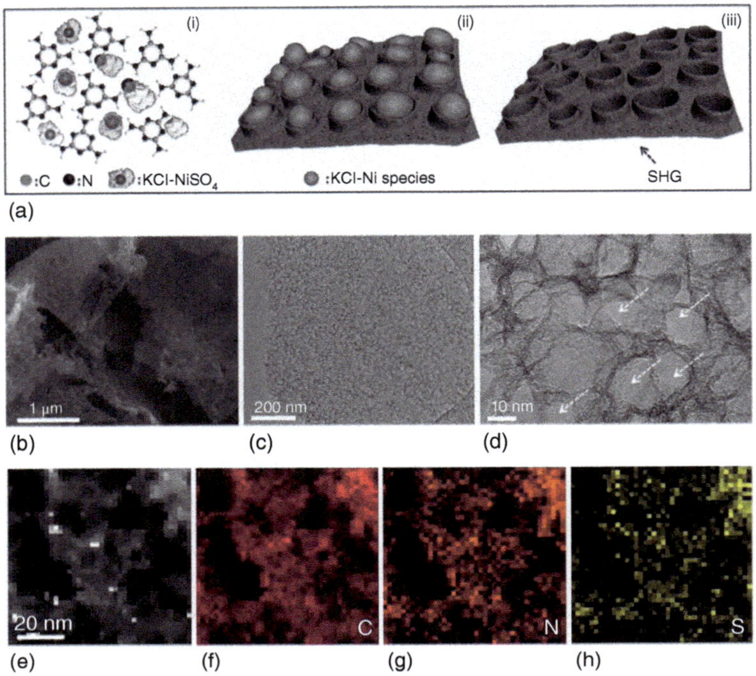

Figure 4.12 (a) Schematic of the preparation process for the SHG. (i) Melamine–nickel sulfate complex mixed well with KCl by ball milling the powder of a melamine–nickel sulfate complex and KCl. (ii) The Ni-KCl@SHG in situ grown from Ni@KCl and KCl seeds were obtained by an annealing process. For clarity, only a piece of the 2D Ni-KCl@SHG is shown. (iii) The structurally well-defined SHG was obtained by etching the Ni@KCl and KCl seeds, followed by additional thermal annealing. (b) SEM image of SHG. (c, d) TEM images under different magnifications. (e) The corresponding high angle annular dark field-scanning transmission electron microscope (HAADF-STEM) image of SHG. (f–h) Energy dispersive spectrometer (EDS) mapping images of C, N, and S elements of (e). Source: Hu and Dai [51]. Reproduced with permission of Wiley-VCH.

revealed the KCl can also facilitate the graphitization and the formation of catalytic active sites in the SHG.

4.2.2 3D Porous Graphene-Based Materials

Compared with 2D graphene, 3D porous graphene can provide continuously interconnected 3D network architecture with higher porosity, larger surface area, higher conductivity, and lower weight density, which make 3D porous graphene and their composites a more promising candidate for a large variety of applications, including energy conversion and storage.

Salt-template procedure is regarded as an ideal pathway for large-scale and low-cost synthesis of 3D porous graphene-based materials. Logically, numerous carbon precursors can be fully miscible with many non-carbonizable inorganic salts to form homogeneous solutions. When the ratio of carbon/template salts reaches a certain value during calcination, the carbon precursors are thermally cross-linked and form a 3D interconnected network around the template salts, which cannot be cross-linked. After the completion of the condensation reaction and the elimination of the template salts by water washing, the resulting 3D graphene-based materials possess a high surface area where the pore size corresponds to salt clusters and their percolation structures, leading to micro- to mesoporous materials [9–14]. In a typical case, 3D porous heteroatom (N, B, or BN dual)-doped graphene was prepared in one step by salt-templating method by choosing ionic liquids as carbon precursors and eutectic mixtures (such as $ZnCl_2$ with LiCl [LiZ], or NaCl (SZ) or KCl (PZ)) as porogen salts. As indicated by X-ray diffraction, X-ray photoelectron spectroscopy, and nitrogen sorption measurements, the washed products were the slightly disordered graphitic heteroatom-doped graphene with surface areas up to $2000\,m^2\,g^{-1}$, higher than any for zeolite or activated carbons and approached the value of single-layer graphene. Microstructure characterizations combined with nitrogen sorption analysis illustrated that the morphologies and porosities of the washed products chiefly relied on the nature of the eutectic salts and not on the ionic liquids (Figure 4.13), thereby validating the templating effect by the porogen salts [9, 10]. It was noted that the pore structure could be tuned by the adjustment of eutectic mixture ratio. Take the $ZnCl_2$/NaCl salts as an example, larger NaCl concentration was more favorable for increasing the mesopore size and introducing additional macropores into the nanocarbons, resulting in a very high maximum total pore volume of $5.2\,cm^3\,g^{-1}$ (specific surface area = $2540\,m^2\,g^{-1}$) [9]. In addition, all the 3D porous nanocarbons produced even using very high salt concentrations could be obtained with a remarkable carbon yield (~35 wt%) [9, 10]. The above results implied that the salt-template method is an effective way to regulate the surface area and porosity of the 3D graphene. Based on this concept, a considerable variety of 3D highly porous graphene-based materials, including heteroatom-doped graphene [61–71] as well as metal (or metal oxide, etc.)-containing graphene composites [72–75], have been successfully constructed through selecting suitable miscellaneous combinations of carbon precursors, porogen salts, and metal (or metal oxide, etc.) precursors. These obtained 3D porous graphene-based materials demonstrated high conductivity, large surface area, and porosity, as well as high stability, thereby showing promise in energy storage and conversion.

In another instance, Wei's group has proposed a "shape fixing via salt recrystallization" route for the preparation of 3D porous N-doped nanocarbons (Figure 4.14) [76]. In the course of synthesis, a 3D polyaniline network produced via a self-assembly process was fixed and fully sealed inside the NaCl crystal through repeated recrystallization of NaCl solution. During pyrolysis, the NaCl crystal functioned as a fully closed nanoreactor to facilitate graphitization and N incorporation, while the gasification in such a closed nanoreactor generated a great many pore in the resultant products. After the removal of NaCl, a 3D porous nanocarbon network with high N doping level and a large number of active sites exposed to the three-phase zones were obtained. This shape fixing via salt recrystallization method provides a powerful approach for the preparation of 3D graphene with high heteroatom doping level and

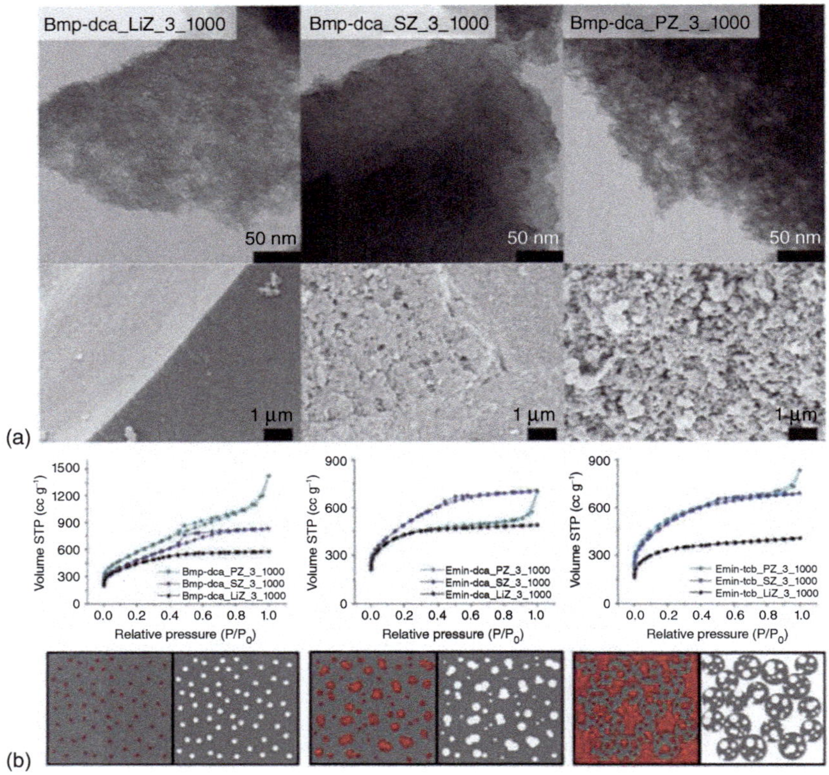

Figure 4.13 (a) TEM (upper row) and SEM pictures (lower row) of N-dCs using Bmp-dca as precursor templated with LiZ (left), SZ (middle), and PZ (right) at equal mass ratios synthesized at 1000 °C. (b) Upper row: Nitrogen sorption isotherms of Bmp-dca (left), Emim-dca (middle), and Emim-tcb (right) derived carbons templated with $LiCl_2/ZnCl_2$ (LiZ), $NaCl/ZnCl_2$ (SZ), and $KCl/ZnCl_2$ (PZ) at equal mass ratios. Lower row: Schematic representation of pore formation for carbons templated with LiZ (left), SZ (middle), and PZ (right) at equal mass ratios. Each left image depicts the carbon (gray)/salt (red) composite and each right the carbon after washing. Source: Fechler et al. [9]. Reproduced with permission of Wiley-VCH.

high utilization of active sites via confined pyrolysis of nanopolymers. Furthermore, this method can be applied to the production of other 3D metal (or metal oxide, metal sulfides, etc.)-containing graphene composites with similar structures for various applications [76].

Although the above 3D porous graphene fabricated by salt-template strategy shows high surface areas and porosity, it usually exhibits disordered morphology, and its pore size distribution is hard to tailor. In contrast, well-ordered 3D porous graphene has the benefits of uniform pore size and high isotropic percolation and thus exhibits great superiority in many applications [1–4]. Very recently, our group has developed a general and scalable pathway, i.e. 3D salt-templating assembly technique, for building well-ordered 3D porous graphene and its composites (as shown in Figure 4.15, take the preparation of MoS_2/graphene composites as an

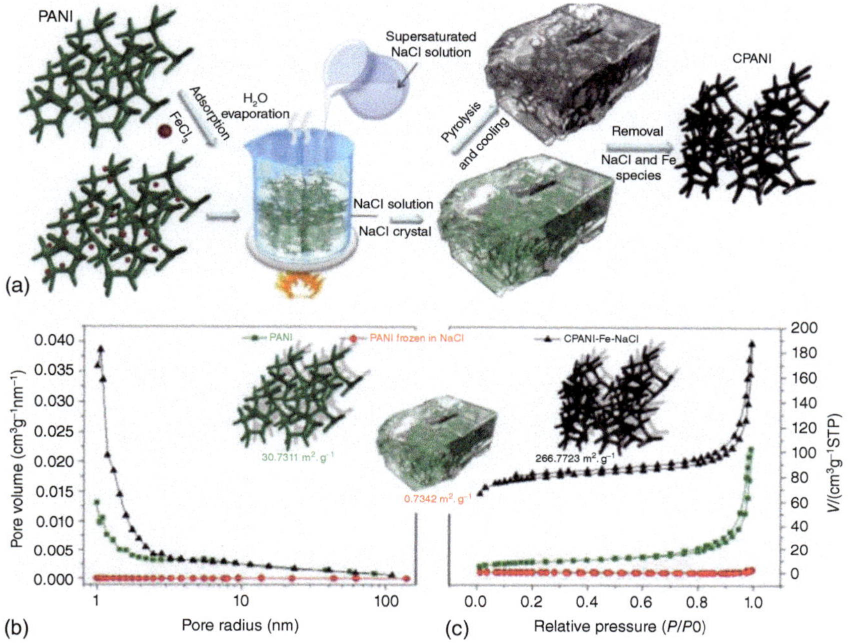

Figure 4.14 (a) Schematic illustration of shape fixing via salt recrystallization method. (b) Pore size distribution for PANI, PANI in NaCl, and CPANI-Fe-NaCl, assessed by BJH. (c) Nitrogen sorption isotherms for PANI, PANI in NaCl, and CPANI-Fe-NaCl. Reproduced with permission. Source: Ding et al. [76]. Reproduced with permission of American Chemical Society.

example) [5, 8, 13, 14, 16, 28, 53–56, 58, 77–81]. Typically, carbon precursors (such as glucose and citric acid), and template salts (such as NaCl), together with or without metal (or metal sulfide, etc.) precursors, were firstly dissolved in distilled water to obtain a homogeneous solution, which was subsequently subjected to freeze-drying treatment to remove water and form dry gel. By modulating the content and crystallization of various salts during the freeze-drying process, multi-scale salts are uniformly coated with an ultrathin complex film of carbon or carbon–metal (or carbon–metal sulfide, etc.) precursors were self-assembled into a regular 3D structure. In the third step, the 3D complex assembly was calcined at high temperature under Ar or Ar/H_2. During carbonization, the carbon or carbon–metal (or carbon–metal sulfide, etc.) precursors within the 2D-confined space between the NaCl surfaces were converted into carbon nanosheets, graphene, or 2D metal (or metal sulfides, etc.)-containing graphene composites. After simple water washing of 3D salt-templating assembly, pure 3D porous graphene-based materials with controllable hierarchical pores, large surface area, and high electron conductivity were generated, including 3D porous (doped) carbon nanosheet networks [16, 53, 77, 78], 3D porous graphene/metal composites [13, 28], and 3D porous graphene (or ultrathin carbon sheet)/metal sulfide composites [14, 79, 80].

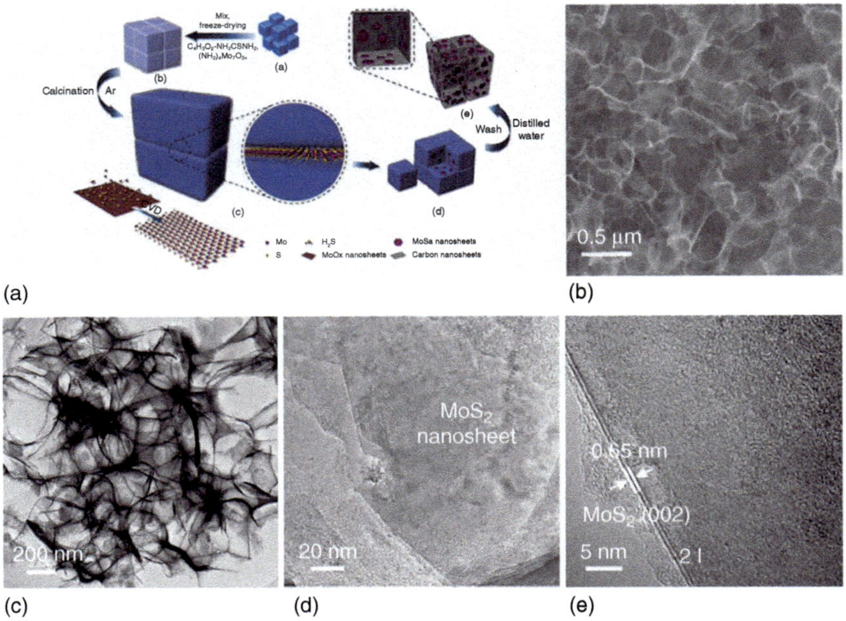

Figure 4.15 (a) Schematic illustration of the 3D NaCl assembly-assisted synthesis of 3D porous MoS_2-containing nanocarbon composites (3D FL-MoS_2@PCNNs). (b) SEM image, (c–e) TEM images of the 3D FL-MoS_2@PCNNs. Source: Zhou et al. [14]. Reproduced with permission of American Chemical Society.

For another example, a top-down strategy toward SnSb in-plane nanoconfined three-dimensional N-doped porous graphene composite microspheres (3D SnSb@N-PG) was synthesized by our group using industrialized spray-drying method, as shown in Figure 4.16 [28]. It can be seen the SnSb nanocrystals with an average diameter of 13 nm were firmly embedded in the N-graphene walls and all-around coated by few-layer graphene (3–5 layers, ~1 nm), indicating a strong interaction between the SnSb nanocrystals and N-graphene. The thickness of N-graphene walls is about 2 nm; nevertheless, the N-graphene wall with thickness less than 1 nm also exists. Through fully validated experimental results, we found out the important factors in the one-step synthesis of SnSb@graphene structure, which can be called as in situ space-confined catalysis. Namely, upon heat treatment, the 3D interconnected cubic NaCl particles acted as the template to guide the formation of interconnected empty nanoboxes structure, and meanwhile offered continuous 2D-confined space between adjacent NaCl surfaces to achieve the in situ generation of SnSb-nanoconfined ultrathin N-doped graphene walls in the nanoboxes. Besides, the 3D salt-templating assembly strategy could also be employed in CVD process, leading to the construction of sandwiched carbon/metal oxide structures or hierarchical carbon nanostructures [55, 56].

Recently, researchers found some special reactions could be used to form or transform 3D salt templates, thereby making further use of these templates. For example, taking advantage of the reaction of Na_2S and $Fe(NO_3)_3$, in situ synthesis

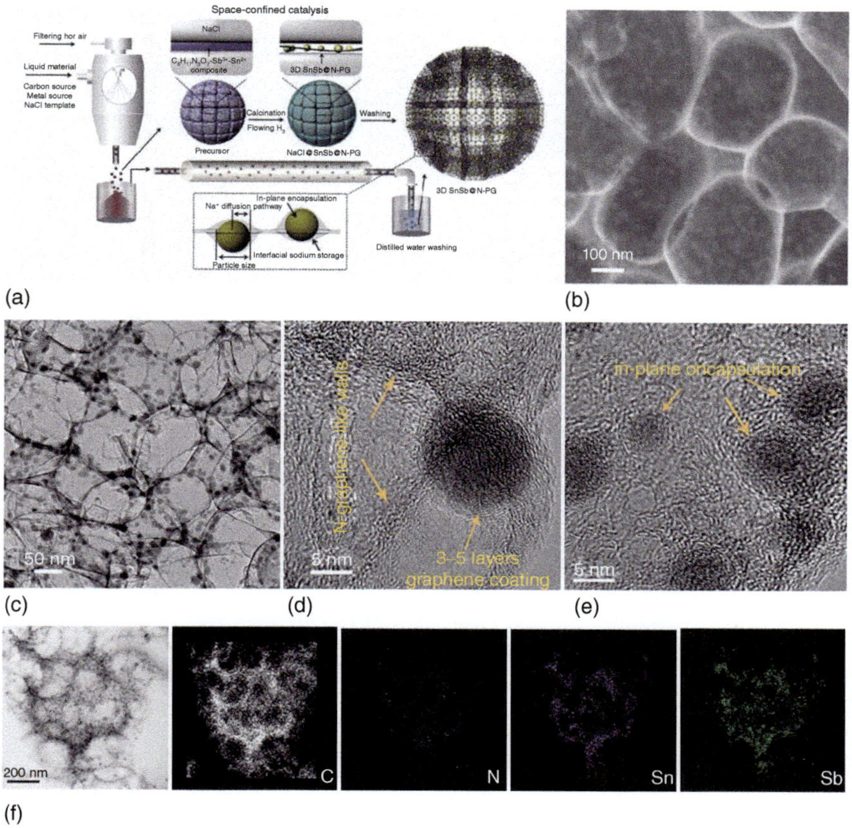

Figure 4.16 (a) Schematic illustration of the top-down in situ synthesis of 3D SnSb@N-PG composite (3D SnSb@N-PG). (b) SEM image, (c–e) TEM images, (f) elemental mapping of the 3D SnSb@N-PG. Source: Qin et al. [28]. Reproduced with permission of Wiley-VCH.

of sulfur nanoparticles into 3D well-ordered porous graphitic carbon (PGC) (designated 3D S@PGC) has been readily realized [20]. As shown in Figure 4.17, this process utilized the self-assembly of water-soluble NaCl and Na_2S crystals as a template for the 3D PGC network and Na_2S (it can react with $Fe(NO_3)_3$ to form sulfur) as the sulfur precursor. Microscopic and spectroscopic techniques confirmed that the in situ formed sulfur nanoparticles were uniformly entrapped in the network walls and covalently bonded to 3D PGC (Figure 4.17b–d). Moreover, the sulfur content in the 3D PGC could be easily adjusted by varying the ratio of Na_2S to glucose. In comparison with the carbon–sulfur composites prepared by conventional methods, these 3D S@PGC composites are endowed with the advantages of carbon–sulfur covalent bonds, a high sulfur content (up to 90 wt%), and nanosized sulfur particles, which highly facilitate the efficient utilization of the loaded sulfur for energy storage (such as a cathode material for Li—S batteries). In another case, utilizing the reaction of $MgCl_2$ and NaOH (it can form $Mg[OH]_2$ and NaCl templates), Wei's group reported a novel heteroatom-doped and

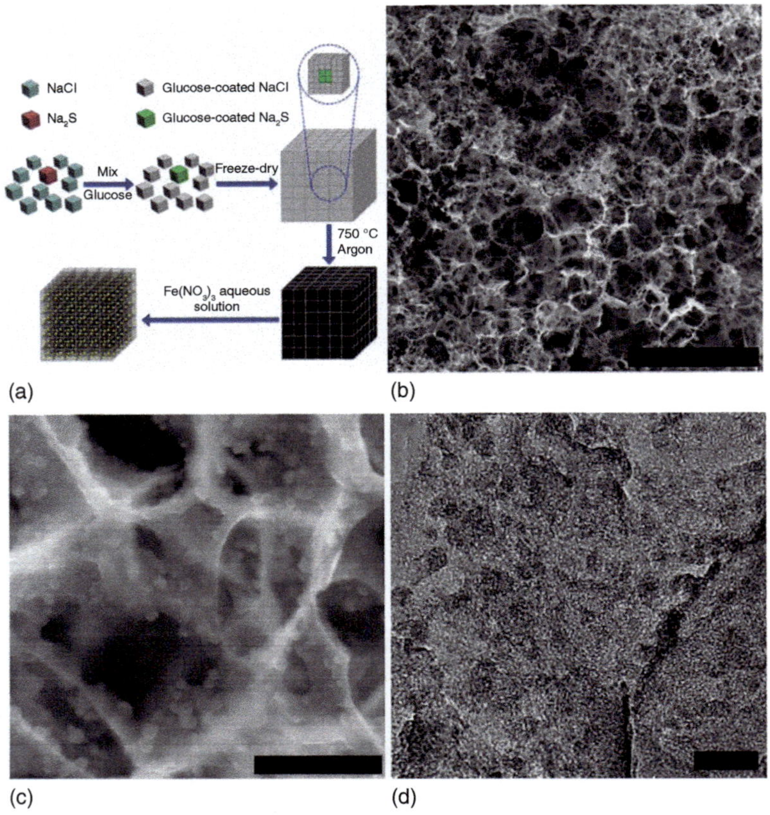

Figure 4.17 (a) Schematic illustration of an in situ strategy for the preparation of 3D S@PGC composites. (b, c) SEM images and (d) TEM image of 3D S@PGC (90% S) composites. Source: Liu et al. [22]. Reproduced with permission of Springer.

edge-rich graphene material (nitrogen-doped graphene mesh) [82], which shows a high specific surface area of 1100 m^2 g^{-1}. It is found the in situ fabricated MgO (calcined product of $Mg[OH]_2$) catalytic templates are critical for effective graphitization of solid carbon/nitrogen sources and the generation of abundant in-plane holes, and the resulting NaCl products can also serve as effective pore-forming agents.

Such 3D salt-templating assembly methods can also be applied to direct synthesis of freestanding graphene-based thin-film materials on current collectors, which can be directly used as the electrodes without additional polymer binders [83–85]. For instance, Wang and coworkers [83] fabricated transparent graphene paper by microwave plasma-enhanced CVD, in which Si wafer served as a substrate, NaCl nanocrystals functioned as a template, and methane acted as a carbon source. The typical synthesizing procedure mainly includes (a) melt of NaCl, (b) formation of NaCl film and NaCl nanocrystal template, and (c) growth of the 3D prism-like graphene building blocks, as shown in Figure 4.18a. The obtained graphene paper

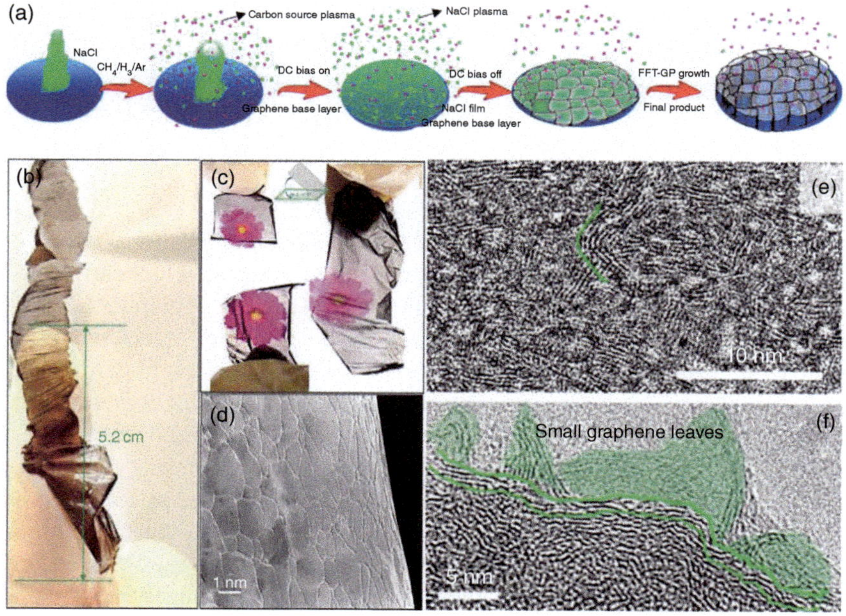

Figure 4.18 (a) Scheme for the synthesis of freestanding, flexible, transparent graphene paper (FFT-GP). (b) Photograph of a self-supporting, 1.2 cm wide, 8.2 cm long graphene sheet that has been hand drawn from a graphene film on a silicon substrate. (c) Photographs, acquired with a 60° angle between the FFT-GP and the underlying logo, illustrating the transparency of the sheet from various angles by demonstrating the visibility of the Sun Yat-Sen University logo placed behind the graphene sheet. (d) Low-resolution SEM micrograph of an FFT-GP sheet. (e) HRTEM micrographs of three- to four-layered small graphene leaves in the center of one surface of the PLG. (f) High-resolution TEM (HRTEM) micrograph of the small graphene leaves at the edge area of PLG. Source: Li et al. [83]. Reproduced with permission of American Chemical Society.

shows both transparency and flexibility in its macrostructure and preserves the advantages of 2D graphene, including ultrathin thickness (~1 nm) and large surface area (~909 $m^2\ g^{-1}$), as shown in Figure 4.18b–f. Another group developed the direct synthesis of vertically aligned 2D N-doped carbon nanosheets embedded with uniform nanosized metal carbides (e.g. $MoC_{0.654}$, WC, TaC, and NbC) (indicated with metal carbide@CNs) on nickel foam by 3D NaCl-templating assembly strategy [86]. It was disclosed that the resulting vertically aligned metal carbide@CNs nanostructures had ultrathin walls, small particle size, porous structure, strong adhesion with the substrate, fine electrical conductivity, and outstanding structural robustness.

In contrast to other template methods reported in literature, this simple and reproducible 3D salt-templating assembly technique can easily control the construction of high-quality graphene into 3D well-ordered network architecture with high conductivity, large surface area, and porosity, yet can achieve an even distribution of active components and robust interfacial bonding independent of their sizes, dimensions, geometries, and compositions, which are very favorable for their applications in catalysis, energy storage and conversion, etc. Moreover, this 3D salt-templating assembly method is also adaptable for the controllable construction of numerous 3D porous inorganic nanostructures for various applications [87].

4.3 Outlook

We have outlined the recent progress of the salt-template methods for producing tailor-made 2D graphene, 3D porous graphene, and their composites for energy conversion and storage. The key to precisely controlling the structure of graphene-based materials by salt-template strategy is to choose proper carbon precursors, template salts, and even metal (or metal oxide, etc.) precursors based on the fully miscible principle, as well as rationally regulate the preparation process (such as mixing and calcination conditions). In the light of the remarkable advances highlighted above, the salt-template approach has been revealed to show tunability, versatility, low cost, and great potential in the large-scale fabrication of graphene-based materials with large surface area and porosity, high electrical conductivity, and robust structural stability for efficient energy storage (e.g. batteries, supercapacitors) and conversion (e.g. fuel cells) applications. However, there are still several challenges that remain to be addressed in this burgeoning area.

First, although many effective salt-template methods have been intensively exploited for fabricating porous graphene-based materials, precise control over their pore morphology, including size and wall thickness, and even periodic ordered pores and hierarchical pores, is still a challenging task. Second, more future efforts are required for illustrating the synergetic effect and template mechanism of multi-scale salt templates during the synthesis process, which can facilitate better control over hierarchical pore structure and surface area of graphene-based materials for energy applications. Third, an important future direction for the salt-templated graphene-based materials consists in developing selection methodology of carbon precursors, template salts, and even metal (or metal oxide, etc.) precursors for linking design, materials, and processing, as well as establishing a general theoretical model for explaining the synthesis mechanism of the resultant products, which are crucial to the controllable production of graphene-based composites with expected morphologies, dimensions, and performance. Fourth, the research into understanding the relationships between salt-templated graphene-based materials and enhanced performance is still in its infancy, and great efforts are urgently required in these further fundamental studies. Fifth, despite great advances gained in the salt-template-directed fabrication of graphene-based materials, their industrial manufacture has not been reported as yet. It is highly desirable that engineers design and develop equipment with compatible scale and function characteristics for achieving continuous preparation of graphene-based materials by salt-template methods. We believe that continued attempts devoted to the salt-template-directed graphene-based materials will lead to their industrial fabrication and a wide range of fruitful applications in energy, catalysis, environment, medicine, etc.

References

1 Bonaccorso, F., Colombo, L., Yu, G. et al. (2015). 2D materials. Graphene, related two-dimensional crystals, and hybrid systems for energy conversion and storage. *Science* 347 (6217): 1246501.

2 Raccichini, R., Varzi, A., Passerini, S. et al. (2015). The role of graphene for electrochemical energy storage. *Nature Materials* 14 (3): 271–279.

3 Shearer, C.J., Cherevan, A., and Eder, D. (2014). Application and future challenges of functional nanocarbon hybrids. *Advanced Materials* 26 (15): 2295–2318.

4 Xiang, Z., Dai, Q., Chen, J.F. et al. (2016). Edge functionalization of graphene and two-dimensional covalent organic polymers for energy conversion and storage. *Advanced Materials* 28 (29): 6253–6261.

5 Nishihara, H. and Kyotani, T. (2012). Templated nanocarbons for energy storage. *Advanced Materials* 24 (33): 4473–4498.

6 Zhao, H., Zhou, M., Wen, L. et al. (2015). Template-directed construction of nanostructure arrays for highly-efficient energy storage and conversion. *Nano Energy* 13: 790–813.

7 Petkovich, N.D. and Stein, A. (2013). Controlling macro- and mesostructures with hierarchical porosity through combined hard and soft templating. *Chemical Society Reviews* 42 (9): 3721–3739.

8 He, C., Zhao, N., Shi, C. et al. (2015). Fabrication of nanocarbon composites using in situ chemical vapor deposition and their applications. *Advanced Materials* 27 (36): 5422–5431.

9 Fechler, N., Fellinger, T.P., and Antonietti, M. (2013). "Salt templating": a simple and sustainable pathway toward highly porous functional carbons from ionic liquids. *Advanced Materials* 25 (1): 75–79.

10 Fellinger, T.P., Thomas, A., Yuan, J. et al. (2013). 25th anniversary article: "cooking carbon with salt": carbon materials and carbonaceous frameworks from ionic liquids and poly(ionic liquid)s. *Advanced Materials* 25 (41): 5838–5854.

11 Jang, B., Park, M., Chae, O.B. et al. (2012). Direct synthesis of self-assembled ferrite/carbon hybrid nanosheets for high performance lithium-ion battery anodes. *Journal of American Chemical Society* 134 (36): 15010–15015.

12 He, C., Wu, S., Zhao, N. et al. (2013). Carbon-encapsulated Fe_3O_4 nanoparticles as a high-rate lithium ion battery anode material. *ACS Nano* 7 (5): 4459–4469.

13 Qin, J., He, C., Zhao, N. et al. (2014). Graphene networks anchored with Sn@graphene as lithium ion battery anode. *ACS Nano* 8 (2): 1728–1738.

14 Zhou, J., Qin, J., Zhang, X. et al. (2015). 2D space-confined synthesis of few-layer MoS_2 anchored on carbon nanosheet for lithium-ion battery anode. *ACS Nano* 9 (4): 3837–3848.

15 Shi, L., Chen, K., Du, R. et al. (2015). Direct synthesis of few-layer graphene on NaCl crystals. *Small* 11 (47): 6302–6308.

16 Zhu, S., Li, J., He, C. et al. (2015). Soluble salt self-assembly-assisted synthesis of three-dimensional hierarchical porous carbon networks for supercapacitors. *Journal of Materials Chemistry A* 3 (44): 22266–22273.

17 Li, S., Cheng, C., Liang, H.W. et al. (2017). 2D porous carbons prepared from layered organic-inorganic hybrids and their use as oxygen-reduction electrocatalysts. *Advanced Materials* 29 (28): 1700707.

18 Fan, X., Yu, C., Yang, J. et al. (2015). A layered-nanospace-confinement strategy for the synthesis of two-dimensional porous carbon nanosheets for high-rate performance supercapacitors. *Advanced Energy Materials* 5: 1401761.

19 Hu, W., Xu, D., Sun, X.N. et al. (2017). Template synthesis of nitrogen-doped carbon nanosheets for high-performance supercapacitors improved by redox additives. *ACS Sustainable Chemistry & Engineering* 5 (10): 8630–8640.

20 Li, G., Sun, J., Hou, W. et al. (2016). Three-dimensional porous carbon composites containing high sulfur nanoparticle content for high-performance lithium-sulfur batteries. *Nature Communications* 7: 10601.

21 Wang, L., Zhu, Q., Zhao, J. et al. (2019). Nitrogen-doped hierarchical porous carbon for supercapacitors with high rate performance. *Microporous and Mesoporous Materials* 279: 439–445.

22 Liu, X., Fechler, N., and Antonietti, M. (2013). Salt melt synthesis of ceramics, semiconductors and carbon nanostructures. *Chemical Society Reviews* 42 (21): 8237–8265.

23 Madaro, F., Sæterli, R., Tolchard, J.R. et al. (2011). Molten salt synthesis of $K_4Nb_6O_{17}$, $K_2Nb_4O_{11}$ and KNb_3O_8 crystals with needle- or plate-like morphology. *CrystEngComm* 13 (5): 1304–1313.

24 Liu, X., Giordano, C., and Antonietti, M. (2014). A facile molten-salt route to graphene synthesis. *Small* 10 (1): 193–200.

25 Liu, X. and Antonietti, M. (2013). Moderating black powder chemistry for the synthesis of doped and highly porous graphene nanoplatelets and their use in electrocatalysis. *Advanced Materials* 25 (43): 6284–6290.

26 Wang, J., Ding, B., Hao, X. et al. (2016). A modified molten-salt method to prepare graphene electrode with high capacitance and low self-discharge rate. *Carbon* 102: 255–261.

27 Zhang, Y., Zai, J., He, K. et al. (2018). Fe_3C nanoparticles encapsulated in highly crystalline porous graphite: salt-template synthesis and enhanced electrocatalytic oxygen evolution activity and stability. *Chemical Communications* 54 (25): 3158–3161.

28 Qin, J., Wang, T., Liu, D. et al. (2018). A top-down strategy toward SnSb in-plane nanoconfined 3D N-doped porous graphene composite microspheres for high performance Na-ion battery anode. *Advanced Materials* 30 (9): 1704670.

29 He, C.N. and Zhao, N.Q. (2012). One-step solid-phase synthesis of ultrasmall homogeneous face-centered tetragonal FePt nanoparticles encapsulated in thin carbon shells. *Journal of Materials Chemistry* 22 (4): 1297–1304.

30 Qin, J., Zhang, X., Zhao, N. et al. (2014). Carbon-coated Ni_3Sn_2 nanoparticles embedded in porous carbon nanosheets as a lithium ion battery anode with outstanding cycling stability. *RSC Advances* 4 (90): 49247–49256.

31 Wu, G., Liang, X., Zhang, H. et al. (2016). Highly stable and sub-3 nm Ni nanoparticles coated with carbon nanosheets as a highly active heterogeneous hydrogenation catalyst. *Catalysis Communications* 79: 63–67.

32 Zhang, J., Zhu, H., Wu, P. et al. (2015). Rational synthesis of Ni nanoparticle-embedded porous graphitic carbon nanosheets with enhanced lithium storage properties. *Nanoscale* 7 (43): 18211–18217.
33 Wu, S., Wang, Z., He, C. et al. (2013). Synthesis of uniform and superparamagnetic Fe_3O_4 nanocrystals embedded in a porous carbon matrix for a superior lithium ion battery anode. *Journal of Materials Chemistry A* 1 (36): 11011–11018.
34 Huang, Y.-G., Pan, Q.-C., Wang, H.-Q. et al. (2016). Sn/SnO embedded in carbon nanosheets as high-performance anode material for lithium ion battery. *Ceramics International* 42 (3): 4586–4593.
35 Wen, F., Hou, H., Xiang, J. et al. (2015). Fabrication of carbon encapsulated Co_3O_4 nanoparticles embedded in porous graphitic carbon nanosheets for microwave absorber. *Carbon* 89: 372–377.
36 Leng, X., Wei, S., Jiang, Z. et al. (2015). Carbon-encapsulated Co_3O_4 nanoparticles as anode materials with super lithium storage performance. *Scientific Reports* 5: 16629.
37 Pan, Q.-C., Huang, Y.-G., Wang, H.-Q. et al. (2016). MoS2/C nanosheets encapsulated Sn@SnOx nanoparticles as high-performance lithium-ion battery anode material. *Electrochimica Acta* 197: 50–57.
38 Huang, Y., Pan, Q., Wang, H. et al. (2016). Preparation of a Sn@SnO2@C@MoS2 composite as a high-performance anode material for lithium-ion batteries. *Journal of Materials Chemistry A* 4 (19): 7185–7189.
39 Chen, L., Wang, Z., He, C. et al. (2013). Porous graphitic carbon nanosheets as a high-rate anode material for lithium-ion batteries. *ACS Applied Materials & Interfaces* 5 (19): 9537–9545.
40 Li, W., Tang, Y., Kang, W. et al. (2015). Core-shell Si/C nanospheres embedded in bubble sheet-like carbon film with enhanced performance as lithium ion battery anodes. *Small* 11 (11): 1345–1351.
41 Fu, G., Cui, Z., Chen, Y. et al. (2017). Ni3Fe-N doped carbon sheets as a bifunctional electrocatalyst for air cathodes. *Advanced Energy Materials* 7 (1): 1601172.
42 Li, S., Wang, Z., Liu, J. et al. (2016). Yolk-Shell Sn@C eggette-like nanostructure: application in lithium-ion and sodium-ion batteries. *ACS Applied Materials & Interfaces* 8 (30): 19438–19445.
43 Wang, H., Pan, Q., Wu, Q. et al. (2017). Ultrasmall MoS_2 embedded in carbon nanosheets-coated Sn/SnOx as anode material for high-rate and long life Li-ion batteries. *Journal of Materials Chemistry A* 5 (9): 4576–4582.
44 Xu, D., Mu, C., Xiang, J. et al. (2016). Carbon-encapsulated Co_3O_4@CoO@Co nanocomposites for multifunctional applications in enhanced long-life lithium storage, supercapacitor and oxygen evolution reaction. *Electrochimica Acta* 220: 322–330.
45 Al-Mamun, M., Yin, H., Liu, P. et al. (2017). Carbon-encapsulated heazlewoodite nanoparticles as highly efficient and durable electrocatalysts for oxygen evolution reactions. *Nano Research* 10 (10): 3522–3533.
46 Chung, D.Y., Kim, M.J., Kang, N. et al. (2017). Low-temperature and gram-scale synthesis of two-dimensional Fe–N–C carbon sheets for robust electrochemical oxygen reduction reaction. *Chemistry of Materials* 29 (7): 2890–2898.

47 Pan, Q., Zheng, F., Ou, X. et al. (2017). MoS2 decorated $Fe_3O_4/Fe_{1-x}S$@C nanosheets as high-performance anode materials for lithium ion and sodium ion batteries. *ACS Sustainable Chemistry & Engineering* 5 (6): 4739–4745.

48 Wu, C., Zhang, Y., Dong, D. et al. (2017). Co_9S_8 nanoparticles anchored on nitrogen and sulfur dual-doped carbon nanosheets as highly efficient bifunctional electrocatalyst for oxygen evolution and reduction reactions. *Nanoscale* 9 (34): 12432–12440.

49 Ren, D., Hu, Y., Jiang, H. et al. (2016). Salt-templating protocol to realize few-layered ultrasmall MoS_2 nanosheets inlayed into carbon frameworks for superior lithium-ion batteries. *ACS Sustainable Chemistry & Engineering* 4 (3): 1148–1153.

50 Li, W., Wang, K., Cheng, S. et al. (2017). A two-dimensional hybrid of SbO_x nanoplates encapsulated by carbon flakes as a high performance sodium storage anode. *Journal of Materials Chemistry A* 5 (3): 1160–1167.

51 Hu, C. and Dai, L. (2017). Multifunctional carbon-based metal-free electrocatalysts for simultaneous oxygen reduction, oxygen evolution, and hydrogen evolution. *Advanced Materials* 29 (9): 1604942.

52 Zheng, X., Cao, X., Li, X. et al. (2017). Biomass lysine-derived nitrogen-doped carbon hollow cubes via a NaCl crystal template: an efficient bifunctional electrocatalyst for oxygen reduction and evolution reactions. *Nanoscale* 9 (3): 1059–1067.

53 Zhu, S., Li, J., Ma, L. et al. (2016). Three-dimensional network of N-doped carbon ultrathin nanosheets with closely packed mesopores: controllable synthesis and application in electrochemical energy storage. *ACS Applied Materials & Interfaces* 8 (18): 11720–11728.

54 Qin, J., Zhang, X., Shi, C. et al. (2015). In situ preparation of interconnected networks constructed by flexible graphene/Sn sandwich nanosheets for high-performance lithium-ion battery anode. *Journal of Materials Chemistry A* 3 (46): 23170–23179.

55 Qin, J., Liu, D., Zhang, X. et al. (2017). One-step synthesis of SnCo nanoconfined in hierarchical carbon nanostructures for lithium ion battery anode. *Nanoscale* 9 (41): 15856.

56 Qin, J., Zhao, N., Shi, C. et al. (2017). Sandwiched C@SnO2@C hollow nanostructures as an ultralong-lifespan high-rate anode material for lithium-ion and sodium-ion batteries. *Journal of Materials Chemistry A* 5 (22): 10946–10956.

57 Chen, B., Qin, H., Li, K. et al. (2019). Yolk-shelled Sb@C nanoconfined nitrogen/sulfur co-doped 3D porous carbon microspheres for sodium-ion battery anode with ultralong high-rate cycling. *Nano Energy* 66: 104133.

58 Zhang, X., Shi, C., Liu, E. et al. (2015). Nitrogen-doped graphene network supported copper nanoparticles encapsulated with graphene shells for surface-enhanced Raman scattering. *Nanoscale* 7 (40): 17079–17087.

59 Liu, X. and Antonietti, M. (2014). Molten salt activation for synthesis of porous carbon nanostructures and carbon sheets. *Carbon* 69: 460–466.

60 He, B., Li, W.-C., and Lu, A.-H. (2015). High nitrogen-content carbon nanosheets formed using the Schiff-base reaction in a molten salt medium as efficient anode materials for lithium-ion batteries. *Journal of Materials Chemistry A* 3 (2): 579–585.

61 Graglia, M., Pampel, J., Hantke, T. et al. (2016). Nitro lignin-derived nitrogen-doped carbon as an efficient and sustainable electrocatalyst for oxygen reduction. *ACS Nano* 10 (4): 4364–4371.

62 Pampel, J. and Fellinger, T.-P. (2016). Opening of bottleneck pores for the improvement of nitrogen doped carbon electrocatalysts. *Advanced Energy Materials* 6 (8): 1502389.

63 Chang, Y., Antonietti, M., and Fellinger, T.P. (2015). Synthesis of nanostructured carbon through ionothermal carbonization of common organic solvents and solutions. *Angewandte Chemie International Edition* 54 (18): 5507–5512.

64 Qian, W., Zhu, J., Zhang, Y. et al. (2015). Condiment-derived 3D architecture porous carbon for electrochemical supercapacitors. *Small* 11 (37): 4959–4969.

65 Fechler, N., Fellinger, T.-P., and Antonietti, M. (2013). One-pot synthesis of nitrogen–sulfur-co-doped carbons with tunable composition using a simple isothiocyanate ionic liquid. *Journal of Materials Chemistry A* 1 (45): 14097–14102.

66 Schipper, F., Vizintin, A., Ren, J. et al. (2015). Biomass-derived heteroatom-doped carbon aerogels from a salt melt sol-gel synthesis and their performance in Li-S batteries. *ChemSusChem* 8 (18): 3077–3083.

67 Elumeeva, K., Fechler, N., Fellinger, T.P. et al. (2014). Metal-free ionic liquid-derived electrocatalyst for high-performance oxygen reduction in acidic and alkaline electrolytes. *Materials Horizons* 1 (6): 588–594.

68 Porada, S., Schipper, F., Aslan, M. et al. (2015). Capacitive deionization using biomass-based microporous salt-templated heteroatom-doped carbons. *ChemSusChem* 8 (11): 1867–1874.

69 Deng, X., Zhao, B., Zhu, L. et al. (2015). Molten salt synthesis of nitrogen-doped carbon with hierarchical pore structures for use as high-performance electrodes in supercapacitors. *Carbon* 93: 48–58.

70 Sakaushi, K., Yang, S.J., Fellinger, T.P. et al. (2015). Impact of large-scale meso- and macropore structures in adenosine-derived affordable noble carbon on efficient reversible oxygen electrocatalytic redox reactions. *Journal of Materials Chemistry A* 3 (22): 11720–11724.

71 Yin, H., Lu, B., Xu, Y. et al. (2014). Harvesting capacitive carbon by carbonization of waste biomass in molten salts. *Environmental Science & Technology* 48 (14): 8101–8108.

72 Li, X., Liang, J., Hou, Z. et al. (2015). A new salt-baked approach for confining selenium in metal complex-derived porous carbon with superior lithium storage properties. *Advanced Functional Materials* 25 (32): 5229–5238.

73 Zhang, C., Antonietti, M., and Fellinger, T.-P. (2014). Blood ties: Co_3O_4 decorated blood derived carbon as a superior bifunctional electrocatalyst. *Advanced Functional Materials* 24 (48): 7655–7665.

74 Elumeeva, K., Ren, J., Antonietti, M. et al. (2015). High surface iron/cobalt-containing nitrogen-doped carbon aerogels as non-precious advanced electrocatalysts for oxygen reduction. *ChemElectroChem* 2 (4): 584–591.

75 Fechler, N., Tiruye, G.A., Marcilla, R. et al. (2014). Vanadium nitride@N-doped carbon nanocomposites: tuning of pore structure and particle size through salt templating and its influence on supercapacitance in ionic liquid media. *RSC Advances* 4 (51): 26981–26989.

76 Ding, W., Li, L., Xiong, K. et al. (2015). Shape fixing via salt recrystallization: a morphology-controlled approach to convert nanostructured polymer to carbon nanomaterial as a highly active catalyst for oxygen reduction reaction. *Journal of American Chemical Society* 137 (16): 5414–5420.

77 Zhu, S., Xu, K., Sui, S. et al. (2017). Synthesis of 2D/3D carbon hybrids by heterogeneous space-confined effect for electrochemical energy storage. *Journal of Materials Chemistry A* 5 (36): 19175–19183.

78 Lu, P., Sun, Y., Xiang, H. et al. (2018). 3D amorphous carbon with controlled porous and disordered structures as a high-rate anode material for sodium-ion batteries. *Advanced Energy Materials* 8 (8): 1702434.

79 Zhao, X., Wang, H.-E., Massé, R.C. et al. (2017). Design of coherent anode materials with 0D Ni_3S_2 nanoparticles self-assembled on 3D interconnected carbon networks for fast and reversible sodium storage. *Journal of Materials Chemistry A* 5 (16): 7394–7402.

80 Xu, Y., Li, W., Zhang, F. et al. (2016). In situ incorporation of FeS nanoparticles/carbon nanosheets composite with an interconnected porous structure as a high-performance anode for lithium ion batteries. *Journal of Materials Chemistry A* 4 (10): 3697–3703.

81 Luo, W., Zhang, P., Wang, X. et al. (2016). Antimony nanoparticles anchored in three-dimensional carbon network as promising sodium-ion battery anode. *Journal of Power Sources* 304: 340–345.

82 Tang, C., Wang, H.F., Chen, X. et al. (2016). Topological defects in metal-free nanocarbon for oxygen electrocatalysis. *Advanced Materials* 28 (32): 6845–6851.

83 Li, N., Yang, G., Sun, Y. et al. (2015). Free-standing and transparent graphene membrane of polyhedron box-shaped basic building units directly grown using a NaCl template for flexible transparent and stretchable solid-state supercapacitors. *Nano Letters* 15 (5): 3195–3203.

84 Li, N., Huang, X., Zhang, H. et al. (2017). NaCl multistage-recrystallization-induced formation of 3D micro-structured ribbon-like graphene based films for high performance flexible/transparent supercapacitors. *Journal of Materials Chemistry A* 5 (28): 14595–14603.

85 Li, N., Huang, X., Zhang, H. et al. (2017). Graphene-hollow-cubes with network-faces assembled a 3D micro-structured transparent and free-standing film for high performance supercapacitors. *Journal of Materials Chemistry A* 5 (32): 16803–16811.

86 Zhu, J., Sakaushi, K., Clavel, G. et al. (2015). A general salt-templating method to fabricate vertically aligned graphitic carbon nanosheets and their metal carbide hybrids for superior lithium ion batteries and water splitting. *Journal of American Chemical Society* 137 (16): 5480–5485.

87 Zhou, J., Qin, J., Zhao, N. et al. (2016). Salt-template-assisted synthesis of robust 3D honeycomb-like structured MoS_2 and its application as a lithium-ion battery anode. *Journal of Materials Chemistry A* 4 (22): 8734–8741.

5

Powder Metallurgy Templates Methods

Junwei Sha[1], Xiaoyu Chu[1], Yuxuan Wang[1], Meixian Li[1], Chunnian He[1,2,3], and Naiqin Zhao[1,2,3]

[1] *Tianjin University, School of Materials Science and Engineering and Tianjin Key Laboratory of Composites and Functional Materials, 135 Yaguan Rd., Jinnan District, Tianjin, 300350, P.R. China*
[2] *Tianjin University, Collaborative Innovation Center of Chemical Science and Engineering, 135 Yaguan Rd., Jinnan District, Tianjin, 300350, P.R. China*
[3] *Tianjin University, Ministry of Education, Key Laboratory of Advanced Ceramics and Machining Technology, 135 Yaguan Rd., Jinnan District, Tianjin, 300350, P.R. China*

As described in the previous chapters, graphene, as a star material which can be used in many fields, can be synthesized through various methods such as mechanical exfoliation [1], arc-discharge [2, 3], epitaxial growth [4], chemical vapor deposition (CVD) [5, 6], and chemical redox [7–10]. Among these methods, CVD method is a simple, effective, and low-cost method for high-quality mono-/multi-layered graphene growth. In CVD method, graphene can be prepared by using Ni or Cu foil as the substrates and hydrocarbon gases (CH_4, C_2H_2, etc.) [11, 12] or carbonaceous organics (glucose, sucrose, etc.) [13] as the carbon sources. However, in some fields, for example, energy-storage- and conversion-related applications, two-dimensional (2D) graphene on Cu or Ni foil often cannot meet the mass or volume requirements [14]. Individual 2D graphene nanosheets, which tend to agglomerate together due to the Van der Waals forces between different sheets, need to be integrated to three-dimensional (3D) macroscopic structure to form 3D graphene monolith (3D GM), which is also called 3D graphene foam (3D GF).

For the synthesis of 3D GM, several template-assisted preparation methods, such as nanoporous metal templates and soluble salt templates, had been introduced previously. Thus, in this chapter, we will introduce a new kind of templates-assisted method, i.e. powder metallurgy templates (PMTs) method.

5.1 Powder Metallurgy

Powder metallurgy is a traditional modern technology for the fabrication of monolith materials from powders, which was also commented "as old as the pyramids and

Templated Fabrication of Graphene-Based Materials for Energy Applications, First Edition.
Edited by Chunnian He, Naiqin Zhao, and Junwei Sha.

as new as a present-day bomber" in S.A. Tsukerman's book in 1965 [15]. Up to now, powder metallurgy method has been widely used for fabricating metal, ceramic, and composite components in modern industries, including automobile, aerospace, and many other civil and military industries.

For the essence [15] of powder metallurgy method, articles will be fabricated through powders with a specific composition. A mixture of specially selected and prepared powders will be prepared firstly through a mechanical or chemical mixing approach. After that, the hybrid powders will be transferred and compressed in a die with a specific size under a high pressure (i.e. 10–100 kg mm^{-2}) to form a monolith. The half-finished monolith will have a strength, which is still not enough for being used, allowing to be transferred for next treatment. Then, the as-obtained monolith will be treated at a high temperature for a certain time to allow the particles to be sintered. The heat-treatment temperature for sintering the half-finished monolith is often below the melting point of the basic component, i.e. usually 66–75% of the melting point. Finally, the sintered monolith will be further treated by other machining and deformation methods, such as hot extrusion, aging, and cutting to meet the final-use demands.

Nowadays, with the development of modern technologies, various new approaches to the powder metallurgy fabrication of monolith materials have been developed, including hot pressing [16], spark plasma sintering (SPS) [17], and hot isostatic pressing (HIP) [18]. Powder metallurgy, which possesses advantages such as designable components, ease of operation, and ease of scale-up, is attracting intense attention and interest of researchers, especially from functional-materials-related fields.

5.2 Powder Metallurgy Templates Methods

PMT method, developed by Sha et al. in 2016 [19], is a simple and scalable method for 3D GM growth, which combines the traditional powder metallurgy technology and CVD. Thus, by using PMT method, the as-obtained 3D GM can possess the advantages of simple synthesis, low cost, free-standing, designable, and high efficiency from PM technology and high quality, mono-/multi-layer, and controllable from CVD method.

5.2.1 Basic Synthesis Procedures of PMT Method

The basic synthesis procedure of PMT method is as follows (see Figure 5.1): the selected metal powders (e.g. Ni powder) and carbonaceous organics (e.g. sucrose) are mixed in a certain solvent in which the carbonaceous organics can be dissolved (for example, DI water). Under mechanical stirring, the mixture is heated by using a hot plate at 80–100 °C to evaporate the DI water. During this process, sucrose will precipitate and coat the surfaces of Ni particles to form a hybrid powder. After evaporating water completely, the hybrid powders will be dried further at 80 °C in a vacuum oven overnight and ground using a mortar and pestle. Then, the Ni/sucrose hybrid

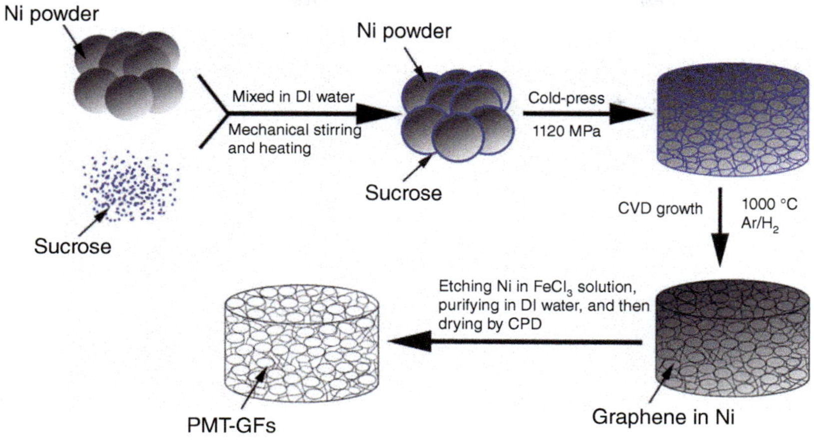

Figure 5.1 Schematics of powder metallurgy templates method with Ni powder as the templates and sucrose as the carbon sources. Source: Sha et al. [19]. Reproduced with permission of American Chemical Society.

powders will be transferred into a die to be cold pressed into a pellet under a high pressure. After that, the Ni/sucrose pellets are loaded into a quartz tube furnace to grow graphene with an atmosphere of Ar/H_2 at a low chamber pressure. The temperature of the furnace will be increased from room temperature to 1000 °C at a heating rate of $10\,°C\,min^{-1}$, and the pellets will be annealed further at 1000 °C for another 30 minutes. During heating, the Ni powders will be sintered to form a skeleton acting as the template and catalyst, and sucrose will be converted into graphene under the catalysis of Ni. The redundant amorphous carbon generated from sucrose decomposition will be absorbed by Ni. Then, the pellets will be moved to the unheated region for rapid cooling to room temperature. Finally, the pellets will be etched in $FeCl_3$ aqueous solution to remove Ni, purified in DI water to remove Ni^{2+}, Fe^{2+}, Fe^{3+}, Cl^-, and any other impurities, and dried by using a critical point dryer (CPD) to obtain a freestanding 3D GM.

5.2.2 The Selection of Metal Templates

In CVD growth of graphene or 3D GM, Cu, Ni, or Ni–Cu alloys are the most commonly used substrates [12, 20]. Due to the different solubility of carbon, which is ~1.3 at.% of carbon in Ni and < 0.001 at.% of carbon in Cu at 1000 °C, the growth mechanism of graphene on Cu or Ni is different. Ruoff and coworkers demonstrated that the CVD growth of graphene occurs through a carbon segregation or precipitation process on Ni while through a surface adsorption process on Cu, as shown in Figure 5.2 [12]. Thus, in PMT method, different metal powder templates will result in different quality and structure of final 3D GM.

Sha et al. chose Ni powders as the templates for 3D GM PMT growth [19]. The Ni-derived 3D GM consisted of particle-like graphene shells, which were connected by 2D graphene sheets, as shown in Figure 5.3. The graphene was multi-layered

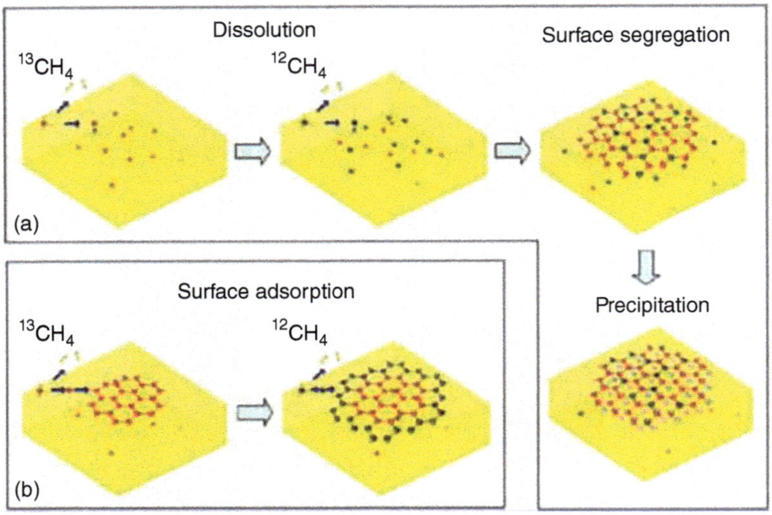

Figure 5.2 Schematics of the possible distribution of C isotopes on (a) Ni and (b) Cu. Source: Li et al. [12]. Reproduced with permission of American Chemical Society.

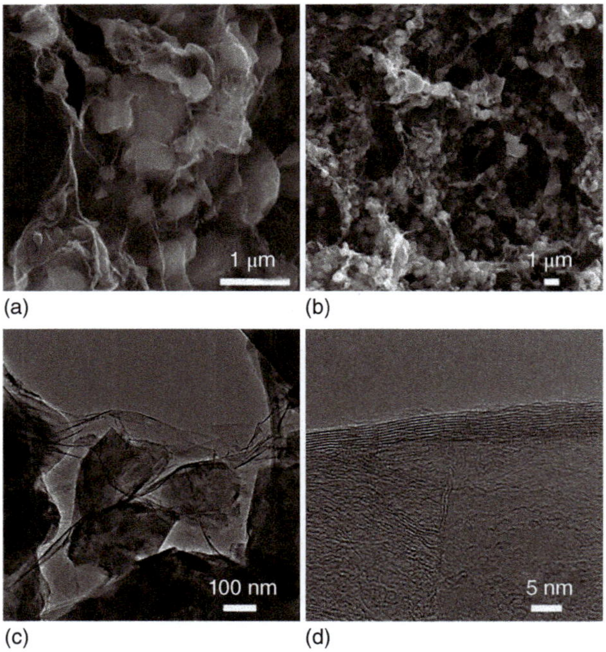

Figure 5.3 (a, b) SEM images and (c, d) TEM images of 3D GM prepared by PMT method with Ni powders as the templates. Source: Sha et al. [19]. Reproduced with permission of American Chemical Society.

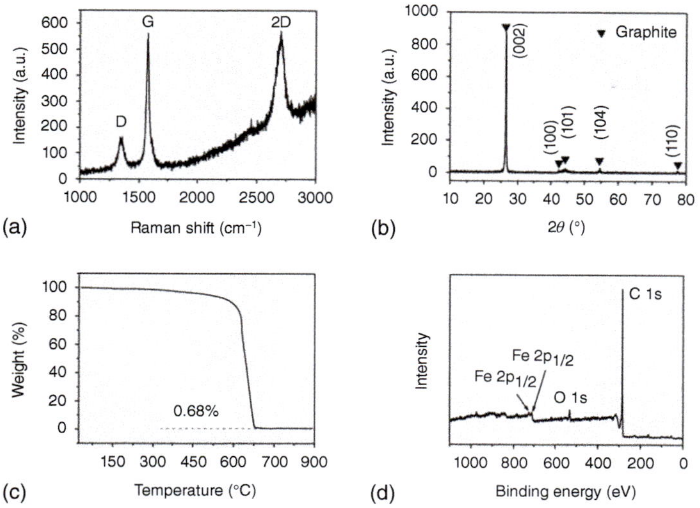

Figure 5.4 (a) Raman spectrum, (b) XRD pattern, (c) TGA curve, and (d) XPS data of 3D GM prepared by PMT method with Ni powders as the templates. Source: Sha et al. [19]. Reproduced with permission of American Chemical Society.

graphene under Ni catalysis. The size of graphene shells is similar with the starting single Ni particles. The size of the holes is also comparable with the Ni particles. Thus, the Ni particles were acting as the templates during graphene growth. The as-obtained 3D GM showed a high quality according to the Raman spectrum (see Figure 5.4), since Ni can absorb redundant amorphous carbon contamination.

It should be noted that the final 3D GM prepared by PMT method would still have some contamination such as Fe or Ni ions, which came from the starting metal powder and etching solution, as shown in the TGA and XPS results in Figure 5.4.

Chen et al. used Cu powder as the templates and PMMA as the solid carbon sources to prepare graphene-reinforced Cu matrix composites by powder metallurgy method, as shown in Figure 5.5 [21]. The as-obtained graphene showed a relatively low quality according to the Raman results compared with that by using Ni templates [19], as shown in Figure 5.6.

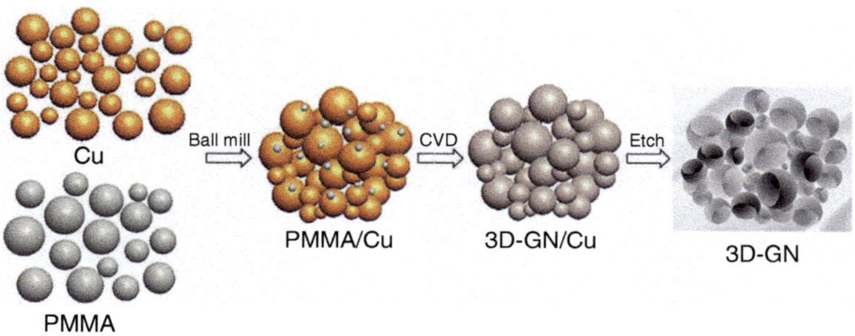

Figure 5.5 Schematic diagram of the formation process of in situ grown 3D-GN. Source: Chen et al. [21]. Reproduced with permission of Journal of Alloys and Compounds.

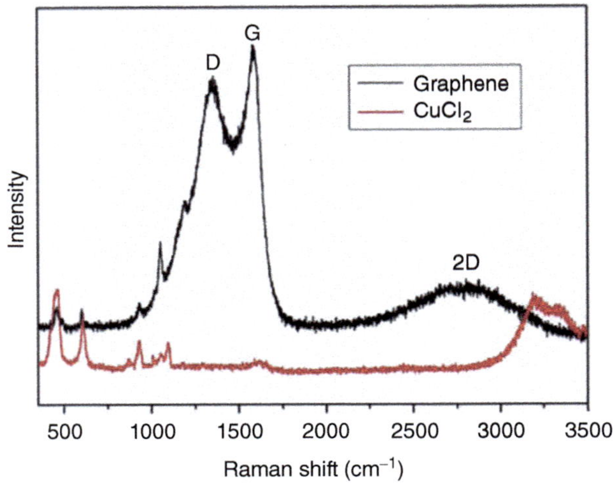

Figure 5.6 Raman spectra of as-obtained graphene by using Cu powder as the templates. Source: Chen et al. [21]. Reproduced with permission of Elsevier.

Besides metal powders like Cu or Ni powders, oxide powders also can be used as the templates in PMT method. Wang et al. reported the facile synthesis of 3D porous cellular carbon by using silica microsphere as the templates, sucrose as the solid carbon source, and $FeCl_3$ as the additive [22]. After removal of Fe nanoparticles and silica templates, a 3D porous cellular carbon framework (PCCF) was achieved, as shown in Figure 5.7.

In addition, the structure and quality of 3D GM can also be controlled by combining different kinds of templates with different functions. Sha et al. prepared three-dimensional bi-functional modular graphene network (3D bmG) by using nickel powder and sodium chloride as dual templates for graphene growth, as shown in Figure 5.8 [23]. 3D bmG-40 consists of thin graphene nanosheets grown

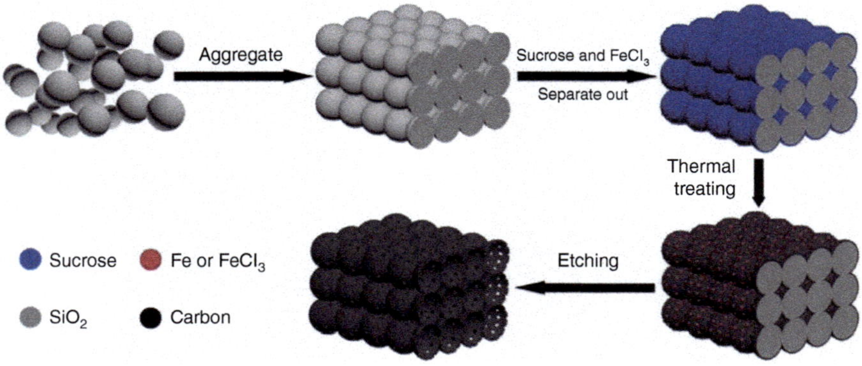

Figure 5.7 The synthesis steps for the preparation of 3D porous cellular carbon framework by using silica as the templates, sucrose as the carbon source, and $FeCl_3$ as the additive. Source: Wang et al. [22]. Reproduced with permission of Elsevier.

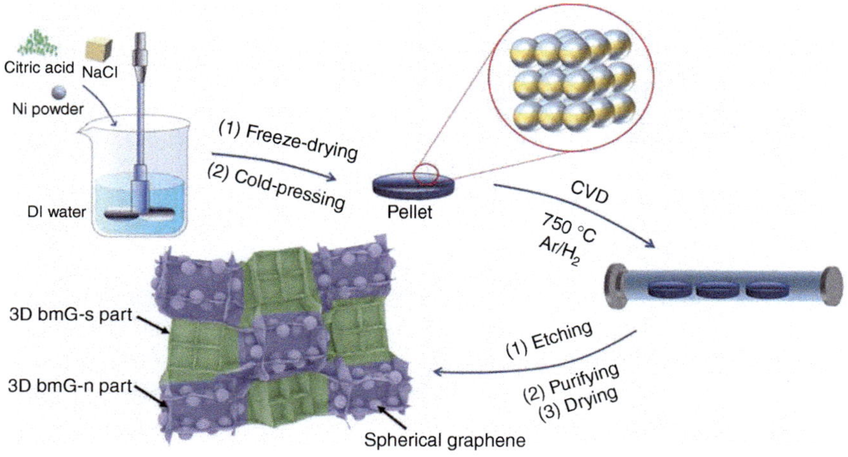

Figure 5.8 Scheme of metal-powder-assisted soluble salt templates method to prepare 3D bmG. The bmG-s part is marked in green, and the bmG-n part is marked in blue. Source: Sha et al. [23]. Reproduced with permission of Elsevier.

Figure 5.9 Raman spectrum surface scanning (I_D/I_G) of 3D bmG. Source: Sha et al. [23]. Reproduced with permission of Elsevier.

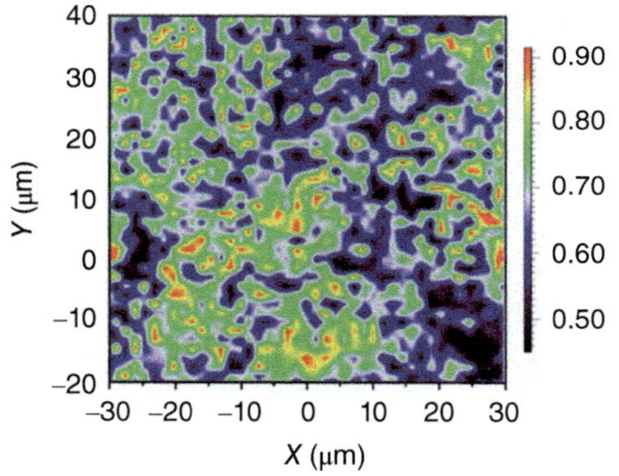

on the surface of salt template and spherical graphene catalyzed by nickel particles. Graphene with two structures presents seamless connection structure. The existence of graphene catalyzed by Ni powder can enhance the structural stability of 3D bmG, inhibit the aggregation of graphene nanosheets, and play a key role in the effective Li^+ transmission and electron transmission of 3D graphene, possibly. Figure 5.9 shows different ratios of I_D/I_G at different regions in 3D bmG. According to the color distribution, graphene with different crystallization degree catalyzed by nickel and salt crystals, respectively, can be distributed evenly, demonstrating the coexistence of two kinds of graphene with different crystallization degree.

Except for the components of the templates materials, which can affect the quality and structure of the final 3D GM products, the size of the templates particles is also a critical factor. The size of the templates can not only affect the size of the pores, but also often determine the sintering temperature. Drieschner et al. presented a facile technique to prepare high-quality GFs with tunable pore size by CVD method [24].

The templates were achieved by sintering Ni or Cu powders. By changing the size of Ni and Cu particles, as well as the corresponding sintering temperatures, the pore size and quality of the final products were controllable.

Thus, by changing the composition and size of the templates, the quality and structure of the final 3D GM are tunable, which is one of the biggest advantages of PMT method for developing new types of 3D materials.

5.2.3 The Selection of Carbon Sources

Except for the components and size of the templates, the component of carbon sources is another key factor determining the performance of the final 3D articles. In 2010, Tour group firstly successfully synthesized high-quality graphene by using solid carbon sources [13]. PMMA, table sugar, and fluorene were chosen to be the carbon sources. Since then, the use of solid carbon sources for growing high-quality carbon materials became a research hotspot.

Sha et al. used sucrose as the carbon source to prepare 3D GF by PMT method [19]. Chen et al. used PMMA as the carbon source to prepare 3D graphene [21]. Wang et al. chose melamine as the carbon source and nitrogen source and Ni particles as the templates to fabricate 3D N-doped GM, as shown in Figure 5.10 [25].

Han et al. reported the development of 3D rebar ultrastiff GF (ruGF) as the 3D conductive fillers for epoxy resin by using Ni powders as the templates, polyacrylonitrile (PAN) as the carbon and nitrogen sources, and CNTs as the reinforcement through a PMT approach [26] (Figure 5.11).

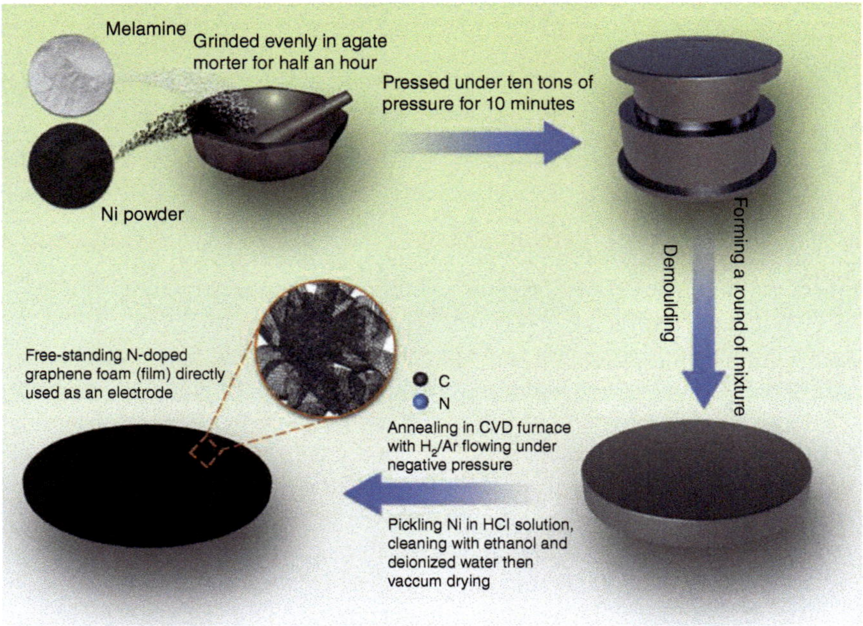

Figure 5.10 Scheme of the synthesis of 3D N-doped GM. Source: Wang et al. [25]. Reproduced with permission of Elsevier.

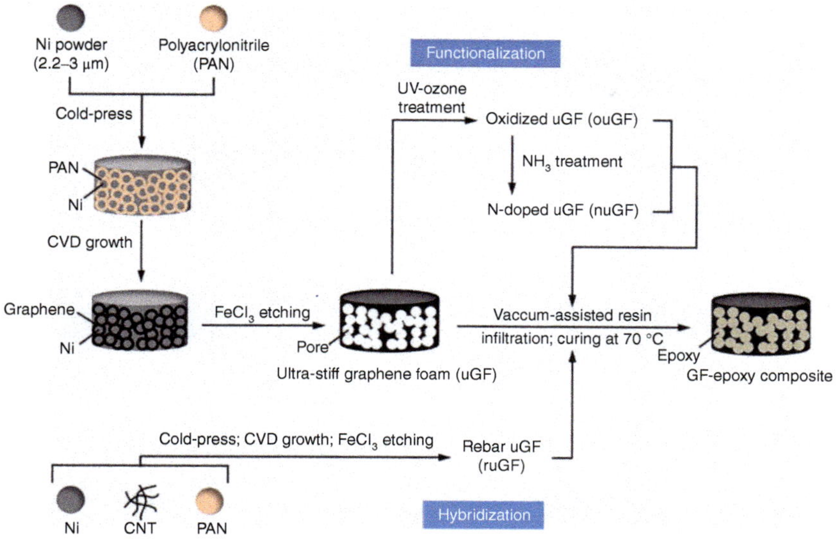

Figure 5.11 Scheme of the preparation of ruGF and ruGF–epoxy composites. Source: Han et al. [26]. Reproduced with permission of American Chemical Society.

There is a prospect that by changing the component of carbon sources, various 3D GMs with elemental doping, or other types of 3D monoliths (e.g. 3D MoS_2, 3D WC_x, and 3D BN) could be developed to meet the requirements of a certain application.

5.2.4 The Influence of Metal Templates/Carbon Sources Ratio

In PMT process, the hybrid powders precursors containing metal templates and carbon sources will be compressed in a die to form a half-finished article, and then sintered at a high temperature to allow the sintering of metal particles and conversion of carbon sources to graphene. During compression, the metal particles are interconnected together by the deformation of metal particles and adhesion of carbon sources. Thus, the ratio of metal templates/carbon sources will have influence on the quality of 3D GM, and formability of the half-articles as well.

Sha used 3 g of Ni powders (2.2–3.0 μm) as the templates, and 0.02, 0.1, 0.3, 0.5, 0.7, and 1.0 g of sucrose as the carbon sources [27]. The Ni particles and sucrose were mixed in 150 ml of DI water, and the half-articles after cold pressing were heated at 1000 °C for 30 minutes to allow the sintering of Ni and growth of graphene. When 1.0 g of sucrose was applied, the Ni powders flowed during high-temperature treatment, indicating that the amount of sucrose was too high to allow the interconnection of metal particles. When the amount of sucrose was decreased to 0.7 g, a 3D GM was obtained. However, the I_D/I_G ratio in Raman spectrum of the sample prepared by using 0.7 g of sucrose was 0.25, which were 0.08–0.1 in the cases of using 0.1–0.5 g of sucrose. It implies that large amounts of amorphous were generated during sucrose decomposition, which could not be absorbed completely by sintered Ni skeleton. When the amount of sucrose was 0.02 g, no graphene signal

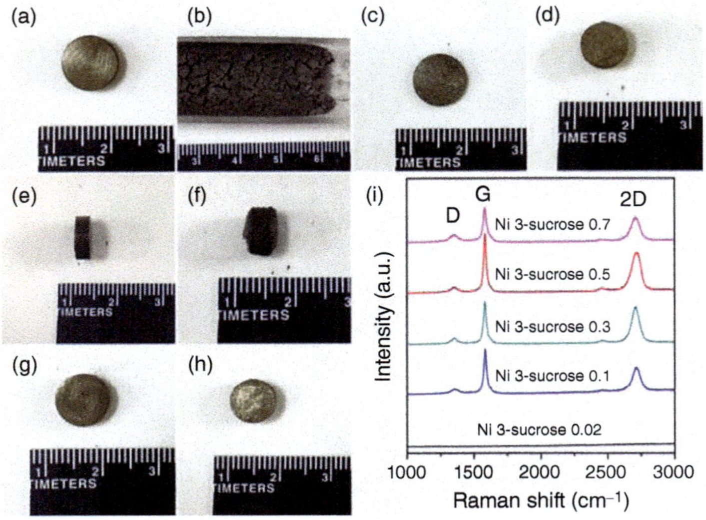

Figure 5.12 Photographs of Ni/sucrose half-articles before and after growth: (a) Ni 3 g/sucrose 1.0 g, before growth; (b) Ni 3 g/sucrose 1.0 g, after growth; (c, e) Ni 3 g/sucrose 0.7 g, before growth; (d, f) Ni 3 g/sucrose 0.7 g, after growth; (g) Ni 3 g/sucrose 0.5 g, before growth; (h) Ni 3 g/sucrose 0.5 g, after growth; (i) Raman spectra of 3D GFs prepared with different sucrose/Ni mass ratios. Source: Sha [27]. Reproduced with permission of Tianjin University.

was detected by Raman microscope since the amount of carbon was too low, i.e. the carbon in Ni could not reach oversaturation to precipitate to form graphene (Figure 5.12).

It should be noted that the ratio of metal/carbon sources should be adjusted according to the component of metal and carbon sources, as well as the size of metal particles. The larger the metal particles are, the smaller the specific surface area of metal particles will be, thus the less the amount of carbon sources could be used. That is, the amount of carbon sources should be determined according to the surface area of the metal templates, as well as the carbon solubility of metal templates.

5.2.5 The Influence of Heating Temperature and Heating Method

The growth temperature and heating method are also the key factors affecting the quality and formability of the articles. Generally, the higher temperature will result in a higher quality of graphene products. By normal CVD method, high-quality graphene usually can be achieved around 1000 °C. When using plasma or microwave-assisted CVD method, the growth temperature can be lower. However, the sintering temperature in PMT method should also be determined according to the size and component of metal templates.

Drieschner et al. investigated the influence of particles size and component on the sintering temperature and the structure of final products. The growth temperature

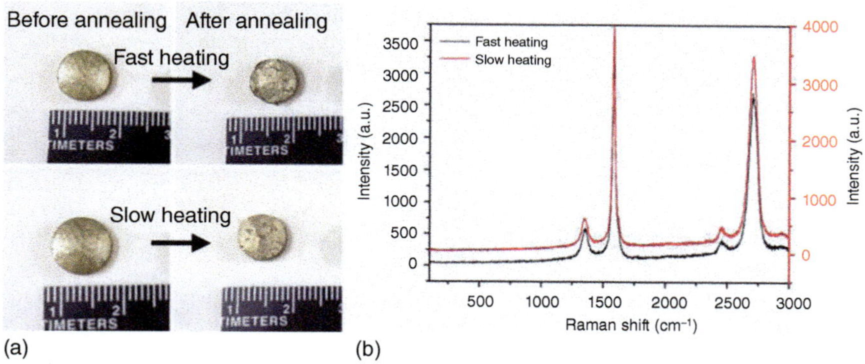

Figure 5.13 (a) Photographs of pellets prepared with 3 g of Ni and 0.5 g of PMMA before and after fast and slow heating. (b) Raman spectra of 3D GMs prepared by fast and slow heating by using 3 g of Ni and 0.5 g of PMMA. Source: Sha [27]. Reproduced with permission of Tianjin University.

ranged from 600 to 1100 °C and varied with the size and component of metal powders [24].

Sha investigated the influence of heating method on the fabrication of 3D GM [27]. The samples were prepared by using 3 g of Ni powder and 1 g of PMMA as the precursors. For a rapid heating, the compressed pellets were directly moved to the hot region (1000 °C) of the tube furnace. For a slow heating, the pellets were put into the tube furnace at room temperature, and then the temperature was increased to 1000 °C at a heating rate of 10 °C min^{-1}. For a rapid cooling, the pellets were pulled to the unheated region directly by using a magnetic extraction boat. In the case of 1 g PMMA/3 g Ni and rapid heating and cooling, the half-finished pellets after heating broke due to the large amount of gases generated. The 3D GM could be obtained only with a slow heating and rapid cooling. When PMMA was decreased to 0.5 g, 3D GM can be obtained with both rapid and slow heating. Thus, the choice of heating method should be adjusted according to the components of compressed pellets. The qualities of the final 3D GM by using rapid and slow heating method were similar since the quality of graphene mainly depends on growth temperature and substrates, generally (Figure 5.13).

5.2.6 The Influence of Cold-Pressing Pressure

Sha also investigated the effects of cold-pressing pressure on the final 3D GM during the PMT process [27]. With 3 g of Ni as the templates and 0.5 g of sucrose as the carbon sources, the half-finished pellets were obtained by cold pressing under 1120, 749, 499, 250, 125, 62, 31, and 10 MPa, respectively. As shown in Figure 5.14, when the pressures were 499–1120 MPa, large amounts of carbon shells can be detected in 3D GM. With the decrease of cold-pressing pressure, the amount of carbon shells decreased, while the amount of 2D graphene sheets increased. When the pressure was 10 MPa, no obvious carbon shell was observed. Thus, the cold-pressing pressure

Figure 5.14 SEM images of 3D GMs prepared under (a, b) 749 MPa, (c, d) 125 MPa, and (e, f) 10 MPa. Source: Sha [27].

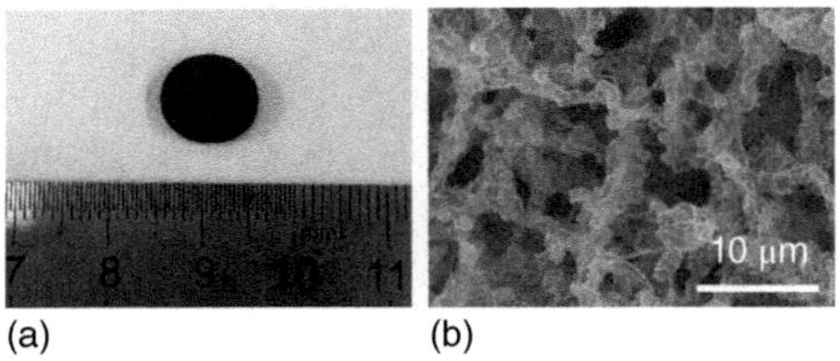

Figure 5.15 (a) Photograph and (b) SEM image of as-prepared LPM-3D rebar GF. Source: Sha [27]. Reproduced with permission of Elsevier.

can affect the structure of 3D GM obviously, suggesting 3D GM's different specific surface areas and other performance.

Tang et al. also prepared 3D rebar GF with Ni/sucrose hybrid powder via a loose PMTs method without cold pressing applied (LPM-3D rebar GF) [28]. The LPM-3D rebar GF presented a macroporous structure with open and interconnected porous pipelines, which consisted of spheroidal graphene connected by 2D graphene sheets and CNTs, as shown in Figure 5.15.

5.3 Mechanism of Powder Metallurgy Templates Method

From the above, the component and size of metal templates, the component of carbonaceous organics, the ratio of metal/organics, the growth temperature and heating

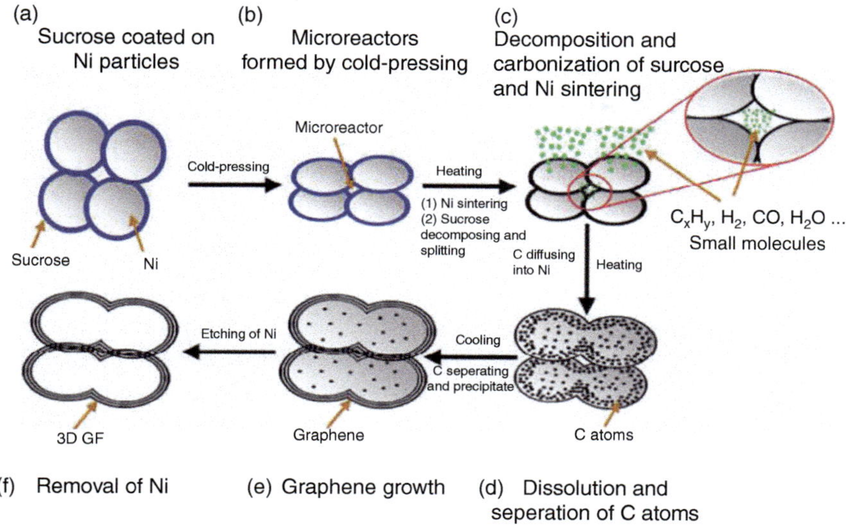

Figure 5.16 Mechanism schematic of preparation of 3D GM by PMT method. (a) Sucrose coated on Ni particles. (b) Microreactors formed by cold-pressing. (c) Decomposition and carbonization of sucrose and Ni sintering. (d) Dissolution and separation of C atoms. (e) Graphene growth. (f) Removal of Ni. Source: Sha [27]. Reproduced with permission of Tianjin University.

method, as well as the cold-pressing pressure will affect the structure and quality of the final 3D GM. Thus, Sha proposed a growth procedure of 3D GM by PMT approach, as shown in Figure 5.16 [27].

The growth procedures of 3D GM mainly include five steps (take the case of using Ni powder and sucrose as an example):

(1) *Coating of organics on the surface of metal templates.* When dissolving in DI water and then precipitating during water evaporation, sucrose will cover the surface of Ni particles, which will act as the carbon source and the binder during cold pressing as well.
(2) *The formation of microreactor.* After cold pressing, the Ni particles will be interconnected together by their deformation and the adhesion by sucrose. The interspace will be a microreactor during the followed heating process for the graphene CVD growth (Figure 5.17). In these microreactors, sucrose will melt firstly, and then decompose to generate hydrocarbon gases or amorphous carbon when increasing the temperature. That is, in essence, the growth of graphene is a normal CVD process in a limited space with generated amorphous carbon as the solid carbon sources or generated hydrocarbon gases as the gaseous carbon source. The volume of the microreactor, as well as the volume and number of channels for hydrocarbon gases release, is tunable by adjusting cold-pressing pressure and the size of the templates.
(3) *The pyrolysis of sucrose and sintering of Ni.* With the increase of the temperature from room temperature, sucrose will melt firstly, and then decompose to

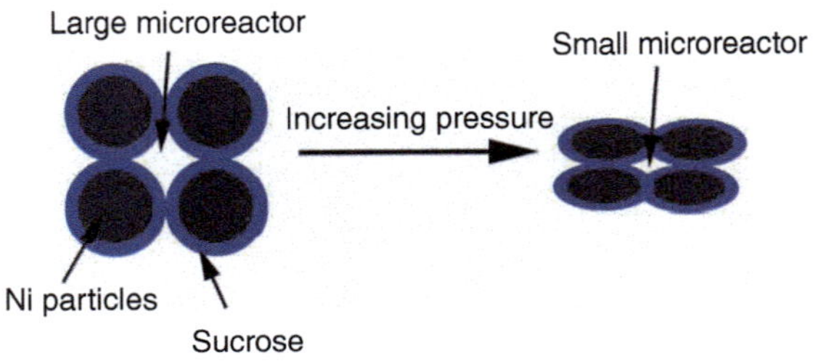

Figure 5.17 Schematic of Ni particle interspace as a microreactor during cold pressing and sintering. Source: Sha [27]. Reproduced with permission of Tianjin University.

generate water vapor, amorphous carbon, and hydrocarbon gases, which will be the carbon sources for graphene growth. Simultaneously, Ni particles will start to migrate and form a bond between the particles below melting point [24]. The hydrocarbon gases will also release through the channels between Ni particles.

(4) *The growth of graphene.* In the microreactor, the generated carbon or hydrocarbon gases will further split and then dissolve into sintered Ni skeleton, and then precipitate to form graphene layers on the surface and interfaces of Ni particles. After removal of the templates, a free-standing 3D GM can be obtained.

5.4 3D GM and Its Composites Prepared by PMT Method

As mentioned above, by changing the component of starting precursors and parameters, various 3D GM and 3D GM-based composites can be developed, which will have prospects in many fields.

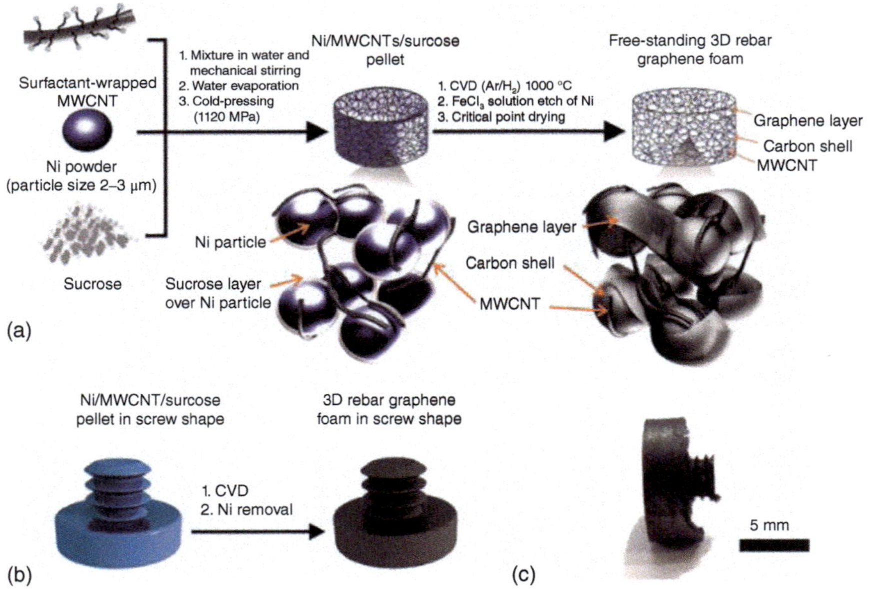

Figure 5.18 (a) Schematic of preparation of 3D rebar GF. (b) Schematic of tuning the shape of the pellet, and (c) photograph of an all-carbon 3D rebar GF screw. Source: Sha et al. [29]. Reproduced with permission of American Chemical Society.

Sha et al. prepared 3D GF through PMT method by using Ni powder as the templates and sucrose as the carbon sources. The 3D GF showed a good electrical conductivity (13.8 S cm^{-1}) and mechanically robust structure [19]. By introducing MWCNTs as the reinforcement, CNTs-reinforced 3D graphene foam (3D rebar GF) was developed [29], as shown in Figure 5.18. The 3D rebar GF showed a good electrical conductivity and excellent mechanical strength. The 3D rebar GF with a mass of 62.8 mg can bear >8500× the foam's weight, and presented a high storage modulus (290.1 kPa) and good structural stability, as shown in Figure 5.19. The 3D rebar GF also showed a good performance as the electrodes of lithium ion capacitor. By changing CNTs to BN nanotubes, 3D BN rebar GFs can be also achieved [29, 30].

Wang et al. prepared 3D N-doped GM by using Ni powder as the templates and melamine as the carbon and nitrogen sources [25]. The resulting 3D N-doped

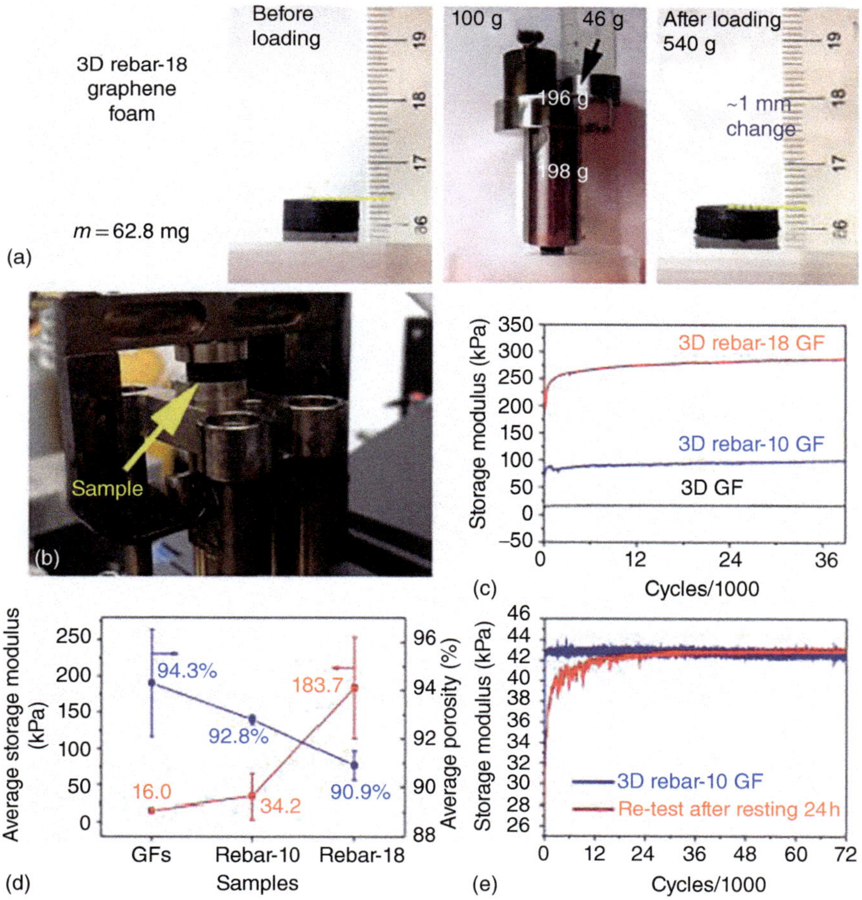

Figure 5.19 (a) Photographs of 3D rebar GF before and after loading 540 g weights. (b–e) Dynamic mechanical analysis (DMA) testing and results of as-prepared 3D rebar GFs. Source: Sha et al. [29]. Reproduced with permission of American Chemical Society.

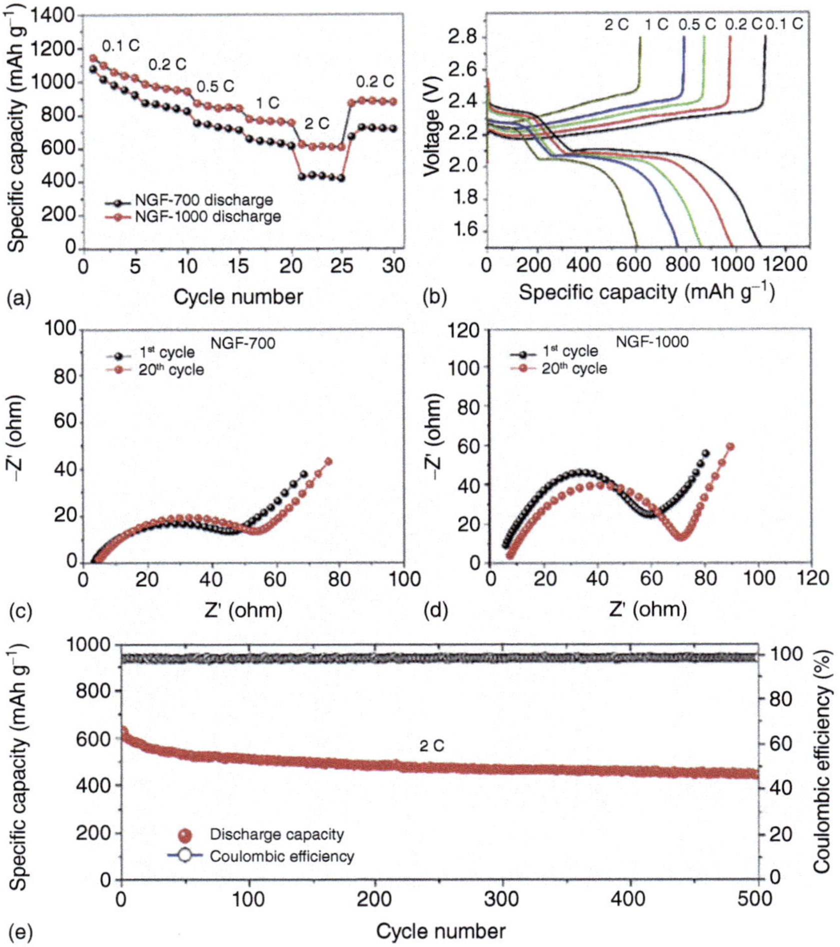

Figure 5.20 (a-e) Performance testing data of 3D N-doped GM as the electrode of Li-S battery. Source: Wang et al. [25]. Reproduced with permission of Elsevier.

GM can be used as binder-free electrode for high-performance lithium/sulfur battery, which delivered a high initial discharge capacity of 987 mAh g^{-1} and can maintain at 819 mAh g^{-1} after 200 cycles at 0.2 C (82.9% capacity retention). Even at a high current density of 2 C, a good long-life-cycle performance was also achieved (Figure 5.20).

The PCCF developed by Wang et al. also presented a good performance as the electrode of Li-S battery [22]. The PCCF/S cathode with a high sulfur loading content (84.75%) exhibited a high initial discharge capacity of 1264.4 mAh g^{-1} at 0.1 C and a good cycling stability (Figure 5.21).

Goh et al. prepared a highly porous N-doped GF by using a commercial Ni foam at the templates (Figure 5.22) [31]. The voids of the commercial Ni foam were filled

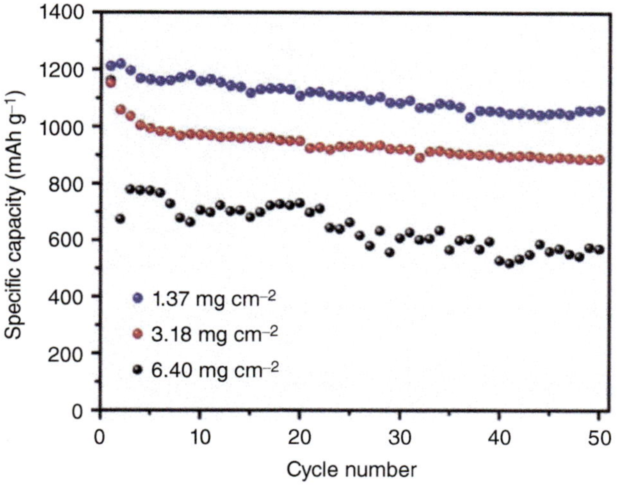

Figure 5.21 The cycling performance of the as-prepared PCCF/S with different sulfur loadings at 0.1 C. Source: Wang et al. [22]. Reproduced with permission of Elsevier.

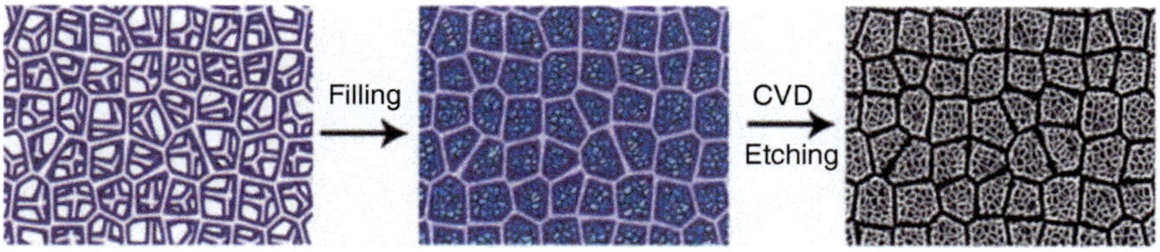

Figure 5.22 Scheme of the preparation process of the 3D GF/graphene "eggshell" composite structure. Source: Goh et al. [31]. Reproduced with permission of American Chemical Society.

using PMMA-coated Ni powders. In essence, that is a PMT process, which was limited inside the voids of Ni foam, resulting in a new 3D GF/graphene "eggshell" composite structure, showing an improved lithium storage performance.

Tang et al. developed a CNTs-reinforced 3D graphene foam via a loose PMTs method without cold-pressing applied (LPM-3D rebar GF) [28]. The introduction of CNTs effectively prevents graphene from agglomeration, increases the nucleation sites, and guides Li electrodeposition, resulting in the superior rate performance and cycle stability of Li/LPM-3D rebar GF anode even at high areal capacity of 30 mAh cm^{-2} (Figure 5.23).

Tynan et al. used dextran/metal salt gel as template to synthesize macroporous 3D graphene materials (Figure 5.24), in which the metal salts are cobalt, nickel, copper, and iron salts. When the metal salt was cobalt salt, the macroporous graphene contained thin graphene sheets of 1–3 layers as well as regions of thicker graphene sheets (>3 layers) [32]. The presence of the thicker regions is thought to help structurally support the 3D foam monolith.

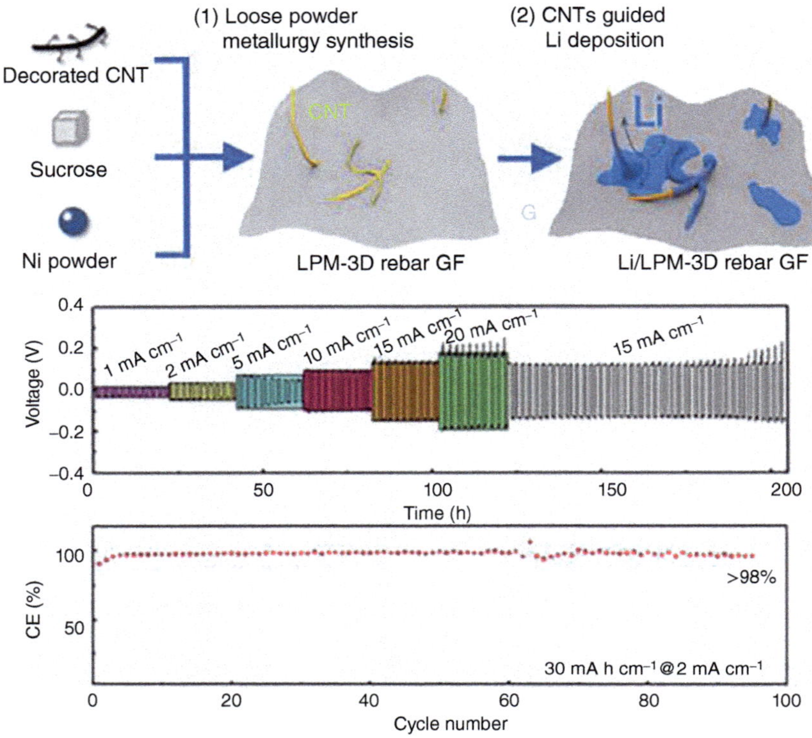

Figure 5.23 Schematic of the preparation of LPM-3D rebar GF as the anode of Li-metal battery, and the performance testing of the samples. Source: Tang et al. [28]. Reproduced with permission of Elsevier.

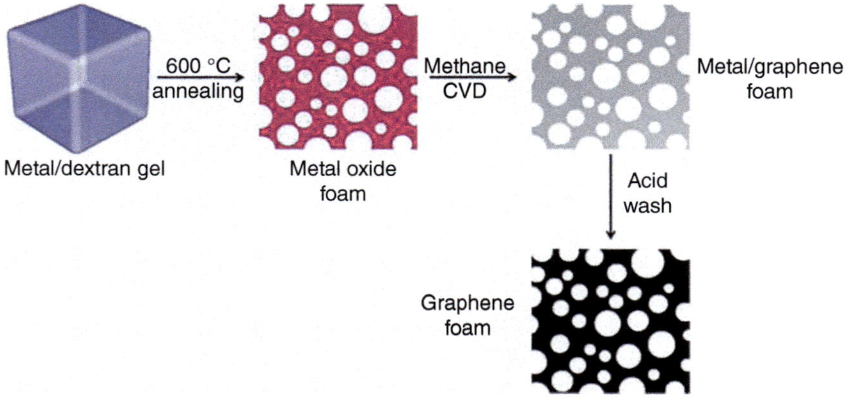

Figure 5.24 Schematic of the synthesis of metal graphene foams. Source: Tynan et al. [32]. Open Access Article.

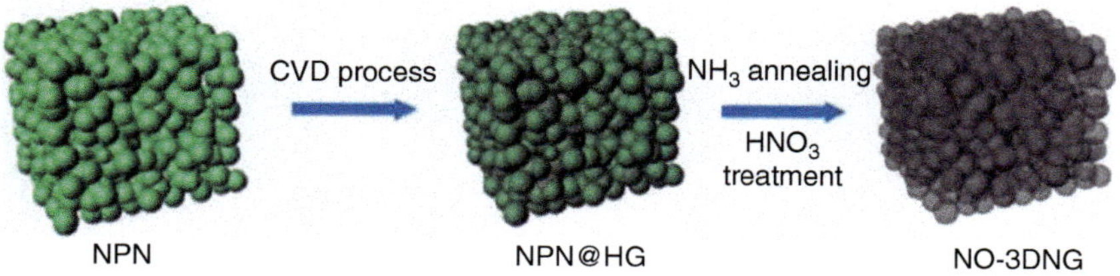

Figure 5.25 Schematic diagram of the fabrication of NO-3DNG film. Source: Wang et al. [35]. Reproduced with permission of Elsevier.

Yamada et al. successfully obtained multilayer graphene by using GaN as template and catalyst and carbon source [33]. The Raman I_D/I_G ratio of the graphene decreased with increasing the annealing temperature. At 700 °C, fine graphene was obtained on the GaN template while maintaining the flatness and smoothness of the GaN surface.

Zhang et al. prepared 3D edge-curled graphene on salt template with the catalysis of nickel [34]. The unique structure not only effectively prevents the compact accumulation and aggregation of graphene, but also significantly improves its lithium storage performance. As the negative electrode of lithium ion battery, it shows excellent rate performance and cycle performance.

Wang et al. prepared high-quality independent 3D nanoporous graphene films using nanoporous nickel as template and catalyst (Figure 5.25) [35]. Compared with macroporous graphene films, nano-porous graphene films show unusually large working windows in neutral, acidic, and alkaline aqueous electrolytes, and have high bulk density.

5.5 Additive Manufacturing

For simply, rapidly, efficiently, and controllably synthesizing 3D GM, additive manufacturing is another alternative method. Recently, additive manufacturing has attracted intense attention both from academia and industry. Additive manufacturing, also known as 3D printing, is a simple and efficient technique for directly producing 3D objects controlled by computer programs. In 3D printing, polymers, ceramics, or metals will be heated and deposited on a substrate layer by layer to build 3D monoliths. Represented by GE company, additive manufacturing has been used widely in automobile and aerospace-related fields. However, very limited work for graphene synthesis by 3D printing has been reported so far.

Wang et al. printed a conductive silver-nanoparticle-decorated graphene nanocomposite film by inkjet printing, showing a high conductivity of 2160 $S\,m^{-1}$ [36]. The resistivity of the as-prepared hybrid film was $4.62\times10^{-4}\,\Omega\,m$ (corresponding to a conductivity of $2.16\times10^{3}\,S\,m^{-1}$) upon 15 printing times (Figure 5.26).

Kim et al. reported the 3D printing to write free-standing reduced GO wires [37], as shown in Figure 5.27. The resistance of GO wire was a function of applied horizontal stain, as shown in Figure 5.28.

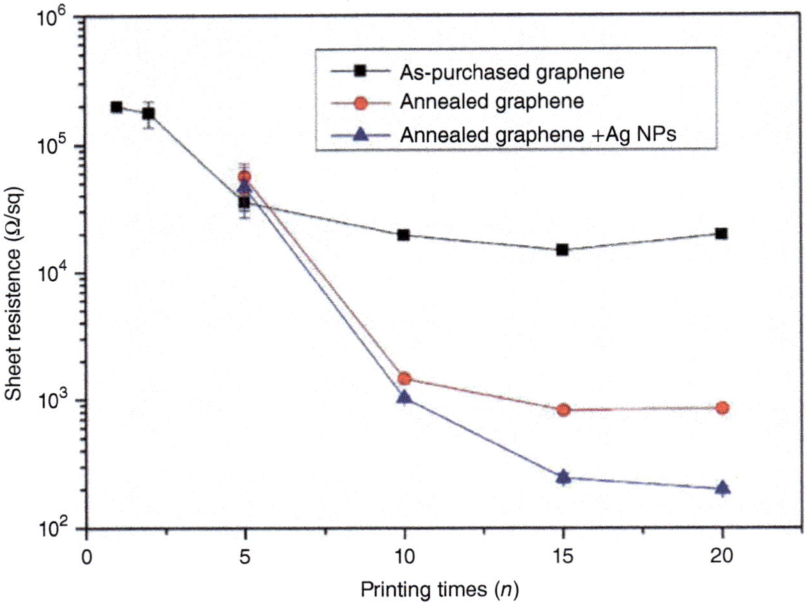

Figure 5.26 The sheet resistance profiles of as-prepared graphene/Ag film, graphene film, and as-purchased graphene. Source: Wang et al. [36]. Reproduced with permission of Elsevier.

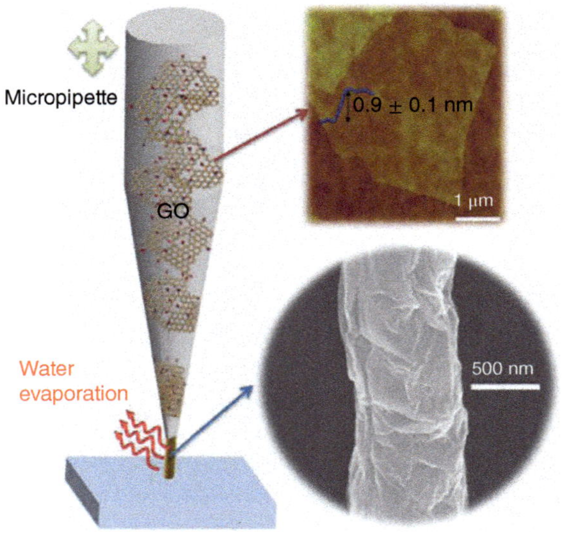

Figure 5.27 Schematic of GO nanowire fabrication. Source: Wang et al. [36]. Reproduced with permission of John Wiley and Sons.

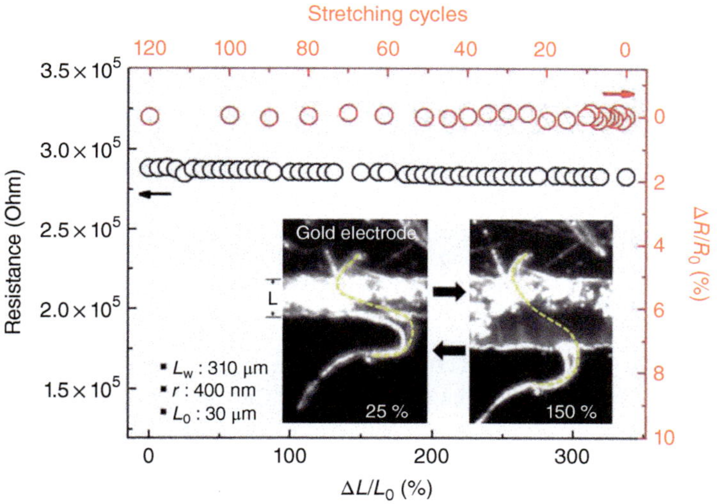

Figure 5.28 Resistance of the fabricated rGO nanowire as a function of strain ($\Delta L/L_0$) (L_w = 310 μm, r = 400 nm, and L_0 = 30 μm). Source: Kim et al. [37]. Reproduced with permission of John Wiley and Sons.

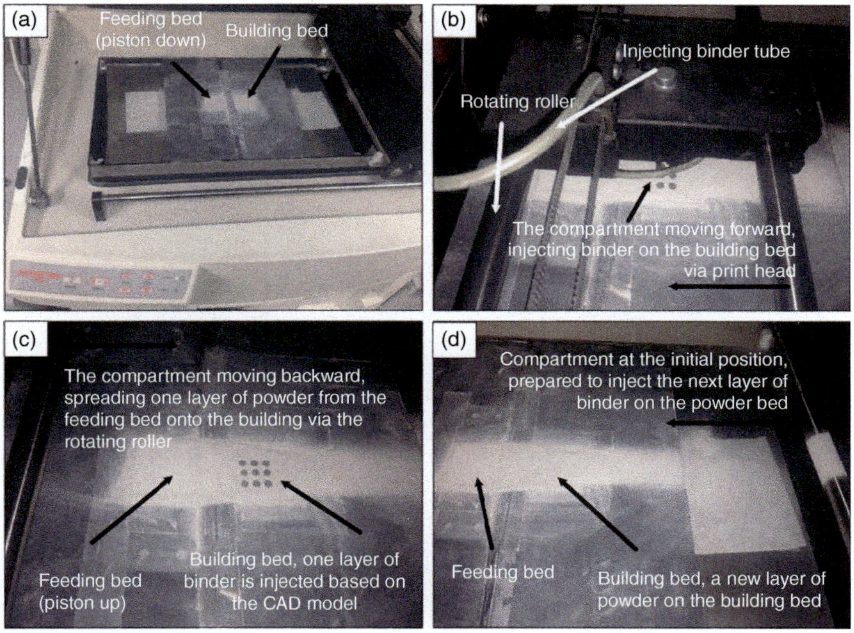

Figure 5.29 (a-d) Fabrication process of 3D printed hydroxyapatite and GO nanoparticles structures by powder-bed additive manufacturing. Source: Kim et al. [37]. Reproduced with permission of John Wiley and Sons.

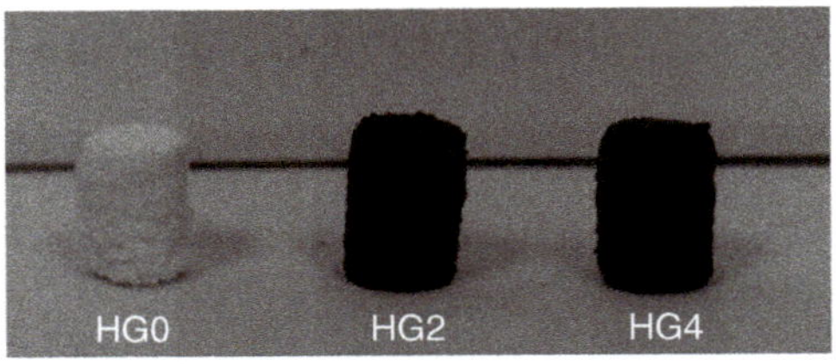

Figure 5.30 Photograph of as-printed samples. HG0 refers to 0 wt% of GO to hydroxyapatite, HG2 refers to 0.2 wt% of GO to hydroxyapatite, and HG4 refers to 0.4 wt% of GO to hydroxyapatite. Source: Kim et al. [37]. Reproduced with permission of John Wiley and Sons.

Azhari et al. developed 3D porous structures using a blend of hydroxyapatite and GO nanoparticles (Figure 5.29) [38]. The photograph of the samples is shown in Figure 5.30.

From the above, the 3D printed graphene-based materials are still limited to micro-sized scales with the precursors of GO contained in inkjet-printable or UV-curable inks. Directly synthesizing 3D GM by 3D printing is still challenge.

Sha et al. developed a 3D printing method for directly synthesizing 3D GM by combining powder-bed 3D printing technique and PMT method [39]. The hybrid powders of Ni/sucrose were used as the precursors. Under the protection of Ar/H_2, the hybrid powders were heated by using a CO_2 laser. Under laser irradiation, Ni particles were sintered, and sucrose was converted to graphene under the catalysis of sintered Ni skeleton, which is similar as the process in PMT method without a tube furnace employed. The 3D printed GM also showed a comparable electrical conductivity ($8.7\,S\,m^{-1}$) and good structural stability. The storage modulus of the 3D printed GM was 11 kPa, which was comparable with the samples prepared by using PMT method. After 70 000 cycles, no collapse was detected.

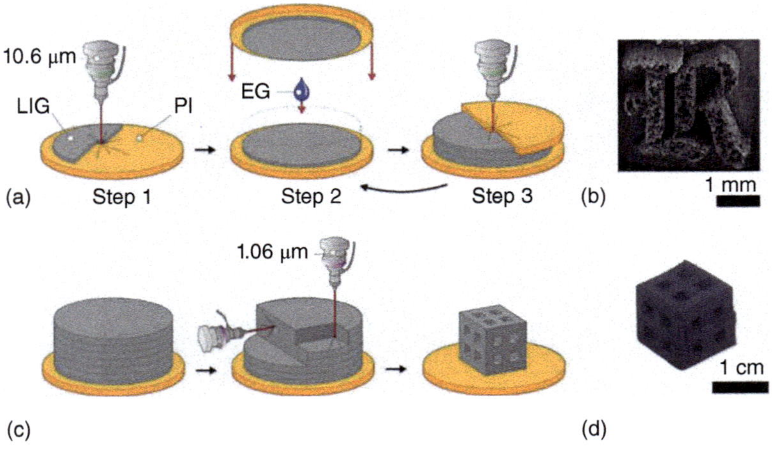

Figure 5.31 (a and c) Schematic of laminated object manufacturing process for fabricating 3D GM. (b and d) SEM images of as-prepared 3D GM. Source: Luong et al. [40]. Reproduced with permission of John Wiley and Sons.

Luong et al. also developed a new laminated object manufacturing process (laminated 3D printing) to fabricate 3D GM [40], as shown in Figure 5.31. PI film substrates were directly converted to graphene by using a CO_2 laser. The 3D GM showed a good electrical conductivity and mechanical strength, as well as a good prospect in energy storage and flexible electronic sensor applications.

Ling et al. applied the laser-induced graphene into the fabrication, and modeling of soft electrothermal actuators, explored their applications in mechanically guided 3D assembly and human–soft actuators interaction, as shown in Figure 5.32 [41]. The LIG had demonstrated excellent physical and chemical properties that can be

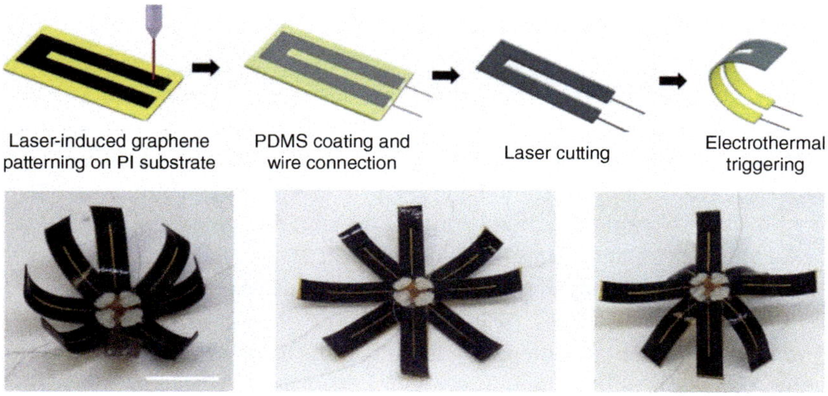

Figure 5.32 The fabrication of soft electrothermal actuators based on laser-induced graphene. Source: Ling et al. [41]. Reproduced with permission of John Wiley and Sons.

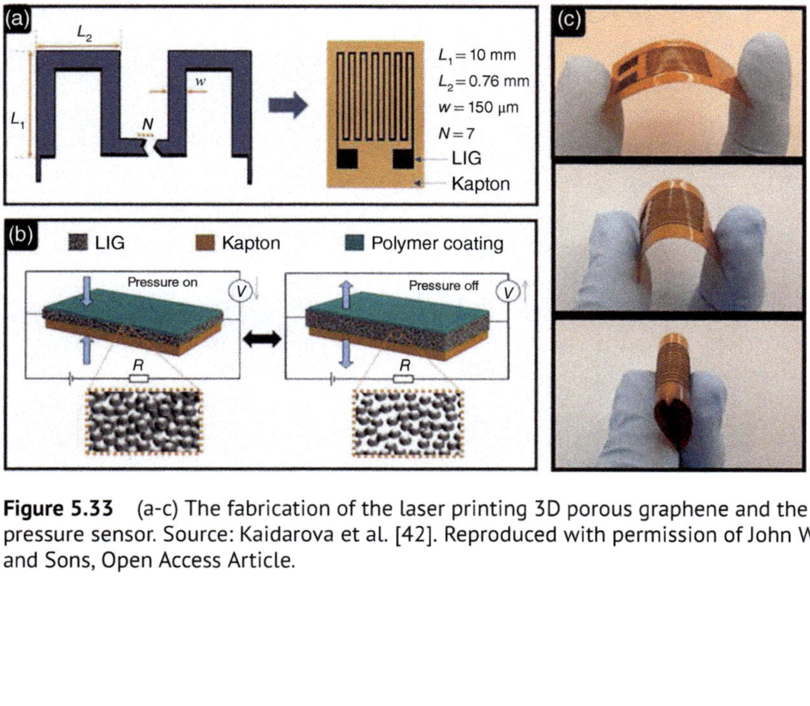

Figure 5.33 (a-c) The fabrication of the laser printing 3D porous graphene and the flexible pressure sensor. Source: Kaidarova et al. [42]. Reproduced with permission of John Wiley and Sons, Open Access Article.

used to manufacture complex 3D architectures, including reconfigurable structures that can reshape among three distinct geometries.

Kaidarova et al. produced 3D porous graphene by laser printing on polyimide films. The piezoresistive effect of 3D porous graphene was utilized to develop the flexible pressure sensor, as shown in Figure 5.33 [42]. The pressure sensors have a sensitivity in the range of 1.23×10^{-3} kPa with a high resolution. Meanwhile, the detection limit was 10 Pa with an extremely wide dynamic range of at least 20 MPa.

5.6 Outlook for PMT and Additive Manufacturing Method

In conclusion, PMT method is a simple and efficient method to fabricate 3D GM and 3D GM-based composites. By changing the components and size of the starting templates and organics, heating treatment parameters, as well as the cold-pressing pressure, various 3D monoliths could be developed. Moreover, by combining PMT method with additive manufacturing, 3D monoliths with different composition and shapes could also be created. Thus, PMT method and 3D printing method are two effective methods for the fabrication of 3D monoliths, which have good prospects in both academia and industry.

However, the investigation of preparing 3D monoliths by using PMT method and 3D printing method is still limited. The cost for fabricating 3D monoliths, especially 3D GM-based materials, is still high compared to the more mature and widely used approaches. The mechanism of PMT and 3D printing 3D GM still needs further study. Therefore, more work and researches are needed to mature these two approaches in the future.

References

1 Novoselov, K.S., Geim, A.K., Morozov, S.V. et al. (2004). Electric field effect in atomically thin carbon films. *Science* 306 (5696): 666–669.

2 Subrahmanyam, K.S., Panchakarla, L.S., Govindaraj, A. et al. (2009). Simple method of preparing graphene flakes by an arc-discharge method. *Journal of Physical Chemistry C* 113 (11): 4257–4259.

3 Wu, Z.S., Ren, W.C., Gao, L.B. et al. (2009). Synthesis of graphene sheets with high electrical conductivity and good thermal stability by hydrogen arc discharge exfoliation. *ACS Nano* 3 (2): 411–417.

4 Emtsev, K.V., Bostwick, A., Horn, K. et al. (2009). Towards wafer-size graphene layers by atmospheric pressure graphitization of silicon carbide. *Nature Materials* 8 (3): 203–207.

5 Reina, A., Jia, X., Ho, J. et al. (2009). Large area, few-layer graphene films on arbitrary substrates by chemical vapor deposition. *Nano Letters* 9 (1): 30–35.

6 Wei, D., Liu, Y., Wang, Y. et al. (2009). Synthesis of N-doped graphene by chemical vapor deposition and its electrical properties. *Nano Letters* 9 (5): 1752–1758.

7 Hummers, W.S. and Offeman, R.E. (1958). Preparation of graphitic oxide. *Journal of the American Chemical Society* 80 (6): 1339–1339.

8 Muzyka, R., Kwoka, M., Smedowski, L. et al. (2017). Oxidation of graphite by different modified Hummers methods. *New Carbon Materials* 32 (1): 15–20.

9 Wu, X., Xu, H., Shen, Y. et al. (2014). Treatment of graphite felt by modified Hummers method for the positive electrode of vanadium redox flow battery. *Electrochimica Acta* 138: 264–269.

10 Marcano, D.C., Kosynkin, D.V., Berlin, J.M. et al. (2010). Improved synthesis of graphene oxide. *ACS Nano* 4 (8): 4806–4814.

11 Sun, Z., Raji, A.-R.O., Zhu, Y. et al. (2012). Large-area Bernal-stacked Bi-, Tr-, and tetralayer graphene. *ACS Nano* 6 (11): 9790–9796.

12 Li, X.S., Cai, W.W., Colombo, L. et al. (2009). Evolution of graphene growth on Ni and Cu by carbon isotope labeling. *Nano Letters* 9 (12): 4268–4272.

13 Sun, Z., Yan, Z., Yao, J. et al. (2010). Growth of graphene from solid carbon sources. *Nature* 468 (7323): 549–552.

14 Geim, A.K. and Novoselov, K.S. (2007). The rise of graphene. *Nature Materials* 6 (3): 183–191.

15 Tsukerman, S.A. (1965). *Powder Metallurgy*, 1e. Elsevier.

16 Tang, F., Anderson, I.E., Gnaupel-Herold, T. et al. (2004). Pure Al matrix composites produced by vacuum hot pressing: tensile properties and strengthening mechanisms. *Materials Science and Engineering a-Structural Materials Properties Microstructure and Processing* 383 (2): 362–373.

17 Bathula, S., Anandani, R.C., Dhar, A. et al. (2012). Microstructural features and mechanical properties of Al 5083/SiCp metal matrix nanocomposites produced by high energy ball milling and spark plasma sintering. *Materials Science and Engineering a-Structural Materials Properties Microstructure and Processing* 545: 97–102.

18 McNeese, M.D., Lagoudas, D.C., and Pollock, T.C. (2000). Processing of TiNi from elemental powders by hot isostatic pressing. *Materials Science and Engineering a-Structural Materials Properties Microstructure and Processing* 280 (2): 334–348.

19 Sha, J., Gao, C., Lee, S.-K. et al. (2016). Preparation of three-dimensional graphene foams using powder metallurgy templates. *ACS Nano* 10 (1): 1411–1416.

20 Wu, Y.P., Chou, H., Ji, H.X. et al. (2012). Growth mechanism and controlled synthesis of AB-stacked bilayer graphene on Cu-Ni alloy foils. *ACS Nano* 6 (9): 7731–7738.

21 Chen, Y., Zhang, X., Liu, E. et al. (2016). Fabrication of three-dimensional graphene/Cu composite by in-situ CVD and its strengthening mechanism. *Journal of Alloys and Compounds* 688: 69–76.

22 Wang, Y.Q., Luo, S.Q., Wang, D.Q. et al. (2018). Facile synthesis of three dimensional porous cellular carbon as sulfur host for enhanced performance lithium sulfur batteries. *Electrochimica Acta* 284: 400–407.

23 Sha, J.W., Chu, X.Y., Xu, T.T. et al. (2021). Bi-functional modular graphene network with high rate and cycling performance. *Journal of Power Sources* 504: 230075.

24 Drieschner, S., Weber, M., Wohlketzetter, J., et al. (2016). High surface area graphene foams by chemical vapor deposition. *2d Materials,* 3 (4): 045013.

25 Wang, Y., Huang, J., Chen, X. et al. (2018). Powder metallurgy template growth of 3D N-doped graphene foam as binder-free cathode for high-performance lithium/sulfur battery. *Carbon* 137: 368–378.

26 Han, X., Wang, T., Owuor, P.S. et al. (2018). Ultra-stiff graphene foams as three-dimensional conductive fillers for epoxy resin. *ACS Nano* 12 (11): 11219–11228.

27 Sha, J. (2017). Controllable preparation of three-Dimensional graphene based hybrid materials and their energy storage related applications. Unpublished Doctoral Thesis, Tianjin University.

28 Tang, Y., Sha, J., Wang, N. et al. (2020). Covalently bonded 3D rebar graphene foam for ultrahigh-areal-capacity lithium-metal anodes by in-situ loose powder metallurgy synthesis. *Carbon* 158: 536–544.

29 Sha, J.W., Salvatierra, R.V., Dong, P. et al. (2017). Three-dimensional rebar graphene. *ACS Applied Materials & Interfaces* 9 (8): 7376–7384.

30 Li, Y.L., Peng, Z.W., Larios, E. et al. (2015). Rebar graphene from functionalized boron nitride nanotubes. *ACS Nano* 9 (1): 532–538.

31 Goh, B.-M., Wang, Y., Reddy, M.V. et al. (2014). Filling the voids of graphene foam with graphene "eggshell" for improved lithium-ion storage. *ACS Applied Materials & Interfaces* 6 (12): 9835–9841.

32 Tynan, M.K., Johnson, D.W., Dobson, B.P. et al. (2016). Formation of 3D graphene foams on soft templated metal monoliths. *Nanoscale* 8 (27): 13303–13310.

33 Yamada, J., Ueda, Y., Maruyama, T., et al. (2020). Precipitation of multilayer graphene directly on gallium nitride template using tungsten capping layer. *Journal of Crystal Growth,* 534. 125493

34 Zhang, J., Ai, Y., Wu, J., et al. (2020). Nickel-catalyzed synthesis of 3D edge-curled graphene for high-performance lithium-ion batteries. *Advanced Functional Materials,* 30 (9). 1904645

35 Wang, L., Qin, K., Li, J. et al. (2018). Doping and controllable pore size enhanced electrochemical performance of free-standing 3D graphene films. *Applied Surface Science* 427: 598–604.

36 Wang, G.K., Wang, Z., Liu, Z.H. et al. (2015). Annealed graphene sheets decorated with silver nanoparticles for inkjet printing. *Chemical Engineering Journal* 260: 582–589.

37 Kim, J.H., Chang, W.S., Kim, D. et al. (2015). 3D printing of reduced graphene oxide nanowires. *Advanced Materials* 27 (1): 157–161.

38 Azhari, A., Toyserkani, E., and Villain, C. (2015). Additive manufacturing of graphene-hydroxyapatite nanocomposite structures. *International Journal of Applied Ceramic Technology* 12 (1): 8–17.

39 Sha, J.W., Li, Y.L., Salvatierra, R.V. et al. (2017). Three-dimensional printed graphene foams. *ACS Nano* 11 (7): 6860–6867.

40 Luong, D.X., Subramanian, A.K., Silva, G.A.L., et al. (2018). Laminated object manufacturing of 3D-printed laser-induced graphene foams. *Advanced Materials,* 30 (28). 1707416

41 Ling, Y., Pang, W., Li, X., et al. (2020). Laser-induced graphene for electrothermally controlled, mechanically guided, 3D assembly and human-soft actuators interaction. *Advanced Materials,* 32 (17). 1908475

42 Kaidarova, A., Alsharif, N., Oliveira, B.N.M., et al. (2020). Laser-printed, flexible graphene pressure sensors. *Global Challenges,* 4 (4). 2000001

6

Graphene-Based Materials for Lithium/Sodium-Ion Batteries

Biao Chen

School of Materials Science and Engineering and Tianjin Key Laboratory of Composite and Functional Materials, Tianjin university, Tianjin, 300350, P.R. China

6.1 Introduction

Lithium-ion batteries (LIBs) have been widely used in consumer electronics, such as mobile phones, digital cameras, personal computers, and electric vehicles [1–4]. Until now, the most successful anode material (AM) is graphite. However, the energy density of LIBs is limited by the low theoretical capacity of graphite (367 mAh g^{-1}) [5–7]. In addition, the low abundance of Li elements on the earth cannot meet the increasing demand. Therefore, developing new energy-storage devices with abundant and cheap sources is highly desirable. Among them, sodium-ion batteries (SIBss) have attracted widespread attention [8–10]. The anode material of SIBs plays a key role in electrochemical performance. Therefore, much attention has been paid to developing new anode materials with high theoretical capacity for both LIBs and SIBs.

The anode materials for both LIBs and SIBs can be divided into three kinds of categories—insertion, alloying, and conversion [11]. As shown in Figure 6.1, the insertion materials have small volume change and strong structural stability, whilst both alloying and conversion materials show high theoretical capacities. However, the electrochemical performance and application of insertion and alloying/conversion materials are, respectively, suffered from a small number of storage sites and large volume change. To address the issues, nanostructure engineering with various nanostructures has been prepared to enhance the electrochemical reaction kinetics. However, the obtained electrochemical performance of single material is still unsatisfactory. Therefore, the rational design of anode composites with complementary effects is highly desirable [12].

The graphene shows superior electrical conductivity, large specific surface area, and low density, resulting in great potential application in lithium/sodium-ion

Templated Fabrication of Graphene-Based Materials for Energy Applications, First Edition.
Edited by Chunnian He, Naiqin Zhao, and Junwei Sha.

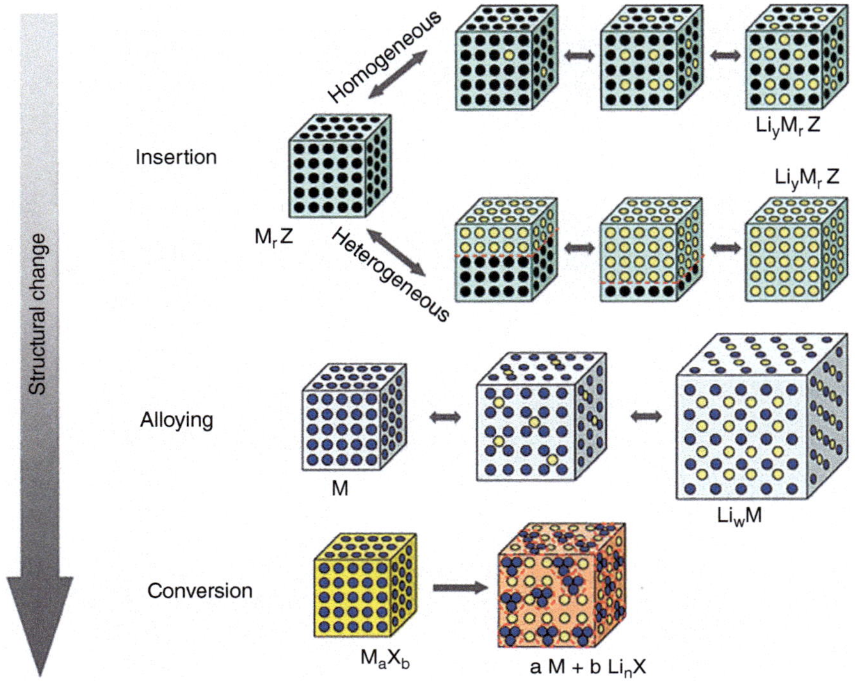

Figure 6.1 A schematic representation of the different reaction mechanisms observed in electrode materials for lithium batteries. Black circles: voids in the crystal structure, blue circles: metal, yellow circles: lithium. Source: Palacin [11]. Reproduced with permission of Royal of Society Chemistry.

batteries [13, 14]. However, it cannot be used directly as anode material due to its ultralow initial coulombic efficiency (CE) and compaction density [15]. Generally, graphene has been widely used as one kind of promising material that can be incorporated with other active materials [16, 17]. In the reported graphene-based composites, the graphene provides a large specific surface to load a high weight content active material and inhibit aggregation, as well as works as high electrical transfer pathway. Especially, templated fabrication graphene matrix always shows porous structure, which can significantly enhance the reaction kinetics, electrical conductivity, and structural stability of the active materials, leading to good cycling and rate performance [18, 19]. Therefore, developing templated fabrication graphene-based anode materials is a promising strategy to prepare high-performance LIBs and SIBs.

Herein, the recent process of graphene-based anode materials for LIBs and SIBs is summarized. The templated preparation, electrochemical performance, and storage mechanism of graphene-based anode materials are systematically discussed. Especially, the role and function of graphene in graphene-based anode materials are highlighted. Sections 6.2–6.4, respectively, introduce the graphene-based insertion-, alloying-, and conversion-types composites for LIBs and SIBs. Section 6.5 gives a brief conclusion and proposes some suggestions for further development.

6.2 Graphene-Based Insertion Composites

6.2.1 TiO_2/Graphene Composites

The titanium-based compounds have been widely used as anode materials for LIBs [20] and SIBs [21]. Among them, TiO_2 has unique physical and chemical properties, including high chemical durability, abundance, low cost, and environmental protection [22]. Therefore, it has been widely used as dual anode materials for both LIBs and SIBs. However, the reversible capacities of TiO_2 are severely limited by its poor conductivity. The incorporation of TiO_2 with templated fabrication graphene can significantly improve its capacity and rate performance.

For application in LIBs, TiO_2/graphene composites with various morphologies have been widely prepared. TiO_2/graphene/TiO_2 sandwich-like structures and 3D porous structures are the typical morphologies. In addition, a free-standing structure was also prepared to optimize the electrochemical properties on electrode level.

For TiO_2/graphene/TiO_2 sandwich-like structures, the TiO_2 display various structures, including 0D nanoparticle [23–25], 1D nanorod [26], 1D nanowire [27], 1D nanoneedle [28], 1D nanobelt [29], 1D nanotube [30], 2D nanosheet [31], 3D nanoflower [32], 3D nanobundle [33], and 3D nanosphere [34]. The TiO_2/graphene/TiO_2 structures show strong interfacial interaction between graphene and TiO_2, which is beneficial to the electrical transfer from graphene to TiO_2. Generally, the TiO_2/graphene/TiO_2 composites are widely prepared by solution-based and sol–gel methods. Taking TiO_2 nanoparticles/graphene as an example, Cao et al. mixed graphene oxide (GO) and tetrabutyl titanate (TBOT) in an autoclave [23]. After hydrothermal process at 160 °C for 12 hours, the TiO_2 nanoparticles anchored on reduced graphene oxide (rGO) composite were obtained. The rGO/TiO_2 nanoparticles showed a larger reversible capacity of 152.6 mAh g^{-1} at 1 A g^{-1} than TiO_2 (9.7 mAh g^{-1}) in a wide voltage window of 0.001–3.0 V. The enhanced reversible capacity is attributed to the highly conductive rGO. In this conventional method, only the oxygen groups on the surface of GO can work as nucleation sites for TiO_2. The obtained TiO_2 nanoparticles display disordered and discontinuous structures to minimize the surface energy. Therefore, many TiO_2 nanoparticles do not directly contact with rGO substrate, while many places on the surface of rGO are not loaded with TiO_2. The obtained reversible capacity is still unsatisfactory. The distribution of the TiO_2 on the surface of rGO also plays an important role in electrochemical performance. Li et al. proposed a sol–gel method to synthesize ultradispersed TiO_2 nanoparticles on rGO composites [24]. The TBOT in pure ethanol shows a slow hydrolysis rate. The nucleation of amorphous TiO_2 on the GO surface is uniform. After calculation, ultradispersed TiO_2 nanoparticles with a diameter size of 5 nm are tightly anchored on the surface of rGO. The obtained ultradispersed TiO_2 nanoparticles on rGO composite displayed a high reversible capacity of 94 mAh g^{-1} at 10 A g^{-1} in a voltage window of 1–3 V. In addition, the effective weight content of TiO_2 nanoparticles in the composites shows important effect. To improve the weight content of TiO_2 nanoparticles, Li et al. proposed a facile strategy to synthesize uniform and continuous TiO_2 nanoparticles network on the surface of rGO, which is different from the conventional one-pot method [25]. The hydrolysis of Ti precursors is controlled at a slow rate. The condensation of Ti oligomers is uniform on the surface of GO. Then the amorphous TiO_2 shell conformably coats on the surface of GO. After calculation, rGO is coated by uniform and continuous TiO_2 nanoparticles network with strong interfacial interaction. The weight content of TiO_2 nanoparticle network is 84%. When evaluated the sandwich-like composite as anode material of LIBs, it showed a large reversible capacity of 237 mAh g^{-1} after 100 cycles as 20 mA g^{-1} in a voltage window of 1–3 V.

The electrical transfer from semiconducting TiO_2 to TiO_2 is hard and slow, which is harmful to the electrochemical reaction kinetics. To address these issues, 1D, 2D, and 3D TiO_2 nanostructures tightly anchored on the surface of graphene were prepared to reduce the interfaces between TiO_2 and TiO_2. For example, He et al. used a conventional method to prepare rGO/TiO_2 nanocrystals [26]. Then the prepared TiO_2 nanocrystals were acted as seeds to prepare nanorods in the subsequent reaction. Finally, sandwich-like TiO_2 nanorods/rGO/TiO_2 nanorods composites were obtained. They displayed a better rate performance than bare TiO_2 nanoparticles and bare TiO_2 nanorods. Li et al. prepared N-doped TiO_2 nanotubes/N-doped graphene composite through one-pot hydrothermal method together with calcination [30]. For comparison, the TiO_2 nanotubes/graphene composite was synthesized. The N-doped TiO_2 nanotubes/N-doped graphene composite exhibited better rate performance than TiO_2 nanotubes/graphene composite and TiO_2 nanotubes, suggesting that the N-doping can enhance the electrical conductivity. Wang et al. used one-pot hydrothermal method to prepare ultrathin anatase TiO_2 nanosheets/rGO composites [31]. When used as anode material of LIBs, from the cyclic voltammogram (CV) curves, a couple of reversible redox peaks were observed at 1.71 and 1.99 V, respectively, attributed to the Li^+ insertion and extraction to anatase TiO_2. The voltage difference (0.28 V) between the two peaks for ultrathin anatase TiO_2 nanosheets/rGO composites is smaller than that for commercial P25, anatase TiO_2 nanosheets, and anatase TiO_2 nanosheets/rGO composites, indicating that the best electrochemical reaction kinetics for ultrathin anatase TiO_2 nanosheets/rGO composites. As a result, the ultrathin anatase TiO_2 nanosheets/rGO composites demonstrated the highest cycling and rate performance. They delivered 174.2 mAh g^{-1} at 167.5 mAh g^{-1} after 200 cycles in a voltage window of 1–3 V.

Although the interfaces of TiO_2/graphene and TiO_2/TiO_2 have been optimized in TiO_2/2D graphene composites, many interfaces between composite particles still exhibit in the anodes, that limit the electrical transfer and electrode stability. To address this issue, templated fabrication of TiO_2/3D graphene composites has been widely reported [35–38]. The reported methods can be divided into three categories. The first is using TiO_2 as template, and then covering 3D graphene networks on its surface. For example, as shown in Figure 6.2a, Yan et al. used microwave solvothermal method to prepare TiO_2 microspheres, then the microsphere was activated in the 3-aminopropyl-trimethoxysilane (APTMS) solution to get positive charges, finally, the negative charges GO was wrapped on the surface of TiO_2 microspheres by electrostatic interactions [35]. The obtained 3D rGO wrapped TiO_2 microspheres

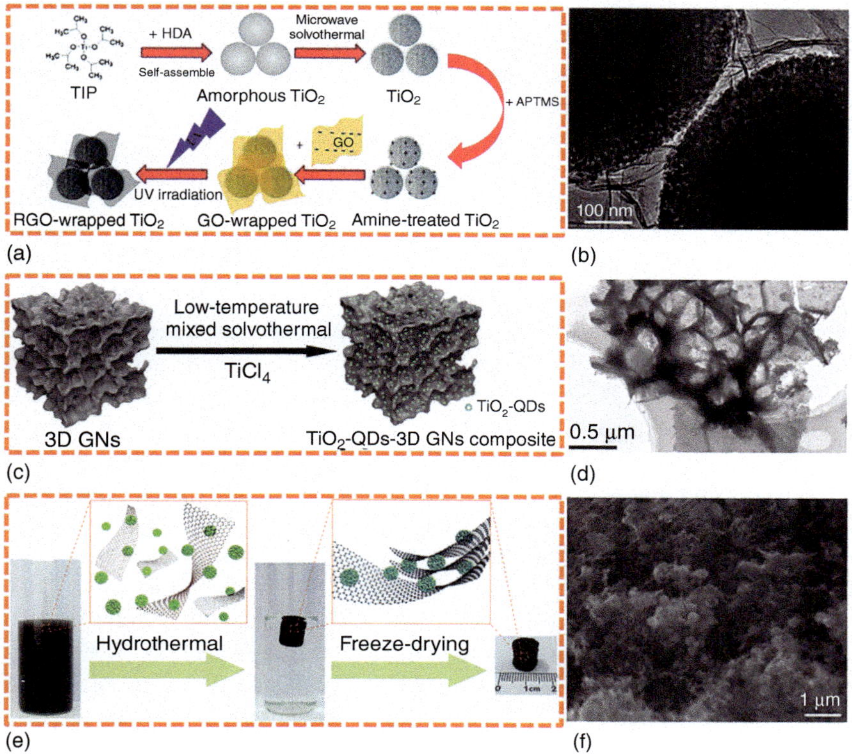

Figure 6.2 (a) Schematic illustration of the formation and (b) TEM image of 3D rGO wrapped TiO_2 microspheres. Source: Yan et al. [35]. Reproduced with permission of Royal of Society Chemistry. (c) Schematic illustration of the formation and (d) SEM image of TiO_2-QDs-3D GNs composites. Source: Li et al. [36]. Reproduced with permission of Elsevier. (e) Schematic illustration of the formation and (f) SEM image of TiO_2/3D rGO aerogel. Source: Yu et al. [38]. Reproduced with permission of Royal of Society Chemistry.

exhibited strong interfacial interaction (Figure 6.2b). The 3D rGO network can not only enhance the electrical conductivity but also prevent the direct contact between TiO_2 and electrolytes. The 3D rGO wrapped TiO_2 microspheres showed significantly enhanced rate performance when compared with TiO_2 microspheres. The second is using 3D graphene as template, and then growing TiO_2 nanostructures on its surface. For example, Li et al. used nickel as template to prepare 3D porous graphene-like networks (3D GNs) [36]. Then TiO_2 quantum dots were decorated on the surface of 3D GNs template by solvothermal method (Figure 6.2c). As shown in Figure 6.2d, the as-prepared TiO_2-QDs-3D GNs composites maintain 3D interconnected porous structure, which is embedded with TiO_2-QDs. The third is using self-assembly of GO during hydrothermal/solvothermal solution to one-step prepare TiO_2/3D rGO composites [37]. For example, as shown in Figure 6.2e, after hydrothermal reaction and freeze-drying treatment, the TiO_2/3D rGO composites were achieved [38]. The SEM image (Figure 6.2f) confirms the 3D rGO network and porous structure. When evaluated the TiO_2, TiO_2/2D rGO composite, and TiO_2/3D rGO composite as anode materials of LIBs, the TiO_2/3D rGO composite exhibited the best cycling and rate performance. Especially, the reversible capacity at high current density for TiO_2/3D rGO composite is significantly larger than those for TiO_2 and TiO_2/2D rGO composite, suggesting the excellent electrical conductivity of TiO_2/3D rGO composite.

To further improve the flexibility and expand the application fields, free-standing TiO_2/3D graphene electrodes, including films and foams without conductive agent and binder were prepared. The free-standing TiO_2/3D graphene film electrodes were prepared by two methods. One is preparing 3D graphene paper in advance, and then decorating TiO_2 into 3D graphene paper template. For example, as shown in Figure 6.3a, Hu et al. used filtration method to fabricate graphene paper, then the TiO_2 nanoparticles were decorated onto the graphene paper through hydrothermal reaction [39]. The obtained TiO_2/graphene films (Figure 6.3b) are free-standing and flexible. Therefore, they can be directly used as anode without the assistance of conductive agents and binders. The other one is mixing TiO_2 or TiO_2 precursor with GO together in advance, and then preparing TiO_2/graphene films via filtration (or spin coating) and post-processing [40, 42, 43]. For example, as shown in Figure 6.3c, Zeng et al. prepared Ti-based MXene/GO films via vacuum filtration [40]. Then the Ti-based MXene/GO film was transformed into TiO_2/rGO or sodium titanate NTO/rGO after annealing and hydrothermal process. The prepared NTO/rGO film shows excellent flexibility (Figure 6.3d) due to the 3D rGO template. In addition, the general synthetic strategy of TiO_2/graphene foams contains two steps—the preparation of free-standing graphene foam and the decoration of TiO_2 on the template. For example, as shown in Figure 6.3e, nickel foam was selected as template to fabricate laterally oriented graphene (LOG) and edge-oriented multilayer graphene (EOG) through plasma-enhanced chemical vapor deposition (PECVD) method. Then the as-prepared EOG foam was used as template to grow TiO_2(B) nanosheets and finally free-standing TiO_2(B)/EOG foam was obtained [41].

For the application in SIBs, the templated fabrication methods of TiO_2/3D graphene composites are similar to those in LIBs. For example, Xiong et al. used TiO_2 nanospheres as template to synthesize 3D rGO coated TiO_2 nanosphere composites via hydrothermal method [44]. The rGO supported TiO_2 nanosphere composites also exhibited significantly larger reversible capacity than bare TiO_2 nanospheres. Zhu et al. prepare TiO_2/3D rGO composites through self-assembly strategy during hydrothermal process [45]. Li et al. [46] and Wu et al. [47] prepared free-standing TiO_2(B) nanofibrils/graphene films and TiO_2/3D graphene foams, respectively, for significantly improving the electrochemical performance of TiO_2 anode of SIBs. The reasons why the incorporation of templated fabrication graphene can significantly improve the capacity, cycling and rate performance of TiO_2 have been studied by density functional theory (DFT). For example, Xu et al. found that graphene incorporation gives rise to interfacial sodium storage [48]. They constructed models of common anatase TiO_2 (101) surface and TiO_2/graphene interface. As shown in Figures 6.4a,b, the two models were inserted of Na coverage

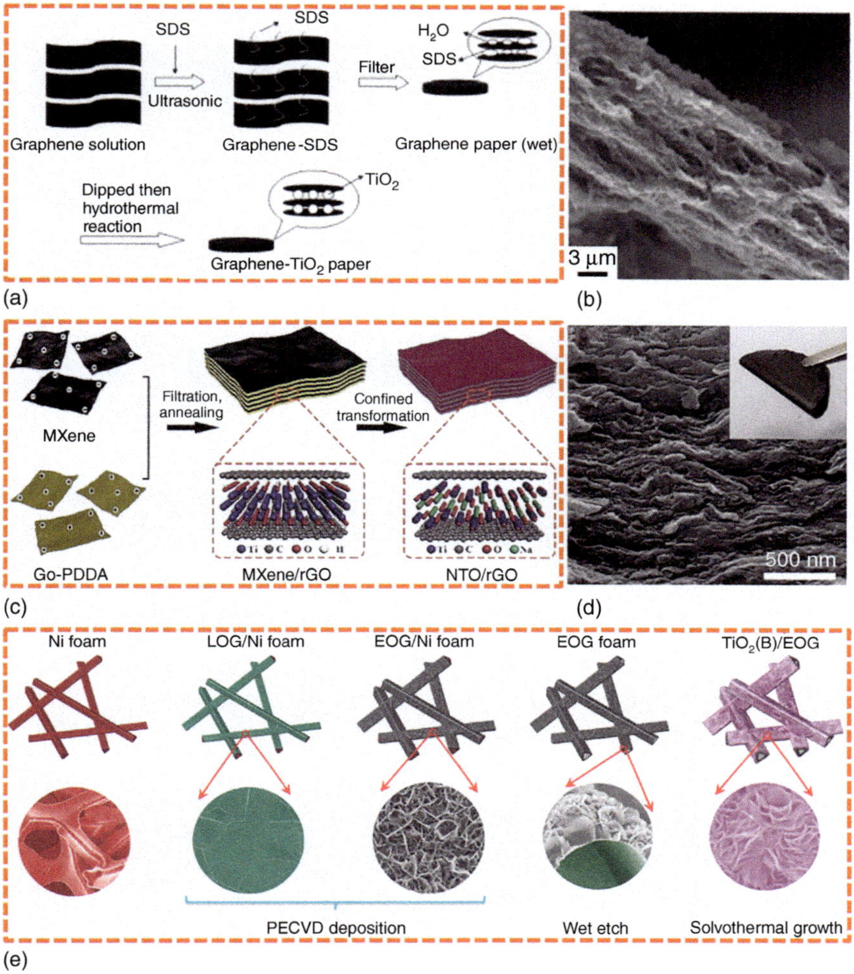

Figure 6.3 (a) Schematic illustration of the formation and (b) SEM image of TiO_2/graphene film. Source: Hu et al. [39]. Reproduced with permission of Elsevier. (c) Schematic illustration of the formation and (d) SEM image of NTO/rGO film. Source: Zeng et al. [40]. Reproduced with permission of WILEY-VCH. (e) Schematic illustration of the process and corresponding SEM images of TiO_2(B)/EOG foam. Source: Ren et al. [41]. Reproduced with permission of Elsevier.

from 0.125 monolayer (ML) to 1 ML. The positive and negative charges changes were obtained and shown in Figures 6.4c,d. The results indicated that graphene layer is beneficial to accept the negative charges to storage additional Na atoms. In addition, graphene can not only improve the electrical conductivity of TiO_2 but also enhance the Na^+ diffusion kinetics in TiO_2. As shown in Figure 6.4e, the Na^+ diffusion energy barrier in graphene-TiO_2(B) (001) interface is significantly smaller than that in bulk TiO_2(B) [49].

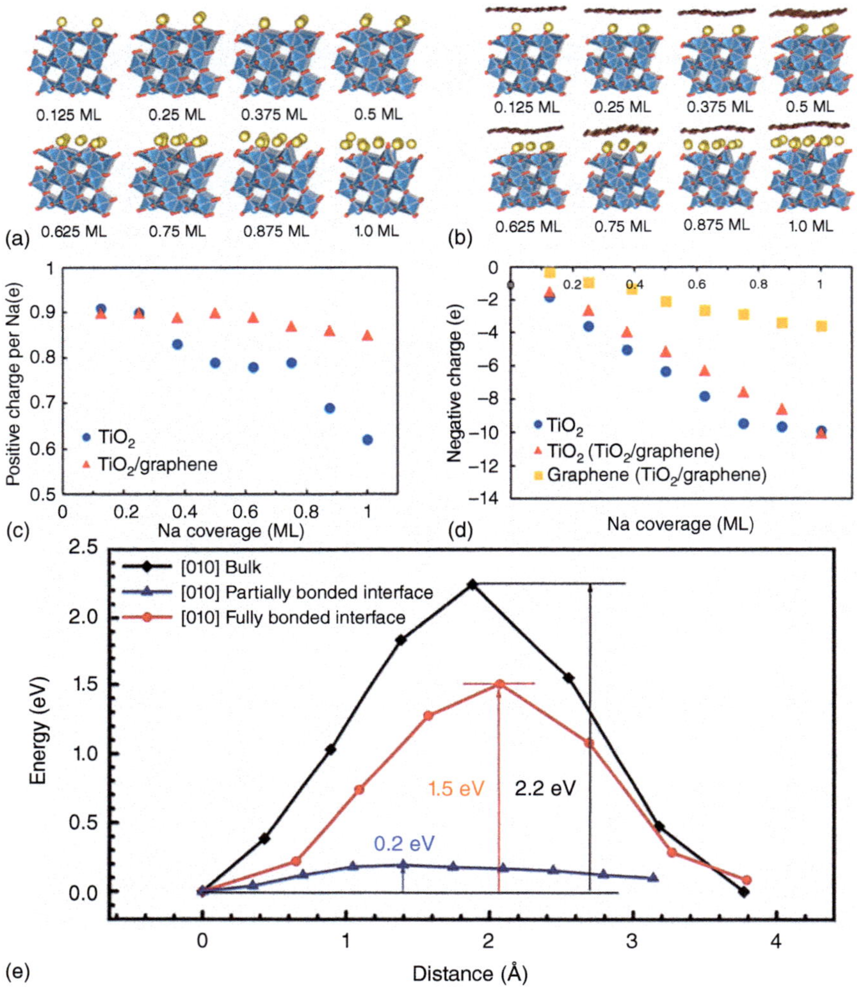

Figure 6.4 Side views of the models with an increase of Na coverage from 0.125 monolayer (ML) to 1 ML on (a) the TiO_2 (101) surface and (b) at the TiO_2/graphene interface. Red, O; cyan, Ti; brown, C; yellow, Na. (c) Positive charges per Na atom with an increase of Na coverage on the TiO_2 (101) surface and at the TiO_2/graphene interface, respectively. (d) Negative charges in TiO_2 and graphene with an increase of Na coverage. Source: Xu et al. [48]. Reproduced with permission of American Chemical Society. (e) Migration activation energy of the Na^+ ion diffusing along the [010] direction in bulk TiO_2(B), fully bonded and partially bonded graphene-TiO_2(B) (001) interface calculated with DFT. Source: Chen et al. [49]. Reproduced with permission of Springer Nature.

6.3 Graphene-Based Alloying-Type Composites

6.3.1 Metal/Graphene Alloy-Type Composites

The group IVA and VA elements have been widely demonstrated to combine with Li^+ and Na^+ to form alloys. Metal alloy anode materials, especially, Sn and Sb have been widely explored in LIBs and SIBs in recent years due to their high theoretical capacities [50, 51]. For example, the Sn can form $Li_{22}Sn_5$ and $Na_{15}Sn_4$ alloys with Li and Na, respectively, resulting in high theoretical capacities of 994 and 847 mAh g^{-1} [50]. However, the accompanying volume changes of Sn anode (260% and 420% for LIBs and SIBs, respectively) cause structural pulverization. Therefore, it is highly urgent to incorporate alloy anode materials with graphene matrix to enhance electrical conductivity and structural stability. This review will discuss the preparation method, structure design, and electrochemical performance of Sn/graphene and Sb/graphene composites.

For Sn/graphene composites, the Sn/graphene/Sn heterostructures were prepared using GO or graphene as templates [52, 53]. For example, Wang et al. used hydrothermal method to prepare Sn nanoparticle decorated graphene composite (Sn/graphene) [52]. The Sn^{2+} can be adsorbed by GO template via electrostatic interaction. Then the Sn^{2+} can be reduced to Sn nanoparticles by $NaBH_4$ during hydrothermal process. Therefore, the Sn nanoparticles anchored on the rGO surface structure were obtained. When compared with bare Sn particles, the Sn/graphene exhibited significantly better cycling performance. However, the Sn/graphene only showed 508 mAh g^{-1} at 55 mA g^{-1} after 100 cycles, which is 62.7% of the first charge capacity. This is still unsatisfactory. Although the Sn nanostructure is anchored on the surface of graphene, it is still exposed to the electrolyte. The point contact between Sn and graphene is harmful to the structural stability. Wan et al. prepared ultrasmall Sn nanoparticles anchored rGO surface with Sn–O–C bonds (Sn–O–G) [53]. The strong Sn–O–C bond is formed with a small amount of $NaBH_4$, while the conventional Sn/G composite is formed with an excess amount of $NaBH_4$. The TEM image shows that the obtained Sn nanoparticles are ultrasmall (<3 nm) in Sn–O–G. The strong bond between G and Sn in Sn–O–G is beneficial to structural stability. As a result, the Sn–O–G displayed a significantly larger reversible capacity and cycling performance than Sn/G composite. This result suggests that the interfacial interaction between Sn and graphene shows important effect on the electrochemical performance of Sn/graphene/Sn structure.

In addition to the Sn/graphene/Sn heterostructures, the graphene/Sn/graphene heterostructures were widely reported due to the dual-sided protection [54–56]. The Sn was encapsulated in graphene, leading to strong structural stability. Generally, during the preparation process of graphene/Sn/graphene, an additional carbon coating layer needs to introduce. For example, as shown in Figure 6.5a, Luo et al. used conventional hydrothermal method to prepare SnO_2/G/SnO_2 heterostructures as template in advance. Then the SnO_2/G/SnO_2 heterostructures were coated with glucose-derived carbon by next hydrothermal method [54]. After calculation, the Sn nanosheets confined in graphene structure were obtained. As shown in

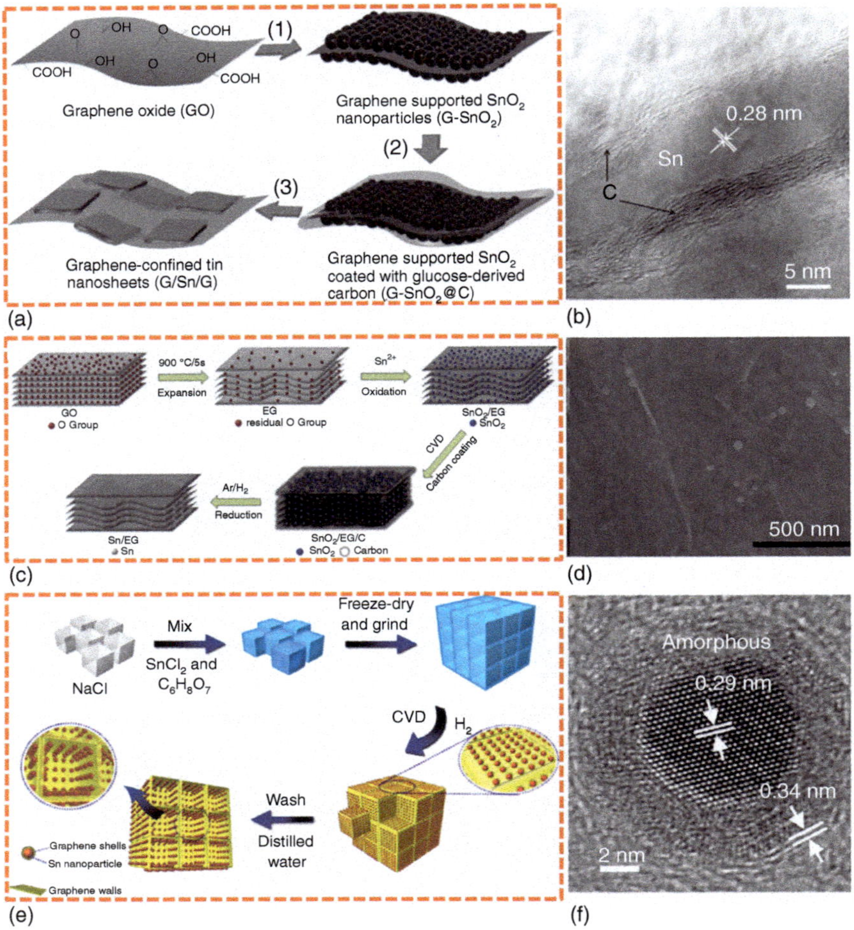

Figure 6.5 (a) Schematic illustration of the formation and (b) TEM image of graphene/Sn/graphene composites. Source: Luo et al. [54]. Reproduced with permission of WILEY-VCH. (c) Schematic illustration of the formation and (d) SEM image of graphene/Sn/graphene composites. Source: Yan et al. [55]. Reproduced with permission of Elsevier. (e) Schematic illustration of the formation and (f) TEM image of Sn@G-PGNWs composites. Source: Qin et al. [57]. Reproduced with permission of American Chemical Society.

Figure 6.5b, the Sn nanosheets are completely coated by graphene nanosheets. The graphene/Sn/graphene composite exhibited excellent structural stability and cycling performance. Another prepared method is shown in Figure 6.5c [55], the SnO_2 (~5 nm) embedded in graphene structure was prepared in advance (SnO_2/EG). Then the SnO_2/EG was coated with carbon by chemical vapor deposition (CVD) method using methylbenzene as C source. The SnO_2/EG/C was further reduced in Sn/G at high temperature with the assistance of H_2. The SEM image (Figure 6.5d) confirmed that the Sn nanoparticles (~30 nm) are completely coated by graphene. However, if the carbon coating layer does not introduce, the obtained

Sn nanoparticles show large size. For example, Zhou et al. calculated the SnO_2/GO composite at 500 °C for two hours in H_2/Ar to prepare Sn@rGO composite [56]. Some Sn nanoparticles show large size of ~200 nm due to the aggregation during the reduction process. Therefore, the obtained electrochemical performance of Sn@rGO is unsatisfactory. As mentioned above, the additional carbon coating layer is essential to the reduction of SnO_2 to Sn without aggregation due to its confinement effect [58].

The 3D structure shows smaller interfacial resistance than 2D structure. Developing 3D graphene/Sn/graphene heterostructures shows great interest [57, 59]. For example, Qin et al. developed a facile salt-template method to prepare 3D porous graphene networks anchored with Sn nanoparticles encapsulated with graphene shells of about 1 nm (Sn@G-PGNWs) [57]. As shown in Figure 6.5e, the 3D self-assembling temple introduces a narrow space between NaCl particles, which significantly limits the growth and aggregation of Sn particles during high temperature. Moreover, the Sn also has catalytic effect for graphene growth under high temperature. The obtained Sn nanoparticles exhibited small size (5–30 nm) and were encapsulated with graphene shells (Figure 6.5f). As a result, the Sn@G-PGNWs displayed promising long-life span cycling performance of 682 mA g^{-1} at 2 A g^{-1} after 1000 cycles.

The free-standing and flexible anodes of Sn/graphene composites are also highly desirable [60–62]. For example, Ji et al. used a mechanical method to prepare Sn nanopillar/graphene film [60]. First, the graphene film was transferred to a Cu collector, then the Sn nanopillars were deposited on the graphene film via thermal evaporation. Repeating the above two steps to prepare multilayer Sn nanopillar/graphene film. The film can be directly used as anode of LIBs. However, the free-standing Sn nanopillar/graphene film did not exhibit satisfactory performance due to the weak interfacial interaction between Sn and graphene. Luo prepared a flexible electrode of Sn ultrasmall nanoparticles embedded in hierarchal carbon. In the preparation process, the Sn precursor encapsulated in one-dimensional carbon nanofibers (CNFs) were synthesized in advance via electrospinning process, then the free-standing film was obtained by vacuum filtration of the nanofibers and GO solution. After calculation, the ultrasmall nanoparticles encapsulated in N, S co-doped CNFs sheathed within rGO scrolls films (Sn/NS-CNFs@rGO) were prepared. These flexible Sn/NS-CNFs@rGO exhibited an ultra-long-life span cycling performance (373 mAh g^{-1} after 5000 cycles at 1 A g^{-1}).

To decrease the volume change during the cycling and aggregation during formation, as well as improve the electrical conductivity of Sn anode. Combing Sn/graphene composites with electrochemically inactive metals, such as Fe, Co, Ni, and Cu, is also an effective strategy [63–66]. For example, Zhu et al. used convention hydrothermal method to prepare Sn nanoparticles decorated rGO composites (Sn–G) [64]. This method can also prepare Sn-Co nanoparticles decorated rGO composites (Sn–Co–G) with the addition of $CoCl_2$ precursor. Because the Co is electrochemically inactive, the initial discharge capacity of Sn–Co–G (947 mAh g^{-1}) is slightly lower than that of Sn–G (960 mAh g^{-1}). However, the Sn–Co–G shows a significantly better cycling performance than Sn–G. The Sn–Co–G exhibited

560 mAh g^{-1} at 0.5 A g^{-1} after 60 cycles, which was larger than Sn–G (430 mAh g^{-1}). This result confirmed that combing Sn with a small amount of electrochemically inactive metals is beneficial to the buffer volume change and enhances electrical conductivity.

For the Sb/graphene composites, both theoretical and experimental results reveal that the incorporation of graphene can significantly enhance the electrical conductivity and structural stability of Sb [67–73]. Moreover, the GO template can provide larger surface for antimony oxides or antimony nucleation and dispersion during synthesis. Therefore, the Sb in Sb/graphene composites always shows small size. The preparation approaches and prepared structures of Sb/graphene composites are similar to those of Sn/graphene composites due to their similar chemical and physical properties. Generally, the $SbCl_3$ and GO are the most used precursors of Sb and graphene, respectively. The preparation process contains the decoration of antimony oxides and reduction to Sb. The Sb/graphene/Sb and graphene/Sb/graphene heterostructures were prepared for application in LIBs and SIBs too.

For Sb/graphene/Sb, Nithya used one-pot solution method to prepare Sb decorated rGO composite (Sb/rGO) with $SbCl_3$ and $NaBH_4$ as Sb resource and reduction agent, respectively [67]. The Sb and rGO were also prepared by the same method. When evaluated the three samples as anode materials of SIBs, the Sb/rGO, Sb, and rGO presented 598, 128, and 172 mAh g^{-1} at 131 mA g^{-1} after 50 cycles, respectively. Moreover, the Sb/rGO showed the largest capacity retention of 93%. Liu et al. used a similar method and prepared Sb/rGO/Sb composites [68]. The Sb/rGO/Sb also showed a good cycling performance of 430 mAh g^{-1} at 100 mA g^{-1} after 200 cycles, corresponding capacity retention of 94%. The good cycling performance is attributed to the formation of strong bonding on Sb–O–C between rGO and Sb. Wan prepared Sb–O–G by the above method and Sb/G by reducing Sb–O–C at high temperature in Ar/H_2 atmosphere to investigate the phenomenon [69]. They used various characterization technologies to confirm the Sb–O–C bond in Sb–O–G composite. When evaluated the Sb, Sb/graphene, and Sb-O-G as anode materials of SIBs, the Sb–O–G showed the largest capacity, the best cycling performance, and the best rate performance (Figures 6.6a,b). As shown in Figure 6.6c, in Sb–O–G composites no aggregation of Sb occurs after deep cycles, whereas in Sb/G composites serious aggregation of Sb occurs. The strong Sb–O–C bonds play a positive role in structural stability. The Sb–O–C bonds in Sb–O–G composites are beneficial to protect the anchored Sb nanoparticles from aggregation.

The graphene/Sb/graphene heterostructures were also prepared to further enhance the structural stability of Sb during cycling process [70, 71]. As shown in Figure 6.6d, the as-prepared G@SbO_x was coated by phenol-formaldehyde (PF) resin before calculation [71]. After calcination G@SbO_x@PF at 550 °C for three hours in H_2/Ar, the SbO_x was reduced in Sb nanoparticles and the PF was carbonized. For comparison, the G@Sb composite was also synthesized under the same condition by calcination of the G@SbO_x. The SEM images of G@Sb (Figure 6.6e) and G@Sb@C (Figure 6.6f) show that small size Sb nanoparticles are obtained with the protection of carbon, whereas the large size and aggregate Sb particles are observed without the protection of carbon. The rare performance

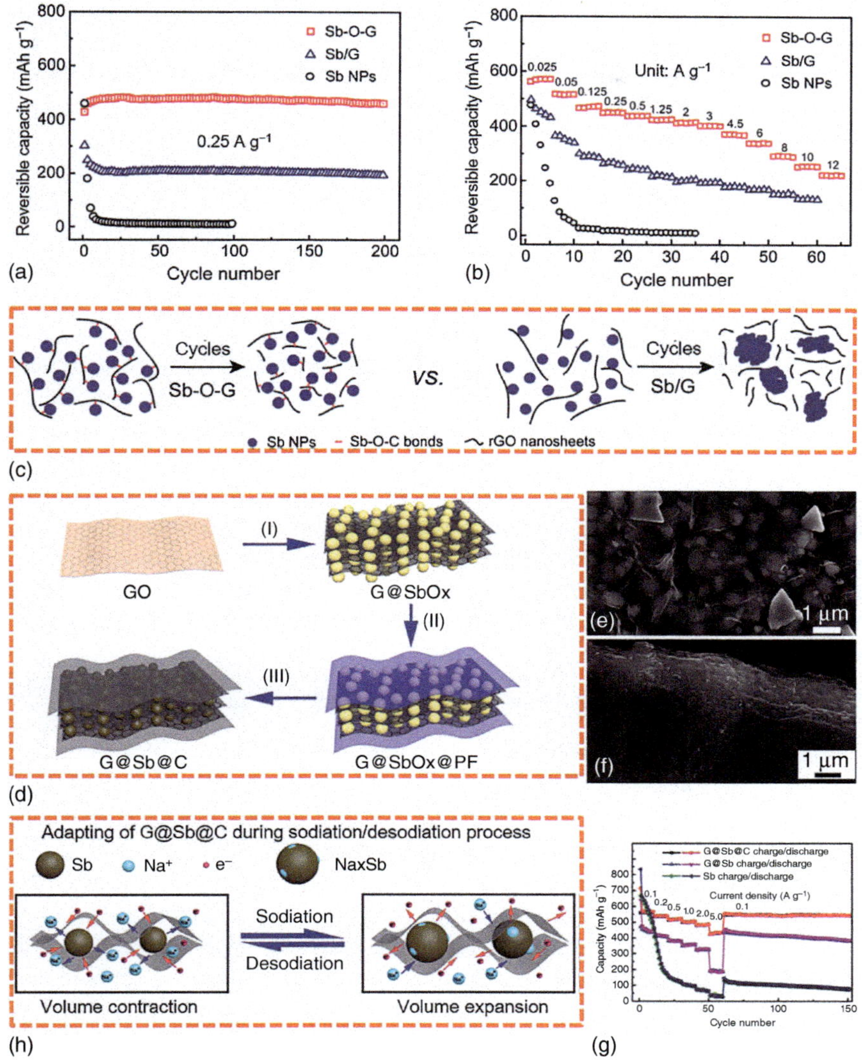

Figure 6.6 The comparison of rate (a) and cycling (b) performances of the Sb–O–G, Sb/G, and pure Sb NPs. (c) The schematic of comparing the structural change of the Sb–O–G and Sb/G materials during cycling when used as anode for an SIB. Source: Wan et al. [69]. Reproduced with permission of American Chemical Society. (d) Schematic illustration of the formation of the sandwich-like G@Sb@C composite. SEM images of (e) G@Sb and (f) G@Sb@C. (g) Rate performance of Sb, G@Sb, G@Sb@C. (h) Schematic illustration of the volume changes of G@Sb@C during the cycling process. Source: (e–h) Wang et al. [71]. Reproduced with permission of Royal of Society Chemistry.

(Figure 6.6g) exhibited that the G@Sb@C shows the largest capacities at all test current densities. This result is attributed to the unique structure of G@Sb@C. The Sb nanoparticles are confined by graphene layer and C layer. The G@Sb@C composites have excellent electrical conductivity. Moreover, the graphene and C protect layers can effectively buffer the volume change during repeated cycles (Figure 6.6h).

The free-standing Sb/graphene anode is also important for the increasing application in flexible devices [72, 73]. For example, Zhang et al. prepared Sb/rGO paper electrode by reducing the peeled SbO_x/GO film [72]. The obtained film is flexible and the Sb nanoparticles are anchored on the rGO template. After cycling at 0.2 A g^{-1} for 100 cycles, the Sb/rGO paper maintained 581 mAh g^{-1}. The rate performance at high current densities for Sb/rGO paper anode may be suffered from the slow electrolyte diffusion perpendicular to 2D rGO. Yi et al. used 3D Ni foam as template to prepare free-standing electrode [73]. First, the Ni can be replaced with Sb^{3+} to form Ni^{2+} and Sb, leading to Sb thin film on the surface of Ni foam. Then the rGO layer was adsorbed on the surface of as-prepared Sb film through electrophoretic deposition process. The Sb weight content is low in the obtained 3D Ni-Sb/rGO foam. The 3D Ni-Sb/rGO foam anode displayed a small reversible capacity. Moreover, the interfacial interaction between Sb and rGO is poor, the cycling performance is also unsatisfactory. Therefore, the flexible Sb/graphene electrode still needs to develop.

6.3.2 Nonmetal/Graphene Alloy-Type Composites

Si and P are the typical nonmetal alloy materials in Group IVA and VA elements, respectively. They also suffer from serious structural pulverization as well as poor electrical conductivity. Therefore, incorporating Si or P with graphene has been considered as a promising strategy to address these problems.

The Si anode materials have attracted great attention in the past decades. The Si/graphene composites have been widely reported by many researchers. The SnO_x and SbO_x can be easily reduced into Sn and Sb, respectively. However, the reduction of SiO_x to Si can happen through magnesium thermal reduction process. Therefore, the preparation strategies for Si/graphene composites are different from metal/Si alloy-type composites. Generally, the synthesis approach of Si/graphene composites can be divided into two methods—one is reduction of SiO_x on GO or graphene templates [74–76], the other one is incorporation of graphene templates with Si [77–85].

In the first method, the magnesium thermal reduction is used to reduce SiO_x. For example, Du et al. used the SiO_2/graphene as template to prepare Si/graphene composite via magnesium thermal reduction [74]. The Si/graphene shows significantly enhanced capacity and cycling performance than Si/graphene mixture and Si. However, the electrochemical performance for Si/graphene is still unsatisfactory, because the anchored Si is still exposed to electrolyte. Chen et al. decorated SiO_2 nanoparticles on the rGO foam (SiO_2@GF) in advance [75]. After magnesium thermal reduction and CVD with C_2H_2, the SiO_2/GF were transformed into C@Si@GF. The CVD process can bring in an additional carbon protecting layer. Therefore, the C@Si@GF

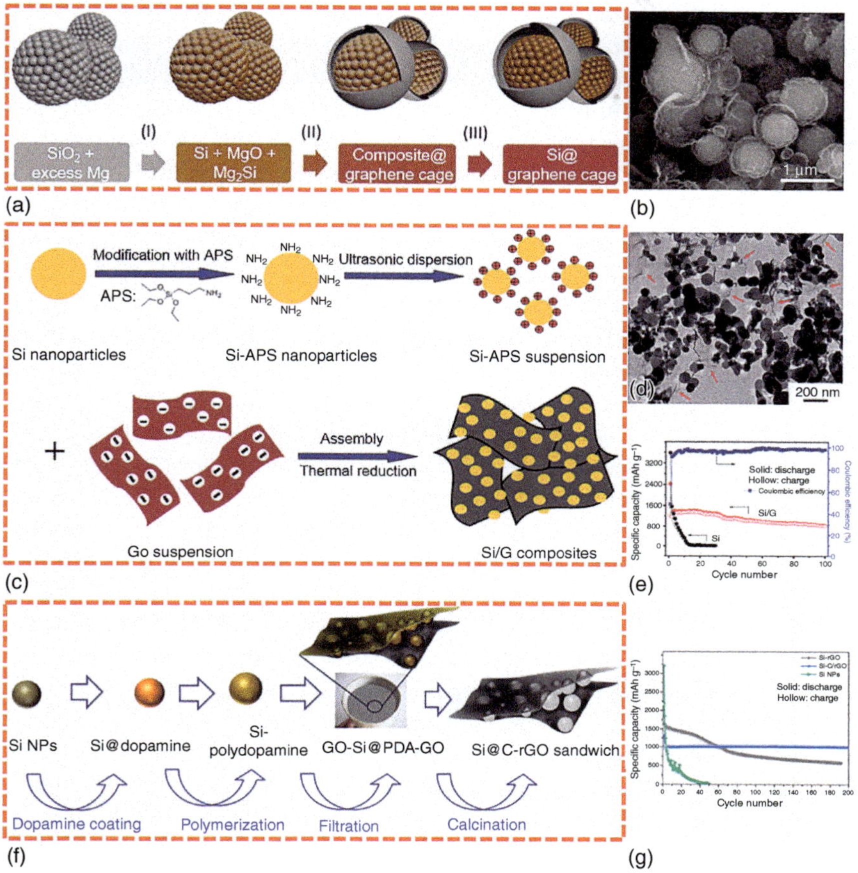

Figure 6.7 (a) Schematic illustration of the formation process and (b) SEM image of Si@graphene cage. Source: Nie et al. [76]. Reproduced with permission of WILEY-VCH. (c) Schematic illustration of the formation process and (d) TEM image of Si–G composites. (e) Cycling performance of Si and Si-G composites. Source: Zhou et al. [78]. Reproduced with permission of Royal of Society Chemistry). (f) Schematic diagram for the synthetic procedure of the sandwich-structured Si@C-rGO composite. (g) Cycling stabilities of Si@C-rGO, Si-rGO, and pristine Si NPs at a current density of 300 mA g^{-1}. Source: Agyeman et al. [81]. Reproduced with permission of WILEY-VCH.

exhibited a good cycling performance. The graphene can also be incorporated after the magnesium thermal reduction. For example, as shown in Figure 6.7a, the SiO_2 spheres were mixed with excess Mg, after magnesium thermal reduction at high temperature, the mixture of mesoporous Si, MgO, and Mg_2Si was obtained [76]. Notably, the MgO can be used as template to grow graphene in the CVD process. Then the MgO and Mg_2Si were removed. Finally, the SEM image (Figure 6.7b) confirmed that the obtained Si@graphene composite shows a yolk–shell cage structure.

In the second method, GO and Si are commonly used precursors. The mixture of Si and graphene template can be obtained by mechanical mixing. For example,

Zhao et al. used simple filtration method to prepare Si/graphene paper electrode [77]. The interfacial interaction between Si and graphene in Si/graphene paper electrode is weak. The cycling and rate performance are still limited. Many efforts have been made on contracting chemical bonding between Si and graphene. Generally, the GO has many functional groups, such as –OH and –COOH, the commercial Si particles always have an amorphous SiO_x layer, which exhibits –OH on their surface. Therefore, both the GO and Si particles always display negative charges. To enhance the electrostatic interaction between GO and Si, turning the negative charge on the Si surface to positive charge has been considered as an effective strategy [78–80]. For example, as shown in Figure 6.7c, Zhou et al. used aminopropyltriethoxysilane (APS) to modify the surface charge of Si nanoparticles [78]. The surface of Si nanoparticles absorbs amino groups ascribed from APS, leading to positive charges. The positive Si-APS nanoparticles and negative GO display strong electrostatic interaction. After thermal reduction, the Si–G composites were prepared. The Si nanoparticles are uniformly distributed on the surface of rGO with strong interfacial interaction (Figure 6.7d). The Si–G composites displayed a good cycling performance of 822 $mAh\,g^{-1}$ after 100 cycles at 0.1 $A\,g^{-1}$, which is significantly larger than that of Si nanoparticles (Figure 6.7e). This strategy is beneficial to prepare Si/rGO composites with strong interfacial interaction. Therefore, the rGO can effectively enhance the electrical conductivity and structural stability of Si. However, the Si is still exposed to the electrolyte, the structural stability is still unsatisfactory. To avoid the direct contact between Si nanoparticles and electrolyte, another strategy is to introduce organic polymers as templates, such as polydopamine (PDA), poly(diallydimethylammonium chloride), polyethylene oxide, polyvinyl alcohol, and polyethylene glycol to modify the surface [81–85]. For example, as shown in Figure. 6.7f, the Si nanoparticles were coated by PDA under stirring. The PDA coating layer serves as a binder to anchor the Si nanoparticles on the surface of GO nanosheets. After calcination at high temperature, Si@C-rGO composites were prepared. The Si is fully coated by carbon and anchored on the surface of rGO [81]. For comparison, the Si-rGO composites were also prepared without addition of PDA. As expected, the Si@C-rGO composites exhibited an excellent cycling performance as anode materials of LIBs, which is greatly better than that of Si particles (Figure. 6.7g). Moreover, the cycling performance of Si@C-rGO is also significantly larger than Si-rGO. Therefore, this strategy shows great improvement in electrical conductivity and structural stability of Si nanoparticles.

The P/graphene composite was also studied as anode materials for LIBs and SIBs. For example, Sun et al. prepared phosphorene/graphene hybrid, in which few phosphorene layers were intercalated between graphene layers [86]. When evaluated the phosphorene/graphene hybrid as anode material of SIBs in a voltage window of 0–1.5 V, it displayed 2440 $mAh\,g^{-1}$ after 100 cycles at 0.05 $A\,g^{-1}$. The excellent reversible capacity and good cycling performance are promising for potential application. However, the P is easily oxidized in air condition. It is hard to prepare stable P/graphene composites. Therefore, the research on P/graphene anode materials needs more attention.

6.4 Graphene-Based Conversion-Type Composites

6.4.1 Transition Metal Oxides/Graphene Composites

The transition metal oxides (MO_x), such as FeO_x, CoO_x, and NiO_x, have been widely studied as anode materials of LIBs and SIBs due to their low cost, good safety, and high theoretical capacity [87]. The general conversion reaction is described as Eq. (6.1):

$$MO_x + 2xLi^+/Na^+ \rightarrow M + xLi_2/Na_2O \tag{6.1}$$

The MO_x materials show large structural change after the discharge process. Therefore, the structural stability is still poor. The obtained Li_2O and Na_2O are insulators. Incorporating the MO_x materials with graphene has been widely reported to enhance structural stability and electrical conductivity [87, 88].

For preparing MO_x/graphene composites, hydro-/solvo-thermal method is mostly used. During the process, Fe, Co, and Ni salts were added into the GO solution, after hydro-/solvo-thermal process and calcination, MO_x/graphene. In addition, the MO_x can also be derived from M metals, $M(OH)_x$, and related MOF. Incorporation of graphene with these precursors is also an approach to prepare MO_x/graphene. Moreover, directly mixing the MO_x and graphene to construct composites is also an effective strategy. The MO_x/graphene/MO_x, MO_x@graphene, MO_x@C/graphene, free-standing MO_x/graphene structures have been widely reported.

For conventional MO_x/graphene/MO_x structures, Zhou et al. in 2010 prepared graphene nanosheets decorated with Fe_3O_4 nanoparticles (Fe_3O_4/GNS) composites [89]. The Fe_3O_4 nanoparticles are uniformly distributed on the surface of graphene. The graphene can enhance the electrical conductivity of Fe_3O_4 and decrease structural pulverization. When evaluated as anode materials of LIBs, Fe_3O_4/GNS showed 1026 mAh g^{-1} after 30 cycles at 35 mA g^{-1}. The reversible capacity and cycling stability of GNS/Fe_3O_4 are significantly larger than commercial Fe_3O_4 and bare Fe_2O_3. The SEM image of GNS/Fe_3O_4 and commercial Fe_3O_4 after 30 cycles were performed to investigate the structural changes. The result confirmed that the GNS/Fe_3O_4 composites show good structural stability due to the assistance of GNS. This structure (FeO_x/graphene/FeO_x) and similar electrochemical results have been also demonstrated in other literature [90]. The CoO_x/graphene/CoO_x structures have also been widely reported to improve the electrochemical performance of CoO_x [91–93]. For example, Wu et al. prepared Co_3O_4 nanoparticles anchored graphene (Co_3O_4/graphene) composites via calcination of $Co(OH)_2$/graphene composites at 450 °C in air [91]. As expected, the Co_3O_4/graphene showed a better cycling performance than Co_3O_4 and graphene. For the NiO_x/graphene/NiO_x structures, several literatures have been reported for application in LIBs and SIBs [94–96]. For example, Cheng's group synthesized NiO nanosheets anchored on the graphene through Ni–O–C bonds composites (NiO NS/graphene) [94]. The strong interfacial chemical bonds not only facilitate the electrical transfer from graphene to NiO but also hinder the structural pulverization of NiO during cycling. As a result, the NiO NS/graphene composites exhibited

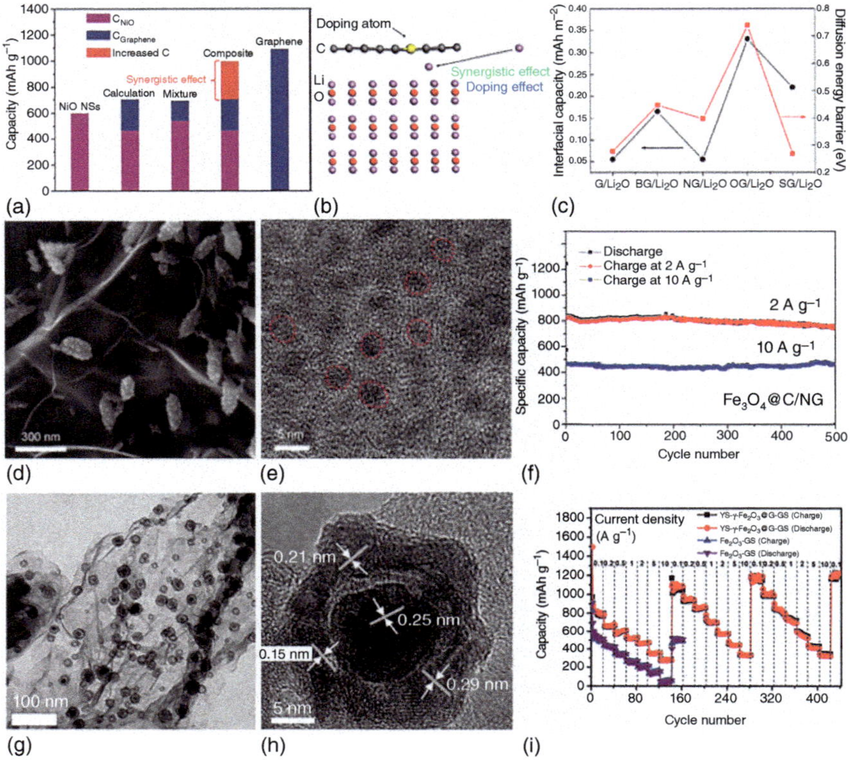

Figure 6.8 (a) First reversible specific capacity of the graphene, NiO NSs, NiO NS/graphene composite, NiO NS-graphene mixture, and the calculated specific capacity based on experimental values. The red rectangle indicates the increased capacity due to the synergistic effect between graphene and NiO NSs. Source: Zhou et al. [94]. Reproduced with permission of American Chemical Society. (b) Side view of atomic structure of the doped graphene/Li_2O (111) interface. (c) Li-ion storage behavior in graphene/Li_2O and doped graphene/Li_2O interfaces. Source: Wang et al. [97]. Reproduced with permission of American Chemical Society. (d) TEM image, (e) HRTEM image, and (f) cycling performance of Fe_3O_4@C/NG composites. Source: Qi et al. [98]. Reproduced with permission of American Chemical Society. (g) TEM image and (h) HRTEM image of YS-γ-Fe_2O_3@G-GS. (i) Rate performance of YS-γ-Fe_2O_3@G-GS and Fe_2O_3−GS composites. Source: Zhang et al. [99]. Reproduced with permission of Royal of Society Chemistry.

better cycling stability than graphene, NiO NS, and NiO NS/graphene mixture. Meanwhile, the reversible capacity showed the same rule. Moreover, the first reversible capacity of NiO NS/graphene composites is larger than the theoretically calculated capacity based on experimental values (Figure 6.8a). This result directly demonstrated that synergistic effect exists between graphene and NiO due to the strong interfacial interaction. A similar phenomenon was also confirmed in other experimental reports [100]. Moreover, Wang et al. used computing method via DFT to study the Li-ion storage behavior in the Li_2O/graphene and doped Li_2O/graphene interfaces [97]. The atomic structures are shown in Figure 6.8b. The obtained result

(Figure 6.8c) revealed that the synergistic effect in Li_2O/graphene interface is attributed to the enhanced interfacial interaction by Li insertion. The doping elements showed different influences on the interfacial synergistic effect.

To avoid the physical contact between MO_x and electrolyte, MO_x@graphene structures were proposed [101–104]. Generally, the preparation of MO_x@graphene structures contains two steps—the first is preparing MO_x in advance and the second is coating graphene on the surface of MO_x. For example, Wang et al. used a one-step P-milling method for 20 hours to prepare Fe_2O_3 particles embedded in few-layer graphene nanosheets composites (P20) [101]. The Fe_2O_3 particles are completely covered by graphene nanosheets. For comparison, other Fe_2O_3/graphene composites were also prepared with different P-milling times. The P5 and P10 samples correspond to 5 and 10 hours, respectively. In the two samples, the Fe_2O_3 particles anchor on the surface of graphene without being completely covered. The cycling performance exhibited that the P20 sample exhibits a high reversible capacity of 758 mAh g^{-1} after 300 cycling at 0.2 A g^{-1}. This reversible capacity is significantly larger than P5, P10, and bare Fe_2O_3. The Fe_2O_3 particles completely covered by graphene nanosheets in P20 are significantly beneficial to enhance the structural stability. Jing et al. prepared graphene embedded Co_3O_4 rose-spheres (G-Co_3O_4) for enhanced performance in LIBs [102]. The authors synthesized Co_3O_4 template in advance. Then the as-prepared Co_3O_4 was modified by APS to change into positive charge. The positively charged Co_3O_4 rose-spheres were easily coated by GO with negative charge through electrostatic interaction. After reduction, G-Co_3O_4 composites were prepared. The G-Co_3O_4 composites displayed smooth surface with exposed graphene coating layer. The cycling performance demonstrated that the G-Co_3O_4 composite presented larger reversible capacity and better cycling stability than Co_3O_4 and graphene. The flower-like NiO coated by rGO composites were also prepared and demonstrated to be show enhanced electrochemical performance [103]. Notably, the transition metals, such as Fe, Co, Ni, have catalytic effects to grow graphene nanosheets on their surface. For example, Zou et al. used Ni-MOF as a precursor to prepare hollow NiO/Ni/graphene composites [104]. The as-prepared Ni-MOF was calcined at 450 °C for 30 minutes in Ar atmosphere. During the process, the hollow structure was formed and the graphene was catalyzed on the surface of Ni nanoparticles. The obtained hollow Ni/graphene composites were further heated to 200 °C in air to oxidize the Ni to NiO. The TEM and high-resolution TEM (HRTEM) images revealed the graphene coating structure in NiO/Ni/graphene nanocomposites. Because the graphene effectively protects the NiO nanoparticles with a coating structure, the NiO/Ni/graphene nanocomposites display long-life span cycling performance in LIBs. However, the SIBs' performance is still unsatisfactory due to the large volume change.

The MO_x@C/graphene structure, which equals to introduction of an additional carbon coating layer in conventional MO_x/graphene/MO_x structure, can be easily prepared by many methods. Among them, in situ preparation of MO_x@C/graphene composites via MOF/graphene templates has been reported [98, 105, 106]. For example, Qi et al. used hydrothermal method to synthesize nitrogen-doped graphene (NG)/MIL-88-Fe-NH_2 in advance with different content. Then the

Fe_3O_4@C/NG composites were prepared by calcination of NG/MIL-88-Fe-NH_2 at 500 °C in N_2 atmosphere for one hour [98]. The TEM image (Figure 6.8d) confirmed that the MOF-derived gap-like structures are tightly anchored on the surface of NG. The HRTEM image (Figure 6.8e) revealed that the Fe_3O_4 nanoparticles are embedded in carbon materials in the gap-like structure. The structural stability of Fe_3O_4 nanoparticles is greatly enhanced by the carbon coating layer, the electrical conductivity is improved by NG. The small size of nanoparticles is favorable to ionic diffusion process. As a result, the Fe_3O_4@C/NG composites exhibited large reversible capacity and cycling stability at high current densities. As shown in Figure 6.8f, the reversible capacity of Fe_3O_4@C/NG remained 458 mAh g^{-1} even after cycling at 10 A g^{-1} for 1000 cycles. This is an outstanding long-life cycling performance among all reported MO_x@C/graphene composites. The other traditional method is coating carbon on MO_x/graphene/MO_x template [99, 107, 108]. For example, Zhang et al. prepared Fe_2O_3/graphene sheets (Fe_2O_3/GS) composites in advance. Then the Fe_2O_3/GS composites were post-treated in CVD process under CH_4 atmosphere. Finally, the sandwiched graphene sheets inserted with yolk–shell γ-Fe_2O_3@graphene (YS-γ-Fe_2O_3@G-GS) composites were obtained [99]. The post CVD process introduces a space between γ-Fe_2O_3 and new-formed graphene shell (Figures 6.8g,h). This composite should achieve excellent electrical conductivity, durably structural stability, and large diffusion kinetics. As a result, the YS-γ-Fe_2O_3@G-GS composites showed a extremely higher rate performance than Fe_2O_3/GS (Figure 6.8i). The YS-γ-Fe_2O_3@G-GS composites exhibited 350 mAh g^{-1} even at 10 A g^{-1}. The YS-γ-Fe_2O_3@G-GS composites also delivered outstanding recovery after three rounds of rate tests.

In addition, the free-standing MO_x/graphene electrodes were also prepared for flexible devices [109–111]. For example, Wang et al. prepared $Co(OH)_2$/GO hybrid films through vacuum filtration of suspension containing $Co(OH)_2$ and GO nanosheets [110]. After calcination, the Co_3O_4/graphene sheets (Co_3O_4/GS) hybrid films were obtained. The films are flexible and can be directly used as anodes of LIBs. Shao et al. prepared NiO microspheres loaded on graphene foam (NiO/GF) films [111]. The obtained films can be directly used as anodes of LIBs. Although the electrical conductivity and structural stability of free-standing films have been significantly enhanced on electrode level, the obtained electrochemical performance, especially long-life span cycling performance of reported free-standing MO_x/graphene electrodes is still unsatisfactory. This is because of the lack of attention focusing on materials and atomic levels.

6.4.2 Transition Metal Sulfides/Graphene Composites

6.4.2.1 Conventional Metal Sulfides/Graphene Composites

The conventional transition metal sulfides (MS_x), such as FeS_x, CoS_x, and NiS_x, have been widely studied as anode materials of LIBs and SIBs [112]. The general conversion reactions in LIBs and SIBs are described as Eq. (6.2):

$$MS_x + 2xLi^+/Na^+ \rightarrow M + xLi_2/Na_2S \tag{6.2}$$

This conversion storage mechanism leads to high theoretical capacity, as well as large structural change. The transformed Li_2S and Na_2S not only show poor electrical conductivity but also exhibit dissolution property with electrolytes. Therefore, incorporation of MS_x materials and graphene has been widely reported to enhance structural stability and electrical conductivity [112].

Three methods have been widely adopted to prepare MS_x/graphene composites—mixing the metal sulfides and graphene [113], preparing MO_x/graphene template followed with sulfuration [114], and one-pot synthesis of MS_x/graphene composites [115–118]. Various MS_x/graphene structures have been designed and reported. The typical structures contain MS_x/graphene/MS_x, MS_x@graphene, MS_x@C/graphene, free-standing MS_x/graphene structures. In the following section, the development of each templated fabrication of structure is briefly introduced.

The MS_x/graphene/MS_x is the most widely reported structure due to its easy preparation. For example, Xu et al. prepared FeS_2 microspheres anchored on rGO (FeS_2/rGO) composites by a one-pot hydrothermal method [115]. The FeS_2 nucleated and grew on the surface of rGO with strong chemical bond. The TEM image confirmed that the FeS_2 microspheres with a diameter of 500 nm are tightly anchored on the surface of rGO. The strong interfacial interaction between rGO and FeS_2 facilitates the electrical transfer from rGO to FeS_2. In addition, the rGO can limit the aggregation of FeS_2 during hydrothermal process. This is beneficial to decrease the ionic diffusion path distance and enhance the reaction contact area. The rGO also shows durable structural stability to accommodate the volume change of FeS_2 during discharge–charge. As a result, the prepared FeS_2/rGO composites possess high electrical conductivity, structural stability, and Li^+ diffusion kinetics. When evaluated bare FeS_2, rGO, and FeS_2/rGO composite as anode materials of LIBs, the FeS_2/rGO composite presented the best cycling stability and reversible capacity. The FeS_2@rGO composite maintained 1131 $mAh\,g^{-1}$ after 100 cycles at 0.1 $A\,g^{-1}$. Xie et al. prepared CoS_2/graphene/CoS_2 heterostructure (CoS_2/G) via one-pot hydrothermal method using $CoCl_2$, thioacetamide, and GO as Co, S, and graphene sources, respectively [116]. The GO provides abundant nucleation sites for CoS_2 growth. Therefore, the size of CoS_2 in CoS_2/G is significantly smaller than that in bare CoS_2. The CoS_2 nanocrystals show a small size (100–150 nm). The CoS_2 nanocrystals are uniformly distributed on the surface of graphene sheets. The graphene can effectively enhance the electrical conductivity and structural stability of CoS_2. Therefore, the CoS_2/G composite displayed better cycling performance than bare CoS_2. Geng et al. also prepared graphene nanosheets decorated with NiS (NiS/GNS) composites by one-pot solvothermal method [117]. The NiS nanosheets are assembled into flower structures without the addition of GO. The addition amount of GNS shows important influence on the morphologies of obtained NiS/GNS composites. The NiS displayed nanoflake and nanorod-assembled flower morphologies in NiS/GNS (4 : 1) and NiS/GNS (3 : 1) composites. The NiS nanorod in flower morphology becomes sharper with increasing the addition amount of GO in the solvothermal process. The nanorod-assembled NiS nanoflowers are uniformly distributed on the surface of graphene in NiS/GNS (1 : 1). The diameter of individual NiS nanorod is around 30–40 nm. Both the cycling stabilities of

NiS/GNS (2 : 1) and NiS/GNS (1 : 1) composites are better than NiS. This result indicates that graphene can significantly enhance the structural stability of NiS. The initial discharge capacities of NiS, NiS/GNS (2 : 1), and NiS/GNS (1 : 1) gradually become smaller due to the decreased weight content of NiS. However, after 60 cycles, the NiS/GNS (1 : 1) exhibited a high reversible capacity of 887 mAh g^{-1}, which is significantly larger than those of NiS/GNS (2 : 1) (458 mAh g^{-1}) and NiS (144 mAh g^{-1}).

For preparing MS_x@graphene structure, the general approach is coating the as-prepared MS_x with graphene. Because the MS_x materials are completely coated by graphene, the MS_x is fully protected from direct contact with electrolytes. Therefore, the MS_x@graphene composites should possess more durable structural stability than conventional MS_x/graphene/MS_x composites. For example, Qiu et al. prepared N-doped graphene-coated yolk–shell CoS_2 spheres (yolk–shell CoS_2@NG) composites [119]. The authors prepared yolk–shell CoS_2 spheres in advance, then the as-prepared CoS_2 spheres were modified by APS to become positively charged. The positively charged CoS_2 particles were mixed with negatively charged GO and ammonia solution to prepare yolk–shell CoS_2@NG composite under hydrothermal reaction. The NG sheets are completely covered the CoS_2 spheres. The yolk–shell CoS_2@NG composite delivered significantly larger reversible capacity than yolk–shell CoS_2 at all tested current densities. Notably, the reversible capacity of yolk–shell CoS_2@NG composite exhibited an excellent recovery when the current density returns to 0.2 C. This implies that the unique structure results in good reversibility. Bi et al. found that the hollow Ni-MOFs spheres can be transformed into hollow NiS_2@graphene (NiS_2@G) composites through calculation with S [120]. The Ni can catalyze the coated carbon to become graphene shell at high temperature. The Ni can further be sulfurized to NiS_2 by S. However, the solid Ni-MOFs spheres would obtain porous NiS_2@C composites. The hollow structure is delivered. The HRTEM image confirmed the core–shell NiS_2@graphene structure. When evaluated the porous NiS_2@C and hollow NiS_2@graphene composites as anode materials of SIBs, the NiS_2@graphene exhibited larger capacity at all tested current densities than porous NiS_2@C.

The MS_x@C/graphene structure is also prepared and studied in LIBs and SIBs due to its dual carbon protecting layers. For example, Yao et al. prepared porous FeS_2 nanoparticles anchored on the surface of rGO and coated by carbon layer (rGO@p-FeS_2@C) composites for SIBs [121]. As shown in Figure 6.9a, the preparation process contains five steps. The SiO_2 serves as template to introduce void between Fe_3O_4 nanoparticles. The TEM image (Figure 6.9b) confirmed the coexistence of FeS_2 nanoparticles, carbon coating layer, rGO interlayer, and

Figure 6.9 (a) Schematic illustration of the formation process and (b) TEM image of rGO@p-FeS_2@C. (c) Cycling performance of rGO@FeS_2, rGO@FeS_2@C, and rGO@p-FeS_2@C. Source: Yao et al. [121]. Reproduced with permission of Elsevier. (d) Schematic illustration of the formation process and (e) SEM image of CoS_2-N-C/3DGN. (f) Cycling performance of CoS_2-N-C and CoS_2-N-C/3DGN. Source: Song et al. [122]. Reproduced with permission of Springer.

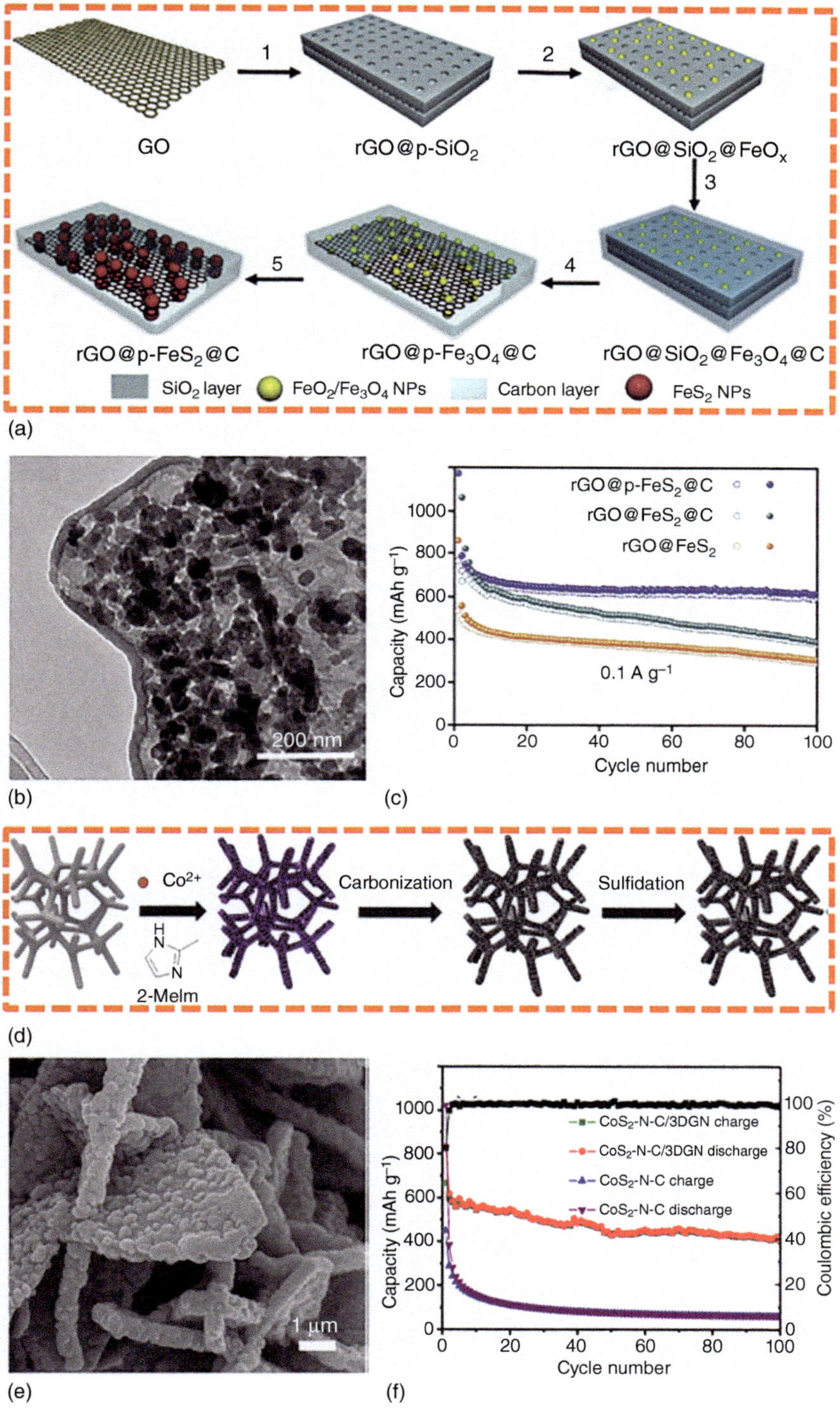
1
2
3
4
5
GO
rGO@p-SiO2
rGO@SiO2@FeOx
rGO@SiO2@Fe3O4@C
rGO@p-Fe3O4@C
rGO@p-FeS2@C
SiO2 layer
FeO2/Fe3O4 NPs
Carbon layer
FeS2 NPs
(a)
200 nm
(b)
rGO@p-FeS2@C
rGO@FeS2@C
rGO@FeS2
0.1 A g−1
Capacity (mAh g−1)
Cycle number
(c)
Co2+
2-MeIm
Carbonization
Sulfidation
(d)
1 μm
(e)
CoS2-N-C/3DGN charge
CoS2-N-C/3DGN discharge
CoS2-N-C charge
CoS2-N-C discharge
Capacity (mAh g−1)
Coulombic efficiency (%)
Cycle number
(f)

porous structure. For comparison, the rGO@FeS_2 and rGO@FeS_2@C composites were also synthesized without SiO_2 loading/carbon coating and SiO_2 loading, respectively. The cycling performance between 0.05 and 2.8 V at 0.1 A g^{-1} was presented in Figure 6.9c. The rGO@p-FeS_2@C exhibited the best cycling stability. It achieved a high reversible capacity of 598 mAh g^{-1} after 100 cycles, which is larger than those of rGO@FeS_2 (293 mAh g^{-1}) and rGO@FeS_2@C (375 mAh g^{-1}). In addition, designing interior void space between metal sulfide precursor and carbon coating is another effective strategy to prepare yolk–shell metal sulfides@carbon/graphene composites. For example, Zhao et al. prepared yolk–shell Fe_3O_4@C/G template by etching Fe_3O_4@C/G sample with 1 m HCl [123]. After sulfurating the template, yolk–shell FeS_2@C/G with void space was prepared. The MOF/GO template derived MS_x@C/graphene method has also been widely reported. For example, yolk–shell FeS_2@carbon microboxes/graphene framework composites [124] and rGO-wrapped Co_9S_8 QD embedded within mesoporous hollow carbon composites [125] were reported.

The free-standing MS_x/graphene electrodes were also reported. For example, He et al. used cauliflower-like FeS_2 anchored on three-dimensional graphene foams (3DGF-FeS_2) as electrodes of LIBs [126]. The 3DGF was fabricated by CVD method using the Ni foams as template. The free-standing 3DGF-FeS_2 shows excellent electrical conductivity at both material and electrode levels. The structural stability on electrode level is also durable. As a result, the 3DGF-FeS_2 exhibited a high capacity of 1080.3 mAh g^{-1} after 100 cycles at 0.178 A g^{-1}, which is 17 times larger than that of conventional FeS_2 electrode (60.9 mAh g^{-1}). Song et al. used the Co-MOF/three-dimensional graphene network (Co-MOF/3DGN) precursor to prepare CoS_2 encapsulated in N-doped carbon/3DGN (CoS_2-N-C/3DGN) composites [122]. Figure 6.9d shows the illustration of the formation process. The 3DGN serves as template to load Co-MOF nanosheet arrays. Under sulfuration, the Co-MOF was transformed into CoS_2-N-C nanosheet arrays. The SEM image (Figure 6.9e) confirmed the CoS_2-N-C nanosheet arrays on 3DGN. The 3DGN provides fast electrical transfer path, while the N-C protects the CoS_2 from pulverization during discharge–charge. As shown in Figure 6.9f, the CoS_2-N-C/3DGN composite exhibited a larger reversible capacity than CoS_2-N-C.

6.4.2.2 2D Metal Disulfides/Graphene Composites

Recently, the 2D metal disulfides (MS_2), such as MoS_2, WS_2, ReS_2, have attracted great attention in LIBs and SIBs due to their typical layer structure stacked by weak van der Walls interaction [127, 128]. The layered structure with large interlayer distance (~0.6 nm) facilitates the Li^+ and Na^+ diffusion. The MS_2 exhibits a conversion storage mechanism in a wide working voltage window (0–3 V). This results in a high theoretical capacity. Whereas, the electrochemical performance of 2D MS_2 is still suffered from the drastic structural change and poor electrical conductivity during discharge–charge [128]. To address this issue, combing 2D MS_2 and 2D graphene is widely reported to enhance electrochemical performance [127]. Among them, four typical structures, including MS_2/graphene/MS_2, graphene/MS_2/graphene, and MS_2@C/graphene, and free-standing MS_2/graphene, have been proposed and prepared.

For MS_2/graphene/MS_2 structure, it is widely prepared by using one-pot hydro-/solvo-thermal method. For example, Chang et al. first prepared MoS_2/graphene composites for application in LIBs in 2011 [129]. They also investigated the influence of molar ratio of MoS_2 and graphene on the morphologies and electrochemical performance. Bare MoS_2, MoS_2/G (1 : 1), MoS_2/G (1 : 2), and MoS_2/G (1 : 4) were designed and prepared. Without the addition of GO in the hydrothermal process, the obtained MoS_2 exhibited stacked structure with multilayers. When the GO was added in the hydrothermal process, it can provide low-energy nucleation site for MoS_2 growth. Therefore, the MoS_2 tends to be anchored on the surface of rGO but not aggregation into stacked structure. The optimized MoS_2/G corresponds to the molar ratio of 1 : 2. As shown in Figure 6.10a, the MoS_2 nanosheets in MoS_2/G (1 : 2) are tightly anchored on the surface of rGO and display few layers. The strong interfacial interaction between MoS_2 (002) fact and G (002) fact improves the electrical conductivity. The few-layer MoS_2 with porous structure facilitates the contact between MoS_2 and electrolyte. Moreover, the graphene substrate can hold the MoS_2 and its conversion products during discharge–charge. As a result, as shown in Figure 6.10b, all MoS_2/G (1 : 1), MoS_2/G (1 : 2), and MoS_2/G (1 : 4) composites exhibited better cycling stability than bare MoS_2. The MoS_2/G (1 : 2) displayed the largest reversible capacity with 1187 mAh g^{-1} after 100 cycles at 0.1 A g^{-1}. Because the GO and M sources, such as MoO_4^-, MoS_4^-, WO_4^{2-}, and ReO_4^- show negative charge, the controllable preparation of MS_2/graphene composites is a major challenge. To address this issue, some researchers introduced cationic surfactants, such as cetyltrimethylammonium bromide (CTAB) and Gemini in the hydro-/solvo-thermal process to modify GO template [133, 134]. The cationic surfactants-modified GO turns to positively charged, which can easily absorb negative M sources by electrostatic interaction. The graphene property plays an important role in electrochemical performance.

Notably, the obtained capacity of MS_2 is larger than that of theoretical capacity. Theoretical calculation revealed that the interface of MS_2 and graphene contributes to an additional pseudocapacitance storage mechanism [135, 136]. The interface engineering between MS_2 and graphene shows an important effect on electrochemical performance. Nonmetal doping graphene, such as N and S doping has been demonstrated to show great impact by both theoretical calculation and experiment [137, 138]. The N and S dopants are beneficial to increase the interfacial storage sites and ionic diffusion kinetics. However, the nonmetal dopants are easily poisoned by alkali metals during cycling. The improvement of interface property is limited by the poor intrinsic activity of nonmental dopants. In this regard, nonmetal and transition metal co-doping graphene was proposed to modify the MoS_2/graphene interface [139]. The Fe-N_4 active site shows high activity for catalyzing the reversible conversion between MoS_2 and Mo/Na_2S.

Although MS_2/graphene/MS_2 composites show large contact area, which is beneficial to structural stability. However, this interface type shows sluggish electrical transfer from graphene to MS_2 due to van der Walls interaction [140]. Recently, MS_2 vertically grown on graphene sheets composites have been reported to address this issue [130, 141, 142]. For example, Liu et al. prepared ReS_2 nanosheets vertically aligned on reduced graphene oxide (v-ReS_2/rGO) composites through

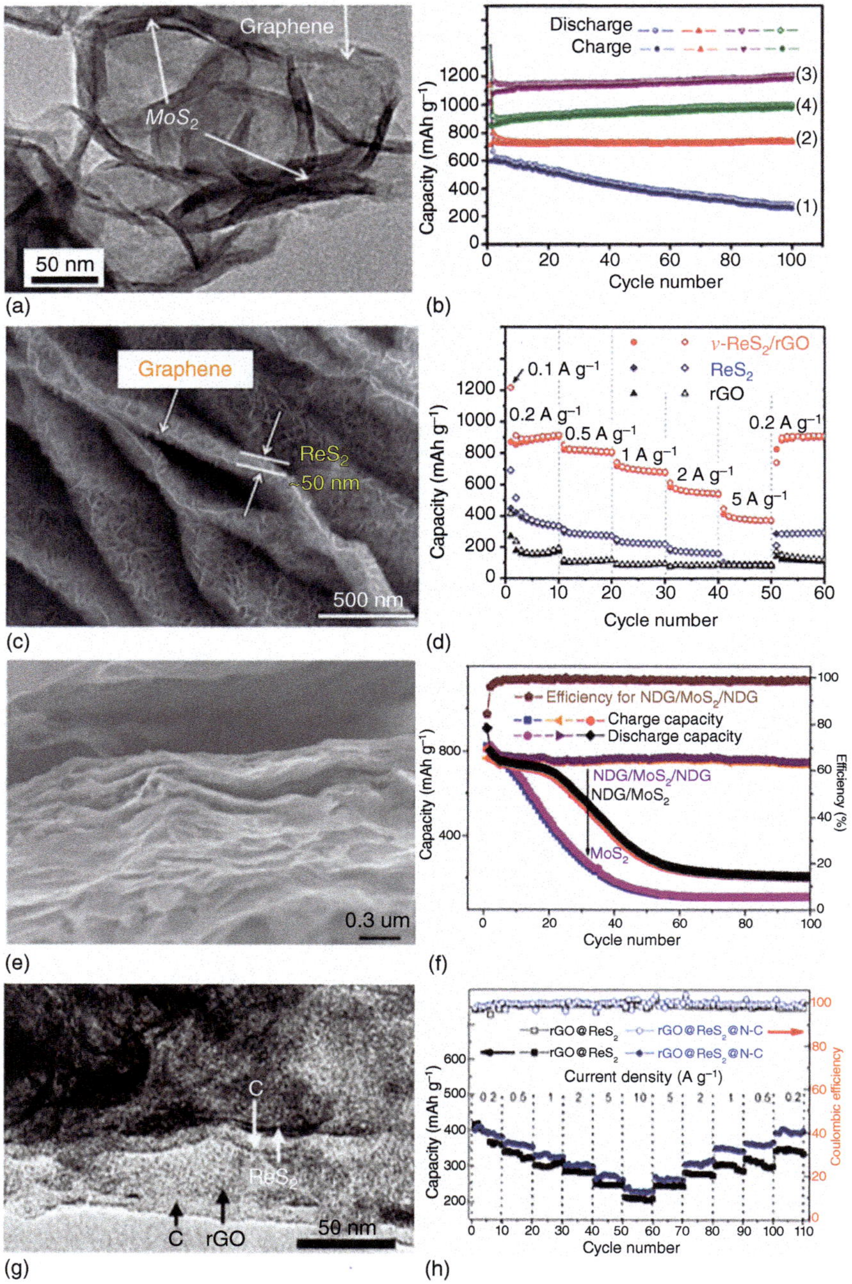
Graphene
MoS2
50 nm
(a)
Discharge
Charge
Capacity (mAh g−1)
Cycle number
(1)
(2)
(3)
(4)
(b)
Graphene
ReS2
~50 nm
500 nm
(c)
0.1 A g−1
0.2 A g−1
0.5 A g−1
1 A g−1
2 A g−1
5 A g−1
0.2 A g−1
v-ReS2/rGO
ReS2
rGO
Capacity (mAh g−1)
Cycle number
(d)
0.3 um
(e)
Efficiency for NDG/MoS2/NDG
Charge capacity
Discharge capacity
NDG/MoS2/NDG
NDG/MoS2
MoS2
Capacity (mAh g−1)
Efficiency (%)
Cycle number
(f)
C
ReS2
C
rGO
50 nm
(g)
rGO@ReS2
rGO@ReS2@N-C
rGO@ReS2
rGO@ReS2@N-C
Current density (A g−1)
0.2
0.5
1
2
5
10
5
2
1
0.5
0.2
Capacity (mAh g−1)
Coulombic efficiency
Cycle number
(h)

Figure 6.10 (a) TEM image of MoS_2/G (1 : 2). (b) Cycling performance of (1) MoS_2, (2) MoS_2/G (1 : 1), (3) MoS_2/G (1 : 2), and (4) MoS_2/G (1 : 4). Source: Chang and Chen [129]. Reproduced with permission of American Chemical Society. (c) SEM image of *v*-ReS_2/rGO. (d) Rate performance of ReS_2, rGO, and *v*-ReS_2/rGO. Source: Liu et al. [130]. Reproduced with permission of Royal Society of Chemistry. (e) Side-view image of NDG/MoS_2/NDG. (f) Cycling performance of MoS_2, NDG/MoS_2, and NDC/MoS_2/NDG. Source: Chen et al. [131]. Reproduced with permission of Elsevier. (g) TEM image of rGO@ReS_2@N-C. (h) Cycling performance of rGO@ReS_2@N-C and rGO@ReS_2. Source: Chen et al. [132]. Reproduced with permission of WILEY-VCH.

hydroxylamine hydrochloride (HAH)-assisted hydrothermal method [130]. The HAH plays a key role in the formation of vertically aligned ReS_2 nanosheets. They found that HAH changes the rGO charge from negative to positive. The HAH-modified GO adsorbs ReO_4^- anions through favorable electrostatic interaction. As a result, the ReS_2 nanosheets vertically and uniformly grow on rGO. As shown in Figure 6.10c, the vertically aligned ReS_2 nanosheets cover the rGO nanosheets uniformly and densely. The vertically aligned ReS_2 nanosheets show a lot of exposed diffusion channels for Li^+ diffusion. In addition, they contact rGO through C–N–Re bonds. This results in fast electrical transfer from rGO to ReS_2. The rate performance confirmed that the *v*-ReS_2/rGO composite shows significantly larger capacity than rGO and ReS_2 and all tested current densities (Figure 6.10d). However, the structural stability of *v*-ReS_2/rGO still needs to improve since the small contact area between rGO and ReS_2.

For graphene/MS_2/graphene structure, the 2D MS_2 is intercalated between 2D graphene sheets [131, 143]. For example, Chen et al. reported a facile thermal decomposition-reduction method to prepare layer-by-layer nitrogen-doped graphene/MoS_2/nitrogen-doped graphene (NDG/MoS_2/NDG) composites [131]. As shown in Figure 6.10e, the side-view image exhibited that the MoS_2 is encapsulated into NDC sheets. For comparison, conventional NDG/MoS_2 composite and bare MoS_2 were prepared with the same method. The cycling performance in Figure 6.10f showed that the NDG/MoS_2/NDG delivers the best cycling stability with a retention of 97.4%. The NDG/MoS_2/NDG exhibited a high capacity of 750 mAh g^{-1} after 100 cycles at 0.1 A g^{-1}, which is larger than those of NDG/MoS_2 and MoS_2.

The MS_2@C/graphene composites with dual carbon protecting layers were also proposed to enhance the electrochemical properties of MS_2 [132, 144–146]. For example, Zhao et al. designed a sandwich-like structure, in which MoS_2 is encapsulated in carbon and coupled on graphene sheets (G@MoS_2-C) [146]. The designed structure is obtained. The G@MoS_2-C shows better cycling stability than MoS_2-C, suggesting the graphene dual carbon protection layers present more advantages. Chen et al. prepared ReS_2 confined in dual carbon layers of rGO and N-doped carbon [132]. As shown in Figure 6.10g, the ReS_2 nanosheets are confined in 2D-honeycombed carbon nanosheets that comprise an rGO interlayer and an N-doped carbon coating layer (rGO@ReS_2@N-C). Therefore, the rGO@ReS_2@N-C shows excellent structural stability and electrical conductivity. As a result, the rGO@ReS_2@N-C displayed larger reversible capacities than rGO@ReS_2 at all tested current densities (Figure 6.10h).

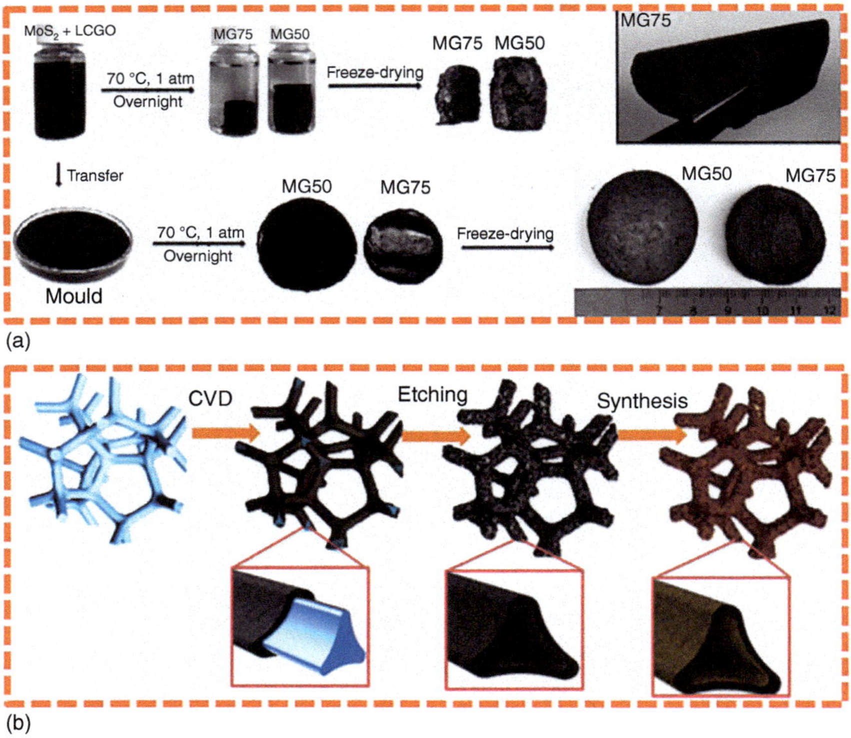

Figure 6.11 Schematic illustration of the formation process of freestanding (a) MoS_2-reduced graphene oxide (MG) film. Source: Chao et al. [149]. Reproduced with permission of WILEY-VCH, and (b) 3D MoS_2-graphene-MoS_2 foam. Source: Geng et al. [150]. Reproduced with permission of WILEY-VCH.

The free-standing MS_2/graphene electrodes without adding conductive agent and binder were reported [147–150]. The self-assembly strategy was also used to prepare free-standing MS_2/graphene films. For example, Chao et al. used MoS_2 dispersion and liquid crystalline graphene oxide (LCGO) dispersion to prepare MoS_2-reduced graphene oxide (MG) film [149]. As shown in Figure 6.11a, the MoS_2 and LCGO dispersion were self-assembled into an MG hydrogel after heating at 70 °C overnight. Then the free-standing MG film was prepared by freeze-drying. The MG75 sample with 75% MoS_2 exhibited a larger reversible capacity than MoS_2, LCGO, and MG50 in LIBs. The MG75 displayed 450 mAh g^{-1} after 500 cycles at 0.4 A g^{-1}. In addition, decorating MS_2 on 3D free-standing graphene substrate was widely used to fabricate free-standing MS_2/graphene foams. For example, as shown in Figure 6.11b, Geng et al. used Ni foam template to prepare 3D graphene foam [150]. Then, the as-prepared free-standing 3D graphene foam was used as template to grow MoS_2 nanosheets on its surface. Finally, the free-standing 3D MoS_2/graphene/MoS_2 was obtained. When evaluated the 3D MoS_2/graphene/MoS_2 as anode of SIBs, it achieved 313 mAh g^{-1} after 200 cycles at 0.05 A g^{-1}, corresponding to a capacity retention of 49.7%. In summary, the graphene substrate significantly improves the electrical conductivity and structural stability of pristine MS_2 nanosheets and the discharged products ($M + A_2S$, $A = Li$, Na), thus the conversion reaction kinetics is improved. However, the high decomposition barrier of A_2S restricts the reaction kinetics during charging, leading to unsatisfied reversibility of conversion reaction. Therefore, designing atomic active site on graphene to catalyze the decomposition of A_2S is highly desirable to improve the reversibility of conversion reaction of MS_2 [151].

6.5 Summary and Outlook

In summary, incorporating graphene with other anode materials (AMs), such as insertion, alloying, and conversion materials, has been widely used to enhance the electrochemical performance of LIBs and SIBs. The graphene with larger surface area hinders the aggregation of AMs. The strong interfacial interaction between AMs and graphene facilitates the electrical transfer from graphene to active materials. In addition, graphene protects the AMs and their transformed materials from pulverization. AMs/graphene/AMs, AMs@graphene, AMs@C/graphene, and free-standing AMs/graphene electrodes have been widely reported for application in LIBs and SIBs.

For AMs/graphene/AMs structure, it is the most widely reported structure due to the easy preparation. The AMs generally nucleate and grow on the surface of GO or graphene. Therefore, the AMs uniformly distribute on the surface of graphene in AMs/graphene/AMs structure. The AMs show large contact area with electrolyte, resulting in high electrochemical reaction kinetics. The graphene interlayer not only serves as high-speed electrical transfer pathway but also works as structural backbone for hindering pulverization. Therefore, the AMs/graphene/AMs can achieve significantly better cycling performance and rate performance than AMs. Although the electrochemical properties of AMs can be significantly improved in AMs/graphene/AMs, the structural stability of AMs is still unsatisfactory. The long-life span cycling performance is rarely reported.

The AMs@graphene structure can prevent direct contact between AMs and graphene. Generally, the composites were prepared by coating GO on the surface of AMs. The AMs were prepared in advance and modified by positive charges. The AMs were tightly covered by rGO through electrostatic interaction. The electrical conductivity and structural stability of AMs are significantly enhanced. However, the as-prepared AMs always show low surface area to hinder self-aggregation. Therefore, the AMs@graphene structure displays low reaction kinetics, the reversible capacity, and rate performance are still unsatisfactory.

AMs@C/graphene composites always possess advantages of both AMs/graphene/AMs and AMs@graphene. The AMs were tightly coated by carbon and anchored on the surface of graphene simultaneously. Therefore, the AMs@C/graphene can present high electrical conductivity, superior structural stability, and high reaction kinetics. Therefore, many reported that AMs@C/graphene composites achieved long-life span cycling performance, which is important for potential application.

However, the exposed carbon-coating layer causes a thick SEI film formation in the first cycle, leading to a low initial coulombic efficiency.

The rapidly increasing development of flexible devices is stimulating the demand for free-standing electrodes. Free-standing AMs/graphene electrodes without conductive agent and binder were prepared and used as flexible LIBs and SIBs. For preparing free-standing AMs/graphene electrodes, mechanical mixing approach, such as spin-coating and filtration, and chemical approach, such as loading AMs on the surface of free-standing graphene foam and GO/AMs self-assembly were widely used. Due to the continuous graphene network in free-standing electrodes, the electrical conductivity and structural stability have been significantly improved on electrode level. However, the AMs on material level in free-standing electrode always lack attention and optimization. Therefore, the electrochemical performance of free-standing AMs/graphene electrodes is always unsatisfactory.

The summarized results exhibit that the graphene material has been widely demonstrated to be robustly structural backbone and fast electrical transfer pathway for enhancing the electrochemical performance of other anode materials. Although great achievement has been made, the structural design on both material and electrode levels needs to be rationally designed further. Numerous efforts need to pay on AMs/graphene composites from lab to products.

References

1 Armand, M. and Tarascon, J.M. (2008). Building better batteries. *Nature* 451 (7179): 652–657.

2 Goodenough, J.B. and Park, K.-S. (2013). The Li-ion rechargeable battery: a perspective. *Journal of the American Chemical Society* 135 (4): 1167–1176.

3 Zhong, Y., Yang, M., Zhou, X. et al. (2015). Structural design for anodes of lithium-ion batteries: emerging horizons from materials to electrodes. *Materials Horizons* 2 (6): 553–566.

4 Li, M., Lu, J., Chen, Z. et al. (2018). 30 years of lithium-ion batteries. *Advanced Materials* 30 (33): 1800561.

5 Ji, L., Lin, Z., Alcoutlabi, M. et al. (2011). Recent developments in nanostructured anode materials for rechargeable lithium-ion batteries. *Energy & Environmental Science* 4 (8): 2682–2699.

6 Mukherjee, R., Krishnan, R., Lu, T.-M. et al. (2012). Nanostructured electrodes for high-power lithium ion batteries. *Nano Energy* 1 (4): 518–533.

7 Roy, P. and Srivastava, S.K. (2015). Nanostructured anode materials for lithium ion batteries. *Journal of Materials Chemistry A* 3 (6): 2454–2484.

8 Hong, S.Y., Kim, Y., Park, Y. et al. (2013). Charge carriers in rechargeable batteries: Na ions vs. Li ions. *Energy & Environmental Science* 6 (7): 2067–2081.

9 Kundu, D., Talaie, E., Duffort, V. et al. (2015). The emerging chemistry of sodium ion batteries for electrochemical energy storage. *Angewandte Chemie International Edition* 54 (11): 3431–3448.

10 Eftekhari, A. and Kim, D.-W. (2018). Sodium-ion batteries: new opportunities beyond energy storage by lithium. *Journal of Power Sources* 395: 336–348.

11 Palacin, M.R. (2009). Recent advances in rechargeable battery materials: a chemist's perspective. *Chemical Society Reviews* 38 (9): 2565–2575.

12 Chen, G., Yan, L., Luo, H. et al. (2016). Nanoscale engineering of heterostructured anode materials for boosting lithium-ion storage. *Advanced Materials* 28 (35): 7580–7602.

13 Geim, A.K. (2009). Graphene: status and prospects. *Science* 324 (5934): 1530–1534.

14 Huang, H., Shi, H., Das, P. et al. (2018). The chemistry and promising applications of graphene and porous graphene materials. *Advanced Functional Materials* 30 (41): 1909035.

15 Wang, G., Shen, X., Yao, J. et al. (2009). Graphene nanosheets for enhanced lithium storage in lithium ion batteries. *Carbon* 47 (8): 2049–2053.

16 Huang, X., Qi, X., Boey, F. et al. (2012). Graphene-based composites. *Chemical Society Reviews* 41 (2): 666–686.

17 Wu, Z.-S., Zhou, G., Yin, L.-C. et al. (2012). Graphene/metal oxide composite electrode materials for energy storage. *Nano Energy* 1 (1): 107–131.

18 Wang, Z., Gao, H., Zhang, Q. et al. (2019). Recent advances in 3D graphene architectures and their composites for energy storage applications. *Small* 15 (3): 1803858.

19 Chen, Z., Jin, L., Hao, W. et al. (2019). Synthesis and applications of three-dimensional graphene network structures. *Materials Today Nano* 5: 100027.

20 Zhu, G.-N., Wang, Y.-G., and Xia, Y.-Y. (2012). Ti-based compounds as anode materials for Li-ion batteries. *Energy & Environmental Science* 5 (5): 6652–6667.

21 Mei, Y., Huang, Y., and Hu, X. (2016). Nanostructured Ti-based anode materials for Na-ion batteries. *Journal of Materials Chemistry A* 4 (31): 12001–12013.

22 Chen, B., Meng, Y., Sha, J. et al. (2018). Preparation of MoS_2/TiO_2 based nanocomposites for photocatalysis and rechargeable batteries: progress, challenges, and perspective. *Nanoscale* 10 (1): 34–68.

23 Cao, H., Li, B., Zhang, J. et al. (2012). Synthesis and superior anode performance of TiO2@reduced graphene oxide nanocomposites for lithium ion batteries. *Journal of Materials Chemistry* 22 (19): 9759–9766.

24 Li, W., Wang, F., Feng, S. et al. (2013). Sol–gel design strategy for ultradispersed TiO_2 nanoparticles on graphene for high-performance lithium ion batteries. *Journal of the American Chemical Society* 135 (49): 18300–18303.

25 Li, W., Wang, F., Liu, Y. et al. (2015). General strategy to synthesize uniform mesoporous TiO_2/graphene/mesoporous TiO_2 sandwich-like nanosheets for highly reversible lithium storage. *Nano Letters* 15 (3): 2186–2193.

26 He, L., Ma, R., Du, N. et al. (2012). Growth of TiO_2 nanorod arrays on reduced graphene oxide with enhanced lithium-ion storage. *Journal of Materials Chemistry* 22 (36): 19061–19066.

27 Yan, X., Li, Y., Li, M. et al. (2015). Ultrafast lithium storage in TiO_2–bronze nanowires/N-doped graphene nanocomposites. *Journal of Materials Chemistry A* 3 (8): 4180–4187.

28 Gan, Y., Zhu, L., Qin, H. et al. (2015). Hybrid nanoarchitecture of rutile TiO_2 nanoneedle/graphene for advanced lithium-ion batteries. *Solid State Ionics* 269: 44–50.

29 Huang, H., Fang, J., Xia, Y. et al. (2013). Construction of sheet–belt hybrid nanostructures from one-dimensional mesoporous TiO_2(B) nanobelts and graphene sheets for advanced lithium-ion batteries. *Journal of Materials Chemistry A* 1 (7): 2495–2500.

30 Li, Y., Wang, Z., and Lv, X.-J. (2014). N-doped TiO_2 nanotubes/N-doped graphene nanosheets composites as high performance anode materials in lithium-ion battery. *Journal of Materials Chemistry A* 2 (37): 15473–15479.

31 Wang, Z., Sha, J., Liu, E. et al. (2014). A large ultrathin anatase TiO_2 nanosheet/reduced graphene oxide composite with enhanced lithium storage capability. *Journal of Materials Chemistry A* 2 (23): 8893–8901.

32 Xin, X., Zhou, X., Wu, J. et al. (2012). Scalable synthesis of TiO_2/graphene nanostructured composite with high-rate performance for lithium ion batteries. *ACS Nano* 6 (12): 11035–11043.

33 Zhen, M., Guo, X., Gao, G. et al. (2014). Rutile TiO_2 nanobundles on reduced graphene oxides as anode materials for Li ion batteries. *Chemical Communications* 50 (80): 11915–11918.

34 Zhou, T., Zheng, Y., Gao, H. et al. (2015). Surface engineering and design strategy for surface-amorphized TiO_2@graphene hybrids for high power Li-ion battery electrodes. *Advanced Science* 2 (9): 1500027.

35 Yan, X., Li, Y., Du, F. et al. (2014). Synthesis and optimizable electrochemical performance of reduced graphene oxide wrapped mesoporous TiO_2 microspheres. *Nanoscale* 6 (8): 4108–4116.

36 Li, Y., Ou, C., Huang, Y. et al. (2017). Towards fast and ultralong-life Li-ion battery anodes: embedding ultradispersed TiO_2 quantum dots into three-dimensional porous graphene-like networks. *Electrochimica Acta* 246: 1183–1192.

37 Luo, R., Ma, Y., Qu, W. et al. (2020). High pseudocapacitance boosts ultrafast, high-capacity sodium storage of 3D graphene foam-encapsulated TiO_2 architecture. *ACS Applied Materials & Interfaces* 12 (21): 23939–23950.

38 Yu, S.X., Yang, L.W., Tian, Y. et al. (2013). Mesoporous anatase TiO_2 submicrospheres embedded in self-assembled three-dimensional reduced graphene oxide networks for enhanced lithium storage. *Journal of Materials Chemistry A* 1 (41): 12750–12758.

39 Hu, T., Sun, X., Sun, H. et al. (2013). Flexible free-standing graphene–TiO_2 hybrid paper for use as lithium ion battery anode materials. *Carbon* 51: 322–326.

40 Zeng, C., Xie, F., Yang, X. et al. (2018). Ultrathin titanate nanosheets/graphene films derived from confined transformation for excellent Na/K ion storage. *Angewandte Chemie International Edition* 57 (28): 8540–8544.

41 Ren, G., Hoque, M.N.F., Liu, J. et al. (2016). Perpendicular edge oriented graphene foam supporting orthogonal TiO_2(B) nanosheets as freestanding electrode for lithium ion battery. *Nano Energy* 21: 162–171.

42 Shi, Y., Yang, D., Yu, R. et al. (2018). Robust binder-free anodes assembled with ultralong mischcrystal TiO_2 nanowires and reduced graphene oxide for high-rate and long cycle life lithium-ion storage. *Journal of Power Sources* 383: 115–123.

43 Luo, J., Li, F., Zhou, Y. et al. (2018). Paper-like TiO_2/graphene-carbon nanotube hybrid electrode with high mass loading: toward high-performance lithium ion battery. *Journal of Alloys and Compounds* 749: 697–704.

44 Xiong, Y., Qian, J., Cao, Y. et al. (2016). Graphene-supported TiO_2 nanospheres as a high-capacity and long-cycle life anode for sodium ion batteries. *Journal of Materials Chemistry A* 4 (29): 11351–11356.

45 Zhu, J., Li, Y., Huang, Y. et al. (2019). General strategy to synthesize highly dense metal oxide quantum dots-anchored nitrogen-rich graphene compact monoliths to enable fast and high-stability volumetric lithium/sodium storage. *ACS Applied Energy Materials* 2 (5): 3500–3512.

46 Li, M., Yang, Y., Guang Choo, E.S. et al. (2018). TiO_2–B nanofibrils reinforced graphene paper for multifunctional flexible electrode. *Journal of Power Sources* 394: 131–139.

47 Wu, G., Chen, J., Guo, Y. et al. (2017). Freestanding sodium-ion batteries electrode using Graphene foam coaxially integrated with TiO_2 nanosheets. *Journal of the Electrochemical Society* 164 (13): A3060–A3067.

48 Xu, G.L., Xiao, L., Sheng, T. et al. (2018). Electrostatic self-assembly enabling integrated bulk and interfacial sodium storage in 3D titania-graphene hybrid. *Nano Letters* 18 (1): 336–346.

49 Chen, C., Wen, Y., Hu, X. et al. (2015). Na+ intercalation pseudocapacitance in graphene-coupled titanium oxide enabling ultra-fast sodium storage and long-term cycling. *Nature Communications* 6: 6929.

50 Ying, H. and Han, W.-Q. (2017). Metallic Sn-based anode materials: application in high-performance lithium-ion and sodium-ion batteries. *Advanced Science* 4 (11): 1700298.

51 Liu, Z., Song, T., and Paik, U. (2018). Sb-based electrode materials for rechargeable batteries. *Journal of Materials Chemistry A* 6 (18): 8159–8193.

52 Wang, G., Wang, B., Wang, X. et al. (2009). Sn/graphene nanocomposite with 3D architecture for enhanced reversible lithium storage in lithium ion batteries. *Journal of Materials Chemistry* 19 (44): 8378–8384.

53 Wan, F., Lü, H.-Y., Wu, X.-L. et al. (2016). Do the bridging oxygen bonds between active Sn nanodots and graphene improve the Li-storage properties? *Energy Storage Materials* 5: 214–222.

54 Luo, B., Wang, B., Li, X. et al. (2012). Graphene-confined Sn nanosheets with enhanced lithium storage capability. *Advanced Materials* 24 (26): 3538–3543.

55 Yan, Y., Ben, L., Zhan, Y. et al. (2016). Nano-Sn embedded in expanded graphite as anode for lithium ion batteries with improved low temperature electrochemical performance. *Electrochimica Acta* 187: 186–192.

56 Zhou, X., Bao, J., Dai, Z. et al. (2013). Tin nanoparticles impregnated in nitrogen-doped graphene for lithium-ion battery anodes. *The Journal of Physical Chemistry C* 117 (48): 25367–25373.

57 Qin, J., He, C., Zhao, N. et al. (2014). Graphene networks anchored with Sn@graphene as lithium ion battery anode. *ACS Nano* 8 (2): 1728–1738.

58 Wang, J., Yang, J., Xiao, Q. et al. (2019). Hierarchical sulfur-doped graphene foam embedded with Sn nanoparticles for superior lithium storage in LiFSI-based electrolyte. *ACS Applied Materials & Interfaces* 11 (33): 30500–30507.

59 Qin, J., Zhang, X., Zhao, N. et al. (2015). In situ preparation of interconnected networks constructed by using flexible graphene/Sn sandwich nanosheets for high-performance lithium-ion battery anodes. *Journal of Materials Chemistry A* 3 (46): 23170–23179.

60 Ji, L., Tan, Z., Kuykendall, T. et al. (2011). Multilayer nanoassembly of Sn-nanopillar arrays sandwiched between graphene layers for high-capacity lithium storage. *Energy & Environmental Science* 4 (9): 3611–3616.

61 Luo, L., Song, J., Song, L. et al. (2019). Flexible conductive anodes based on 3D hierarchical Sn/NS-CNFs@rGO network for sodium-ion batteries. *Nano-Micro Letters* 11 (1): 63.

62 Li, N., Song, H., Cui, H. et al. (2014). Sn@graphene grown on vertically aligned graphene for high-capacity, high-rate, and long-life lithium storage. *Nano Energy* 3: 102–112.

63 Xin, F.-X., Tian, H.-J., Wang, X.-L. et al. (2015). Enhanced electrochemical performance of Fe0.74Sn5@reduced graphene oxide nanocomposite anodes for both Li-ion and Na-ion batteries. *ACS Applied Materials & Interfaces* 7 (15): 7912–7919.

64 Zhu, J., Wang, D., Liu, T. et al. (2014). Preparation of Sn-Co-graphene composites with superior lithium storage capability. *Electrochimica Acta* 125: 347–353.

65 Zhang, H., Zhang, M., Zhang, M. et al. (2017). Hybrid aerogel-derived Sn–Ni alloy immobilized within porous carbon/graphene dual matrices for high-performance lithium storage. *Journal of Colloid and Interface Science* 501: 267–272.

66 Zhu, J., Hu, G., and Zhang, J. (2016). Preparation of Sn-Cu-graphene nanocomposites with superior reversible lithium ion storage. *Materials Letters* 185: 565–568.

67 Nithya, C. and Gopukumar, S. (2014). rGO/nano Sb composite: a high performance anode material for Na+ ion batteries and evidence for the formation of nanoribbons from the nano rGO sheet during galvanostatic cycling. *Journal of Materials Chemistry A* 2 (27): 10516–10525.

68 Liu, X., Gao, M., Yang, H. et al. (2017). 2D sandwich-like nanosheets of ultrafine Sb nanoparticles anchored to graphene for high-efficiency sodium storage. *Nano Research* 10 (12): 4360–4367.

69 Wan, F., Guo, J.-Z., Zhang, X.-H. et al. (2016). In situ binding Sb nanospheres on graphene via oxygen bonds as superior anode for ultrafast sodium-ion batteries. *ACS Applied Materials & Interfaces* 8 (12): 7790–7799.

70 Park, J.-S. and Kang, Y.C. (2019). Uniquely structured Sb nanoparticle-embedded carbon/reduced graphene oxide composite shell with empty voids for high performance sodium-ion storage. *Chemical Engineering Journal* 373: 227–237.

71 Wang, J., Yang, J., Yin, W. et al. (2017). Carbon-coated graphene/antimony composite with a sandwich-like structure for enhanced sodium storage. *Journal of Materials Chemistry A* 5 (39): 20623–20630.

72 Zhang, W., Liu, Y., Chen, C. et al. (2015). Flexible and binder-free electrodes of Sb/rGO and $Na_3V_2(PO_4)_3$/rGO nanocomposites for sodium-ion batteries. *Small* 11 (31): 3822–3829.

73 Yi, Z., Han, Q., Cheng, Y. et al. (2016). A novel strategy to prepare Sb thin film sandwiched between the reduced graphene oxide and Ni foam as binder-free anode material for lithium-ion batteries. *Electrochimica Acta* 190: 804–810.

74 Du, Y., Zhu, G., Wang, K. et al. (2013). Si/graphene composite prepared by magnesium thermal reduction of SiO_2 as anode material for lithium-ion batteries. *Electrochemistry Communications* 36: 107–110.

75 Chen, S., Bao, P., Huang, X. et al. (2014). Hierarchical 3D mesoporous silicon@graphene nanoarchitectures for lithium ion batteries with superior performance. *Nano Research* 7 (1): 85–94.

76 Nie, P., Le, Z., Chen, G. et al. (2018). Graphene caging silicon particles for high-performance lithium-ion batteries. *Small* 14 (25): 1800635.

77 Zhao, X., Hayner, C.M., Kung, M.C. et al. (2011). In-plane vacancy-enabled high-power Si–graphene composite electrode for lithium-ion batteries. *Advanced Energy Materials* 1 (6): 1079–1084.

78 Zhou, M., Pu, F., Wang, Z. et al. (2013). Facile synthesis of novel Si nanoparticles–graphene composites as high-performance anode materials for Li-ion batteries. *Physical Chemistry Chemical Physics* 15 (27): 11394–11401.

79 Wen, Y., Zhu, Y., Langrock, A. et al. (2013). Graphene-bonded and -encapsulated Si nanoparticles for lithium ion battery anodes. *Small* 9 (16): 2810–2816.

80 Ji, J., Ji, H., Zhang, L.L. et al. (2013). Graphene-encapsulated Si on ultrathin-graphite foam as anode for high capacity lithium-ion batteries. *Advanced Materials* 25 (33): 4673–4677.

81 Agyeman, D.A., Song, K., Lee, G.-H. et al. (2016). Carbon-coated Si nanoparticles anchored between reduced graphene oxides as an extremely reversible anode material for high energy-density Li-ion battery. *Advanced Energy Materials* 6 (20): 1600904.

82 Zhou, X., Yin, Y.-X., Wan, L.-J. et al. (2012). Self-assembled nanocomposite of silicon nanoparticles encapsulated in graphene through electrostatic attraction for lithium-ion batteries. *Advanced Energy Materials* 2 (9): 1086–1090.

83 Zhou, X. and Guo, Y.-G. (2013). A PEO-assisted electrospun silicon–graphene composite as an anode material for lithium-ion batteries. *Journal of Materials Chemistry A* 1 (32): 9019–9023.

84 Tao, H., Xiong, L., Zhu, S. et al. (2017). Porous Si/C/reduced graphene oxide microspheres by spray drying as anode for Li-ion batteries. *Journal of Electroanalytical Chemistry* 797: 16–22.

85 Kim, M.K., Shin, W.H., and Jeong, H.M. (2019). Protective carbon-coated silicon nanoparticles with graphene buffer layers for high performance anodes in lithium-ion batteries. *Applied Surface Science* 467-468: 926–931.

86 Sun, J., Lee, H.-W., Pasta, M. et al. (2015). A phosphorene–graphene hybrid material as a high-capacity anode for sodium-ion batteries. *Nature Nanotechnology* 10: 980.

87 Wang, L., Wei, Z., Mao, M. et al. (2019). Metal oxide/graphene composite anode materials for sodium-ion batteries. *Energy Storage Materials* 16: 434–454.

88 Keppeler, M., Shen, N., Nageswaran, S. et al. (2016). Synthesis of α-Fe_2O_3/carbon nanocomposites as high capacity electrodes for next generation lithium ion batteries: a review. *Journal of Materials Chemistry A* 4 (47): 18223–18239.

89 Zhou, G., Wang, D.-W., Li, F. et al. (2010). Graphene-wrapped Fe_3O_4 anode material with improved reversible capacity and cyclic stability for lithium ion batteries. *Chemistry of Materials* 22 (18): 5306–5313.

90 Xu, B., Guan, X., Zhang, L.Y. et al. (2018). A simple route to preparing γ-Fe_2O_3/RGO composite electrode materials for lithium ion batteries. *Journal of Materials Chemistry A* 6 (9): 4048–4054.

91 Wu, Z.-S., Ren, W., Wen, L. et al. (2010). Graphene anchored with Co_3O_4 nanoparticles as anode of lithium ion batteries with enhanced reversible capacity and cyclic performance. *ACS Nano* 4 (6): 3187–3194.

92 Wu, D., Wang, C., Wu, H. et al. (2020). Synthesis of hollow Co_3O_4 nanocrystals in situ anchored on holey graphene for high rate lithium-ion batteries. *Carbon* 163: 137–144.

93 Zhu, J., Tu, W., Pan, H. et al. (2020). Self-templating synthesis of hollow Co_3O_4 nanoparticles embedded in N,S-dual-doped reduced graphene oxide for lithium ion batteries. *ACS Nano* 14 (5): 5780–5787.

94 Zhou, G., Wang, D.-W., Yin, L.-C. et al. (2012). Oxygen bridges between NiO nanosheets and graphene for improvement of lithium storage. *ACS Nano* 6 (4): 3214–3223.

95 Shi, W., Zhang, Y., Key, J. et al. (2018). Three-dimensional graphene sheets with NiO nanobelt outgrowths for enhanced capacity and long term high rate cycling Li-ion battery anode material. *Journal of Power Sources* 379: 362–370.

96 Ma, L., Pei, X.-Y., Mo, D.-C. et al. (2019). Facile fabrication of NiO flakes and reduced graphene oxide (NiO/RGO) composite as anode material for lithium-ion batteries. *Journal of Materials Science: Materials in Electronics* 30 (6): 5874–5880.

97 Wang, T., Zhao, N., Shi, C. et al. (2017). Interface and doping effects on Li ion storage behavior of graphene/Li_2O. *The Journal of Physical Chemistry C* 121 (36): 19559–19567.

98 Qi, L., Xin, Y., Zuo, Z. et al. (2016). Grape-like Fe_3O_4 agglomerates grown on graphene nanosheets for ultrafast and stable lithium storage. *ACS Applied Materials & Interfaces* 8 (27): 17245–17252.

99 Zhang, M., Liu, E., Cao, T. et al. (2017). Sandwiched graphene inserted with graphene-encapsulated yolk-shell [gamma]-Fe_2O_3 nanoparticles for efficient lithium ion storage. *Journal of Materials Chemistry A* 5 (15): 7035–7042.

100 Zhang, Z., Zhao, J., Zhou, J. et al. (2017). Interfacial engineering of metal oxide/graphene nanoscrolls with remarkable performance for lithium ion batteries. *Energy Storage Materials* 8: 35–41.

101 Wang, Y., Yang, L., Hu, R. et al. (2015). A stable and high-capacity anode for lithium-ion battery: Fe_2O_3 wrapped by few layered graphene. *Journal of Power Sources* 288: 314–319.

102 Jing, M., Zhou, M., Li, G. et al. (2017). Graphene-embedded Co_3O_4 rose-spheres for enhanced performance in lithium ion batteries. *ACS Applied Materials & Interfaces* 9 (11): 9662–9668.

103 Li, X., Fan, L., Li, X. et al. (2018). Enhanced anode performance of flower-like NiO/RGO nanocomposites for lithium-ion batteries. *Materials Chemistry and Physics* 217: 547–552.

104 Zou, F., Chen, Y.-M., Liu, K. et al. (2016). Metal organic frameworks derived hierarchical hollow NiO/Ni/Graphene composites for lithium and sodium storage. *ACS Nano* 10 (1): 377–386.

105 Qu, Q., Gao, T., Zheng, H. et al. (2015). Graphene oxides-guided growth of ultrafine Co_3O_4 nanocrystallites from MOFs as high-performance anode of Li-ion batteries. *Carbon* 92: 119–125.

106 Sui, Z.-Y., Zhang, P.-Y., Xu, M.-Y. et al. (2017). Metal–organic framework-derived metal oxide embedded in nitrogen-doped graphene network for high-performance lithium-ion batteries. *ACS Applied Materials & Interfaces* 9 (49): 43171–43178.

107 Wang, X., Zhang, L., Zhang, Z. et al. (2016). Growth of 3D hierarchical porous NiO@carbon nanoflakes on graphene sheets for high-performance lithium-ion batteries. *Physical Chemistry Chemical Physics* 18 (5): 3893–3899.

108 Ding, C., Zhou, W., Wang, X. et al. (2018). Hybrid aerogel-derived carbon/porous reduced graphene oxide dual-functionalized NiO for high-performance lithium storage. *Chemical Engineering Journal* 332: 479–485.

109 Jiang, T., Bu, F., Feng, X. et al. (2017). Porous Fe_2O_3 nanoframeworks encapsulated within three-dimensional graphene as high-performance flexible anode for lithium-ion battery. *ACS Nano* 11 (5): 5140–5147.

110 Wang, R., Xu, C., Sun, J. et al. (2013). Free-standing and binder-free lithium-ion electrodes based on robust layered assembly of graphene and Co_3O_4 nanosheets. *Nanoscale* 5 (15): 6960–6967.

111 Shao, J., Zhou, H., Feng, J. et al. (2019). Facile synthesis of MOF-derived hollow NiO microspheres integrated with graphene foam for improved lithium-storage properties. *Journal of Alloys and Compounds* 784: 869–876.

112 Geng, P., Zheng, S., Tang, H. et al. (2018). Transition metal sulfides based on graphene for electrochemical energy storage. *Advanced Energy Materials* 8 (15): 1703259.

113 Chen, W., Qi, S., Guan, L. et al. (2017). Pyrite FeS_2 microspheres anchoring on reduced graphene oxide aerogel as an enhanced electrode material for sodium-ion batteries. *Journal of Materials Chemistry A* 5 (11): 5332–5341.

114 Zhao, Y., Wang, J., Ma, C. et al. (2019). Interconnected graphene nanosheets with confined FeS_2/FeS binary nanoparticles as anode material of sodium-ion batteries. *Chemical Engineering Journal* 378: 122168.

115 Xu, Q.-T., Xue, H.-G., and Guo, S.-P. (2018). FeS_2 walnut-like microspheres wrapped with rGO as anode material for high-capacity and long-cycle lithium-ion batteries. *Electrochimica Acta* 292: 1–9.

116 Xie, J., Liu, S., Cao, G. et al. (2013). Self-assembly of CoS_2/graphene nanoarchitecture by a facile one-pot route and its improved electrochemical Li-storage properties. *Nano Energy* 2 (1): 49–56.

117 Geng, H., Kong, S.F., and Wang, Y. (2014). NiS nanorod-assembled nanoflowers grown on graphene: morphology evolution and Li-ion storage applications. *Journal of Materials Chemistry A* 2 (36): 15152–15158.

118 Pi, W., Mei, T., Li, J. et al. (2018). Durian-like NiS_2@rGO nanocomposites and their enhanced rate performance. *Chemical Engineering Journal* 335: 275–281.

119 Qiu, W., Jiao, J., Xia, J. et al. (2015). A self-standing and flexible electrode of yolk–shell CoS_2 spheres encapsulated with nitrogen-doped graphene for high-performance lithium-ion batteries. *Chemistry – A European Journal* 21 (11): 4359–4367.

120 Bi, R., Zeng, C., Huang, H. et al. (2018). Metal–organic frameworks derived hollow NiS_2 spheres encased in graphene layers for enhanced sodium-ion storage. *Journal of Materials Chemistry A* 6 (29): 14077–14082.

121 Yao, Q., Zhang, J., Shi, X. et al. (2019). Rational synthesis of two-dimensional G@porous FeS2@C composite as high-rate anode materials for sodium/potassium ion batteries. *Electrochimica Acta* 307: 118–128.

122 Song, J., Zhang, C., Zhang, J. et al. (2019). Construction of CoS_2-N-C sheets anchored on 3D graphene network for lithium storage performances. *Journal of Nanoparticle Research* 21 (5): 90.

123 Zhao, Y., Zhu, J., Ong, S.J.H. et al. (2018). High-rate and ultralong cycle-life potassium ion batteries enabled by in situ engineering of yolk–shell FeS2@C structure on graphene matrix. *Advanced Energy Materials* 8 (36): 1802565.

124 Jing, P., Wang, Q., Wang, B. et al. (2020). Encapsulating yolk-shell FeS2@carbon microboxes into interconnected graphene framework for ultrafast lithium/sodium storage. *Carbon* 159: 366–377.

125 Chen, Z., Wu, R., Liu, M. et al. (2017). General synthesis of dual carbon-confined metal sulfides quantum dots toward high-performance anodes for sodium-ion batteries. *Advanced Functional Materials* 27 (38): 1702046.

126 He, J., Li, Q., Chen, Y. et al. (2017). Self-assembled cauliflower-like FeS_2 anchored into graphene foam as free-standing anode for high-performance lithium-ion batteries. *Carbon* 114: 111–116.

127 Chen, D., Chen, W., Ma, L. et al. (2014). Graphene-like layered metal dichalcogenide/graphene composites: synthesis and applications in energy storage and conversion. *Materials Today* 17 (4): 184–193.

128 Chen, B., Chao, D., Liu, E. et al. (2020). Transition metal dichalcogenides for alkali metal ion batteries: engineering strategies at the atomic level. *Energy & Environmental Science* 13 (4): 1096–1131.

129 Chang, K. and Chen, W. (2011). L-cysteine-assisted synthesis of layered MoS_2/graphene composites with excellent electrochemical performances for lithium ion batteries. *ACS Nano* 5 (6): 4720–4728.

130 Liu, S., Liu, Y., Lei, W. et al. (2018). Few-layered ReS_2 nanosheets vertically aligned on reduced graphene oxide for superior lithium and sodium storage. *Journal of Materials Chemistry A* 6 (41): 20267–20276.

131 Chen, B., Meng, Y., He, F. et al. (2017). Thermal decomposition-reduced layer-by-layer nitrogen-doped graphene/MoS_2/nitrogen-doped graphene heterostructure for promising lithium-ion batteries. *Nano Energy* 41: 154–163.

132 Chen, B., Li, H., Liu, H. et al. (2019). 1T′-ReS_2 Confined in 2D-Honeycombed Carbon Nanosheets as New Anode Materials for High-Performance Sodium-Ion Batteries. *Advanced Energy Materials* 9 (30): 1901146.

133 Huang, G., Chen, T., Chen, W. et al. (2013). Graphene-like MoS_2/graphene composites: cationic surfactant-assisted hydrothermal synthesis and electrochemical reversible storage of lithium. *Small* 9 (21): 3693–3703.

134 Ma, L., Ye, J., Chen, W. et al. (2014). Gemini surfactant assisted hydrothermal synthesis of nanotile-like MoS_2/graphene hybrid with enhanced lithium storage performance. *Nano Energy* 10: 144–152.

135 Lv, Y., Chen, B., Zhao, N. et al. (2016). Interfacial effect on the electrochemical properties of the layered graphene/metal sulfide composites as anode materials for Li-ion batteries. *Surface Science* 651: 10–15.

136 Su, L., Zhao, N., Shi, C. et al. (2018). Hetero-structure effect on Na adsorption and diffusion in two dimensional composites. *Electrochimica Acta* 285: 309–316.

137 Guo, L., Li, J., Wang, H. et al. (2018). Dopant-modulating mechanism of lithium adsorption and diffusion at the graphene/Li_2S interface. *Physical Review Applied* 9 (2): 024010.

138 Li, G., Luo, D., Wang, X. et al. (2017). Enhanced reversible sodium-ion intercalation by synergistic coupling of few-layered MoS_2 and S-doped graphene. *Advanced Functional Materials* 27 (40): 1702562.

139 Chen, B., Wang, T., Zhao, S. et al. (2021). Efficient reversible conversion between MoS_2 and Mo/Na_2S enabled by graphene-supported single atom catalysts. *Advanced Materials* 33 (12): 2007090.

140 Shokhen, V., Miroshnikov, Y., Gershinsky, G. et al. (2017). On the impact of vertical alignment of MoS_2 for efficient lithium storage. *Scientific Reports* 7 (1): 3280.

141 Teng, Y., Zhao, H., Zhang, Z. et al. (2016). MoS_2 nanosheets vertically grown on graphene sheets for lithium-ion battery anodes. *ACS Nano* 10 (9): 8526–8535.

142 Wang, G., Zhang, J., Yang, S. et al. (2018). Vertically aligned MoS_2 nanosheets patterned on electrochemically exfoliated graphene for high-performance lithium and sodium storage. *Advanced Energy Materials* 8 (8): 1702254.

143 Sun, D., Ye, D., Liu, P. et al. (2018). MoS_2/graphene nanosheets from commercial bulky MoS_2 and graphite as anode materials for high rate sodium-ion batteries. *Advanced Energy Materials* 8 (10): 1702383.

144 Xiao, J., Wang, X., Yang, X.-Q. et al. (2011). Electrochemically induced high capacity displacement reaction of PEO/MoS_2/graphene nanocomposites with lithium. *Advanced Functional Materials* 21 (15): 2840–2846.

145 Teng, Y., Zhao, H., Zhang, Z. et al. (2017). MoS_2 nanosheets vertically grown on reduced graphene oxide via oxygen bonds with carbon coating as ultrafast sodium ion batteries anodes. *Carbon* 119: 91–100.

146 Zhao, C., Yu, C., Qiu, B. et al. (2018). Ultrahigh rate and long-life sodium-ion batteries enabled by engineered surface and near-surface reactions. *Advanced Materials* 30 (7): 1702486.

147 Shan, T.-T., Xin, S., You, Y. et al. (2016). Combining nitrogen-doped graphene sheets and MoS_2: a unique film–foam–film structure for enhanced lithium storage. *Angewandte Chemie International Edition* 55 (41): 12783–12788.

148 Wang, Y., Chen, B., Seo, D.H. et al. (2016). MoS_2-coated vertical graphene nanosheet for high-performance rechargeable lithium-ion batteries and hydrogen production. *NPG Asia Materials* 8: e268.

149 Chao, Y., Jalili, R., Ge, Y. et al. (2017). Self-assembly of flexible free-standing 3D porous MoS_2-reduced graphene oxide structure for high-performance lithium-ion batteries. *Advanced Functional Materials* 27 (22): 1700234.

150 Geng, X., Jiao, Y., Han, Y. et al. (2017). Freestanding metallic 1T MoS_2 with dual ion diffusion paths as high rate anode for sodium-ion batteries. *Advanced Functional Materials* 27 (40): 1702998.

151 Chen, B., Zhong, X., Zhou, G. et al. (2022). Graphene-supported atomically dispersed metals as bifunctional catalysts for next-generation batteries based on conversion reactions. *Advanced Materials* 34 (5): 2105812.

7

Graphene-Based Materials for Lithium-Metal Batteries

Rui Zhang

Tianjin University, School of Materials Science and Engineering and Tianjin Key Laboratory of Composites and Functional Materials, 135 Yaguan Rd., Jinnan Dist., Tianjin, 300350, P.R. China

The rapid consumption of fossil fuels and the increasing demand for energy advocate clean, secure, and renewable energy sources, such as solar and wind power. Therefore, more and more attention has focused on the harvest and conversion, storage, and efficient usage of energy. Batteries can play a crucial role because they represent an efficient means of storing and releasing energy electrochemically. Since the introduction of commercial Li-ion batteries in 1991, they have served for decades and achieved tremendous development in portable electronics and electric vehicles [1]. However, Li-ion batteries using graphite anodes (~370 mAh g^{-1}) have almost reached their theoretical specific energy density (~350 Wh kg^{-1}), which are unable to meet the high-energy demands of electric vehicles and grid energy storage [2–6].

Li-metal anode (LMA), which has been investigated for rechargeable lithium-metal batteries (LMBs) since 1962, is the essential anode material for high-energy-density batteries due to its lightweight (0.53 g cm^{-3}), low electrochemical potential (−3.04 V vs. standard hydrogen electrode), and high specific capacity (3860 mAh g^{-1}) [7–11]. The comparison of specific energy density and volumetric energy density of different secondary batteries and petrol (gasoline) is given in Figure 7.1, the specific energy density of Li-ion batteries is much lower than the practical value of petrol. Once the Li-metal anodes are used in the rechargeable batteries, the Li-LMO (LMO is lithium transition-metal oxide) batteries can reach a specific energy density of ~440 Wh kg^{-1}. As for Li–S and Li–air batteries, the specific energy density even can reach ~650 and ~950 Wh kg^{-1}, respectively. According to the volumetric energy density, it also can offer a high energy density of more than 1000 Wh l^{-1}, by using LMAs, comparable to that of petrol [12].

However, the practical applications of LMAs are terribly impeded by some serious difficulties, which are as follows:

(1) High chemical reactivity of metallic Li [12]. The metallic Li is thermodynamically unstable due to its ultralow electrochemical potential. Once the fresh Li is exposed to electrolyte, the irreversible and continuous reactions between Li and

Templated Fabrication of Graphene-Based Materials for Energy Applications, First Edition.
Edited by Chunnian He, Naiqin Zhao, and Junwei Sha.

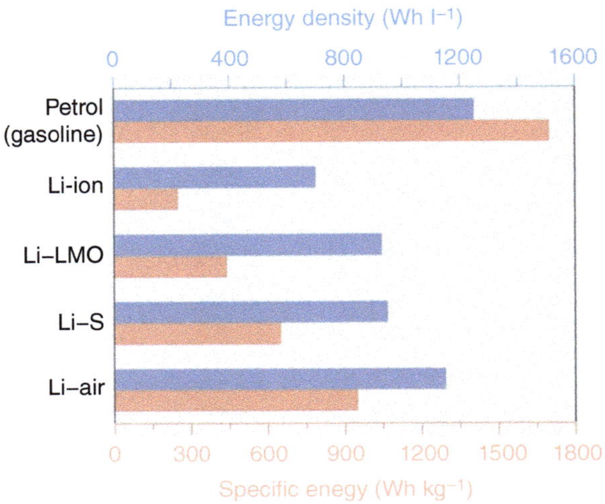

Figure 7.1 Bar chart of the practical specific energy density (pink) and energy density (blue) of the petrol (gasoline), the state-of-the-art Li-ion battery, Li metal | LMO (LMO is a lithium transition-metal oxide) battery, Li–S and Li–air battery. Source: Lin et al. [12]. Reproduced with permission of Springer Nature.

electrolyte will happen and lead to thick solid-electrolyte interphase (SEI) layers on the Li surface.

(2) Uncontrollable lithium dendrite growth. Dendritic electrodeposition is a common phenomenon, the inhomogeneous distribution of Li-ion drives uncontrollable Li dendrite growth during the Li plating process. Li dendrites will react with electrolytes easily and consume active materials, leading to an unstable SEI and reducing the coulombic efficiency (CE) and life span of Li-metal batteries (Figure 7.2) [13]. Besides, the Li dendrites may puncture the separator, causing short circuit, which will result in serious safety issues. Meanwhile, the Li dendrites will lose their electronic contact with current collector during the stripping process, generating loose "dead Li," accordingly leading to capacity fading.

(3) The infinite volume change during the cycling [14]. As we all know, all electrode materials undergo volume change during cycling. The huge volume changes may lead to internal pressure changes and interfacial fluctuations, which will destroy the integrality of the LMAs, thus causing capacity fading. In addition, due to the "hostless" nature of the LMAs, the infinite volume change will lead to a large change in roughness and morphology. Besides, the huge volume change may lead to the fracture of the SEI, further causing chemical side reactions, dendritic Li growth, and capacity fading.

The fatal obstacles of LMAs are summarized in Figure 7.3 [7].

With the urgent need for high-energy-density and high-safe LMAs, various strategies have been proposed to restrain the Li dendrites, stabilize the SEI, and improve the cycling performance of LMAs from different perspectives. To stabilize the SEI, various electrolyte additives, such as Cs^+ and Rb^+ ions, Li_2S_8, $LiNO_3$, and LiF [15–17], have been used. Zhang et al. summarized the formation mechanisms of SEI layer and models of SEI films on LMA [18]. Introducing nanoscale interfacial layers (such as interconnected carbon nanospheres [19], hexagonal boron nitride

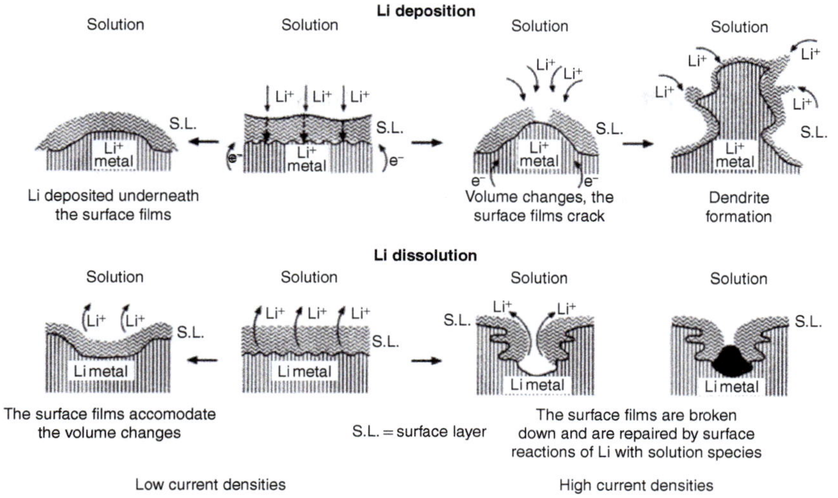

Figure 7.2 An illustration of the morphological phenomena and failure mechanism of LMAs during Li deposition and dissolution: the beginning of Li dendrites growth, nonuniform Li dissolution and the breakdown and reparation of the SEI. Source: Yaron et al. [13]. Reproduced with permission of American Chemical Society.

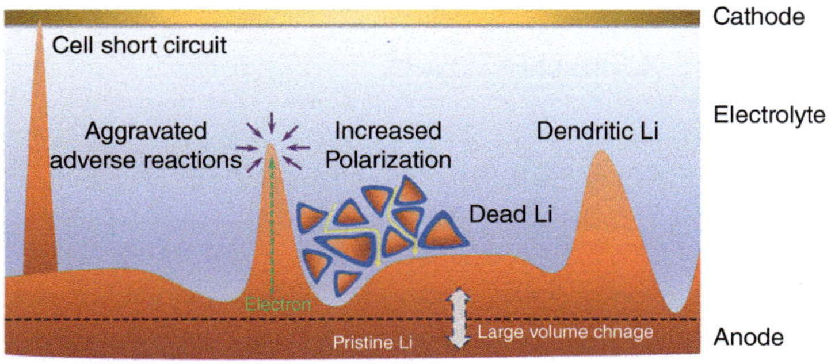

Figure 7.3 Scheme of dilemma for LMA in rechargeable batteries. Source: Cheng et al. [7]. Reproduced with permission of American Chemical Society.

[20, 21], and glass fiber [22]) on the metallic Li surface can prevent the formation of dendrites and improve CE effectively, and more importantly, regulate the Li-ion distribution on the electrode surface to realize uniform Li deposition. Besides, solid electrolyte and gel polymer electrolytes with high modulus are introduced to the LMBs to suppress the Li dendrites growth and improve battery safety [23–26]. To stabilize the interface between the solid electrolyte and Li anodes and reduce the interface resistance, ZnO, Sn, and Mg are used to modify the surface of solid electrolytes [27–29]. 3D Li anodes were also used to suppress the Li dendrites in recent works, such as 3D nickel foam host, 3D porous copper current collectors, and free-standing 3D carbon fibers [30–38]. The metallic Li could be restricted in the porous structure, thus suppressing dendritic Li growth and diminishing volume

change. Moreover, the electroactive surface areas could be significantly increased by using 3D Li anodes, which could promote dendrite-free Li plating due to the low local current density.

According to the above discussion, various materials have been used to improve the cycling performance of LMAs. Among these materials, graphene has shown great potential due to its variety of excellent performance. First, graphene with high flexibility, strength, and various functional groups on the surface could be used to construct a protective layer or a buffer layer that allows transport of Li^+ and enable uniform Li^+ flux. As a result, more homogenous metallic Li deposition and high CE could be achieved. Second, graphene with high chemical stability, good conductivity, and high specific surface area (SSA) can be used as 3D hosts for Li deposition, adjusting the volume change and reducing the local current density for high-performance LMAs. Third, heteroatom-doped graphene, such as N-doped graphene and S-doped graphene, can promote uniform Li nucleation and microscale homogeneous Li deposition, thus suppressing the dendritic Li growth. Last, graphene can be easily modified to form hybrids with other "lithiophilic" materials, such as Zn, Al, Ag, and Au, which could serve as "seeds" to guide a uniform Li deposition.

Here, this chapter summarizes recent key progress on graphene-based materials for LMAs and highlights the advantages and roles of graphene-based materials.

7.1 Graphene-Based Nanoscale Layers

Spatial inhomogeneous distribution of Li-ion on electrode surface contributes to Li dendritic growth directly. During the plating process, the stacked cathodes and separators with inhomogeneous porous inevitably lead to nonuniform distributions of Li-ion at anode/separator interface, leading to concentrated polarization [39]. Thus, the rational design of nanoscale buffer layers to promote more uniform Li-ion flux has been explored in recent research [40–42]. Generally, buffer layers consist of porous substrate with polar surface functional groups and high SSA. During the plating process, the Li-ion near the electrode surface will transfer to the deposition hot spots, such as Li dendrites and the accumulation area of electrons. However, the strong interaction between the Li-ion and polar surface functional groups could slow the diffusion of Li-ion to the deposition hot spots, thereby promoting more homogeneous Li deposition [22].

Graphene oxide (GO) with high flexibility and various functional groups on the surface can be used to construct a buffer layer, which allows transport of Li^+ and enables uniform Li^+ flux. For example, stacked graphene (SG), with higher Li/Na nucleation overpotential than that on Cu foil, was synthesized from GO and was used to orientate and filtrate Li^+/Na^+ flux for uniform Li deposition (Figure 7.4) [43]. When 1 mAh cm^{-2} of Li plating on Cu foil coated with SG, metallic Li deposition occurred underneath the SG layer, proving preferential Li plating on the Cu substrate (Figure 7.4b). By introducing the investigated SG on the inner surface of nanostructured Cu scaffold (NCS-SG), the uniform nucleation and dendrite-free

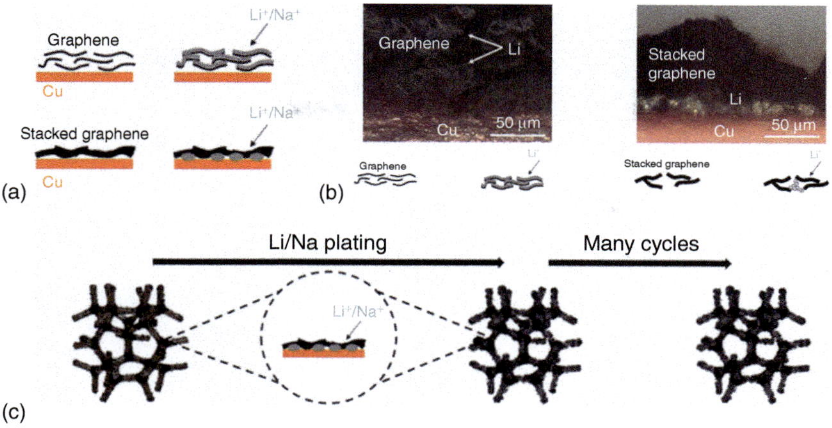

Figure 7.4 (a) Illustration of Li/Na plating on the Cu foil coated with conventional graphene sheets and stacked graphene (SG). (b) Digital photos of the conventional graphene and SG sheets at a Li plating capacity of 1 mAh cm^{-2}. (c) Schematic representation of uniform Li/Na depositions on the stacked graphene coated nanostructured Cu scaffold (NCS-SG) structures. Source: Ren et al. [43]. Reproduced with permission of Elsevier.

deposition of metallic Li on the inner surface of the substrate could be realized (Figure 7.4c). The CEs of NCS-SG electrodes could remain stable at ~97.1% and 93.3% for Li- and Na-metal anodes, respectively. In another study, LMA coated with rGO (reduced graphene oxide) layer was prepared via solvent-evaporation-assisted self-assembly (SEASA) [44]. During the plating/stripping process, the rGO layer cracked into mosaic-like flakes, meanwhile, the interconnected gaps served as the channel for Li-ion diffusion, promoting uniform lithium deposition. As a result, the symmetric cells composed of the rGO-coated Li (gLi) exhibited excellent electrochemical performance with a stable overpotential of 87.9 mV at 5 mA cm^{-2} for more than 600 cycles. Meanwhile, Li–S batteries with a high and stable capacity of 1200 mAh g^{-1} at 1 C for nearly 1000 cycles were prepared to demonstrate the practical application of gLi.

Besides, uniform Li-ion flux could be achieved by using the modified separator with graphene-based material coating. Wu et al. prepared functional porous bilayer composite separators by simply blade-coating polyacrylamide-grafted graphene oxide molecular brushes (GO-g-PAM) onto one side of commercial polypropylene separators (Figure 7.5) [45]. This method is highly convenient for large-scale production and practical application. The GO, which was used as backbone for PAM, was firm enough to improve the mechanical strength of the composite separators. Moreover, the hairy PAM chains on the GO surfaces with a large number of polar groups, such as C=O and N—H bonds, can facilitate uniform and fast Li-ion flux on the surfaces of electrodes, thus uniform and dense Li deposition without dendrite formation can be realized (Figure 7.5b). The morphologies of LMAs after 100 cycles are investigated to further clarify the effect of composite separators on the suppression of Li dendrites (Figure 7.5c–h). As a result, LMAs could reach unprecedented stability of more than 1900 hours even at a high current density of 20 mA cm^{-2}

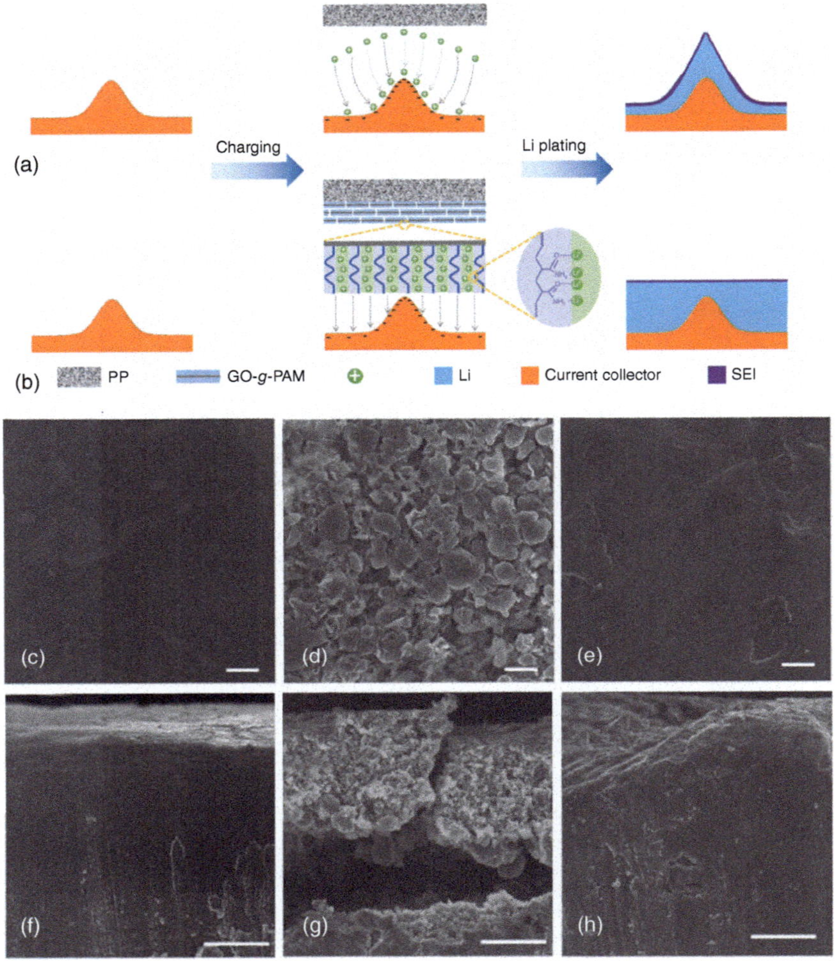

Figure 7.5 (a) Illustration of Li deposition on an electrode with a PP separator. (b) Schematics showing Li deposition on an electrode with a GO-*g*-PAM@PP separator. (c–e) Top-view and (f–h) cross-section SEM images of (c, f) a fresh LMA and LMAs assembled with (d, g) PP and (e, h) GO-*g*-PAM@PP separators after 100 cycles. Scale bars: (c–e) 10 μm and (f–h) 50 μm. Source: Li et al. [45]. Reproduced with permission of Springer Nature.

by using composite separators. Other graphene-based buffer layer materials have also been studied: nitrogen and sulfur co-doped graphene (NSG) layers coated on the separator could provide uniform ionic flux on the Li-metal surface and also act as a physical barrier to restrain dendritic growth (Figure 7.6a) [46]; 3D buffer layer for homogeneous Li-metal deposition was prepared by coating GOn on the graphene framework (GF) membrane (Figure 7.6b–d) [47]; pentaerythritol tetraacrylate and poly(ethylene glycol)diacrylate were incorporated with graphene oxide as multifunctional artificial protective layer [48].

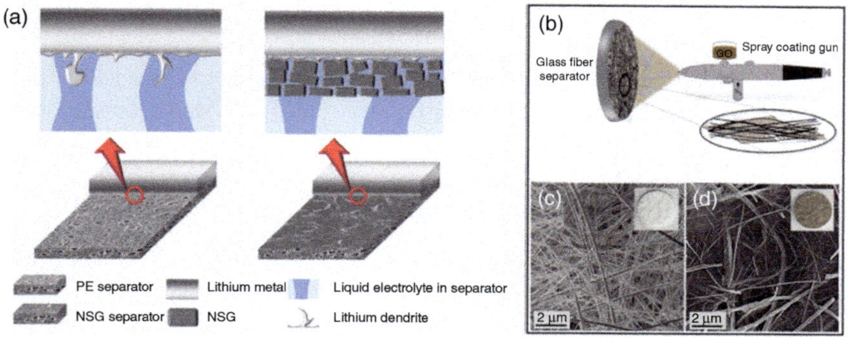

Figure 7.6 (a) Illustration of Li deposition on the lithium metal electrode with the PE separator (left) and NSG-coated PE separator (right) [46]. Source: Shin et al. [46], pp. 23700–23707 Reproduced with permission of American Chemical Society. (b) Schematic illustration of the fabrication of GO–GF separator. (c) SEM image of bare glass fiber separator. (d) SEM image of a GO-modified separator. Source: Foroozan et al. [47]. Reproduced with permission of Wiley.

Based on the above discussion, graphene-based material with high flexibility and various functional groups on the surface can be used to construct a protective layer or a buffer layer that allows transport of Li^+ and enables uniform Li^+ flux. On account of the Li dendrites growth and infinite volume changes of LMAs, such a layer must have an optimum thickness and porosity for the diffusion of Li-ion and also need enough strength to ensure its integrity during the plating/stripping process. Therefore, the porous size and structure, strength, and thickness of the graphene-based layer should be further optimized. Furthermore, it is still difficult to realize the uniform Li nucleation and deposition under an ultrahigh current density and the thickness fluctuation of LMAs still exist during the cycling. To overcome these limitations, three-dimensional (3D) porous hosts were employed to accommodate Li deposition in recent studies.

7.2 Graphene-Based Hosts for Li Storage

Based on the nature of Li-ion and electrons empirically, "Sand's time" model can be used to demonstrate the influences of different factors on the dendritic Li growth:

$$\tau = \pi \mathrm{D} \frac{eC_0\left(\mu_\mathrm{a} + \mu_{\mathrm{Li}^+}\right)^2}{2J\mu_\mathrm{a}}$$

where τ is the time when Li dendrites start to grow, D is the diffusion coefficient, e is the charge, C_0 is the initial concentration of Li salt, μ_a and μ_{Li^+} are, respectively, the anionic and Li^+ mobility, and J is the effective electrode current density. According to the equation, the current density (J) has direct influence on Li dendrite growth; a large Sand's time τ could be achieved by reducing the effective electrode current density (J). Therefore, 3D electrode with high SSA, which can effectively reduce the effective electrode current density, was used to suppress dendrite growth for LMAs in recent works, such as 3D Cu current collectors, 3D Ni foam host, and free-standing 3D carbon fiber host. Furthermore, the Li deposition could be restricted in the 3D structure of the electrode, thus adjusting the volume changes of LMAs. In addition, the 3D electrode affords opportunity for getting dendrites-free LMAs even at a high current density.

7.2.1 Graphene-Based Hosts with High SSA

Graphene with high chemical stability, high SSA, and good conductivity can be used as 3D hosts for Li deposition, adjusting the volume change and reducing the local current density for high-performance LMAs. For example, Zhang's group prepared graphene framework (GF) by hydrothermal reduction of graphene oxide [49]. GF has a high SSA, reducing the real local current density and suppressing the Li dendrite's growth. During the cycling, dual-phase layered structure composed of an in situ formed solid-electrolyte interphase and graphene framework (denoted as SEI-coated graphene, SCG) could promote a uniform Li deposition and dissolution. Besides, GF with porous and flexible structure could store Li deposition and adjust the volume change of Li during the plating/stripping process. As a result, an improved CE of 97% for more than 100 cycles could be realized by using SCG anode. Meanwhile, the stable and efficient Li deposition was maintained for 2000 cycles.

To improve the conductivity and SSA of graphene-based hosts for LMAs, Zhang's group prepared an unstacked graphene framework with a very large SSA (1666 m^2 g^{-1}), pore volume (1.65 cm^3 g^{-1}), and electrical conductivity (435 S cm^{-1}) [50]. Thus, an ultralow local current density could be achieved by using this electrode during the cycling. Due to the low effective current density of graphene framework, during the Li electroplating process, Li-ion migrated through the SEI film and deposited on the substrate homogeneously, creating sandwich-like graphene with core–shell structure, the outer shell of Li, and the outermost shell of SEI (Figure 7.7). Meanwhile, this structure could suppress the Li dendrite's growth due to the decrease of real local current density and maintain a stable, flexible, and compact SEI layer. Thus, a high CE of 93% with a high areal capacity of 5.0 mAh cm^{-2} at a high current density of 2.0 mA cm^{-2} could be achieved. In another study, Luo's group reported a crumpled paper ball-like graphene particle, which can readily assemble to yield a scaffold for LMAs [51]. The ball-like graphene with high SSA is conducting, lightweight, lithiophilic, strong, and mechanically stable. High CE of 97.5% over 750 cycles and high areal capacity of 12 mAh cm^{-2} without dendrite growth were achieved.

7.2.2 Free-Standing 3D Graphene-Based Hosts

Li metal with lightweight (0.53 g cm^{-3}), high specific capacity (3860 mAh g^{-1}), and low electrochemical potential (−3.04 V) has been considered as an essential anode material for high-energy-density batteries. Free-standing 3D graphene-based hosts with porous structures can store plenty of metallic Li for LMA, meanwhile, the lightweight of graphene-based material can contribute to realizing an ultrahigh-specific capacity of the whole electrode.

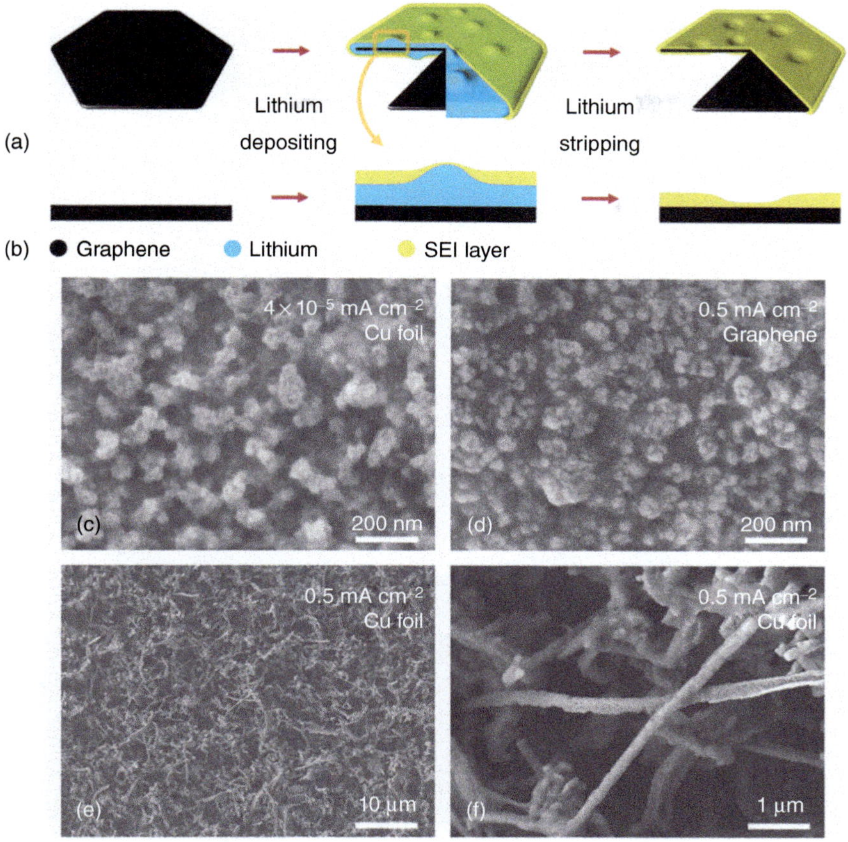

Figure 7.7 (a, b) Schematic diagrams of Li depositing/stripping process on one graphene flake; (c) Li deposition morphology on Cu foil at an ultralow current density of 4×10^{-5} mA cm^{-2}; (d) Li deposition morphology on graphene flake at a current density of 0.5 mA cm^{-2}. (e, f) Li dendrites on Cu foil at an increased current density of 0.5 mA cm^{-2}. Source: Zhang et al. [50]. Reproduced with permission of Wiley.

Recently, Cui's group reported a composite Li–rGO membrane electrode [52]. The Li–rGO with layered structure was fabricated via molten Li infusion into the rGO film. rGO films with layered structure and uniform nanogaps were prepared by a "spark" reaction, the as-prepared GO films were partially put into contact with molten Li, a spark reaction between the molten Li and the GO happened immediately across the whole rGO and generated abundant and uniform layered structure for Li deposition, and the GO was reduced to rGO. Meanwhile, molten Li infusion into rGO film could be realized with the effects of the superior lithiophilic surface of GO and the capillary force of the layered structure (Figure 7.8a,b). The Li–rGO composite electrode exhibited excellent flexibility and twistability due to the flexible nature of rGO and the layered structure (Figure 7.8c). Besides, the composite LMA exhibited low dimension variation during cycling with stable SEI (Figure 7.8d–f). With the benefits of such a structure, the symmetric batteries with

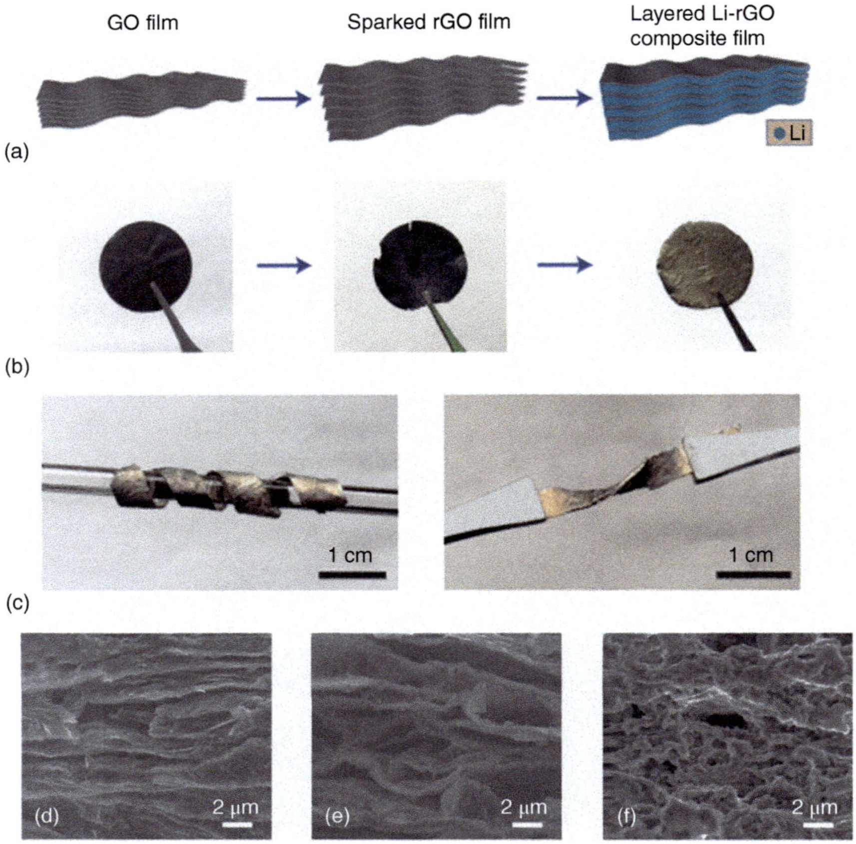

Figure 7.8 (a, b) Fabrication process of layered Li–rGO composite LMA; From left to right: GO film, sparked rGO film, and layered Li–rGO composite film. (c) Mechanical properties testing of Li–rGO electrode to show the good flexibility and twistability. (a–c) Cross-section SEM images of the Li–rGO electrode (a) before and (b) after Li stripping and (c) after 1 cycle. Source: Lin et al. [52]. Reproduced with permission of Springer Nature.

Li–rGO electrode exhibited stable voltage profiles with small voltage hysteresis even at a high areal current density of $3\,mA\,cm^{-2}$. Notably, an ultrahigh-specific capacity of $3390\,mAh\,g^{-1}$ (based on the weight of the whole electrode) could be realized, which could be attributed to the lightweight and porous structure of sparked rGO.

Besides, Zhao et al. reported a carbon nanofiber-stabilized graphene aerogel film (G-CNF film) fabricated by blade coating strategy inspired by constructional engineering [53]. The G-CNF film with large SSA, porous structure, and robust skeleton could sufficiently reduce the local current density, thus suppressing the Li dendrite growth. Besides, as a host material for Li deposition, G-CNF film also could solve the problems associated with large volume change during the plating/stripping process. As a result, G-CNF electrode can keep high CE of nearly 99% for more than 700 hours with a high areal capacity of $10\,mAh\,cm^{-2}$. Notably, an ultrahigh-specific

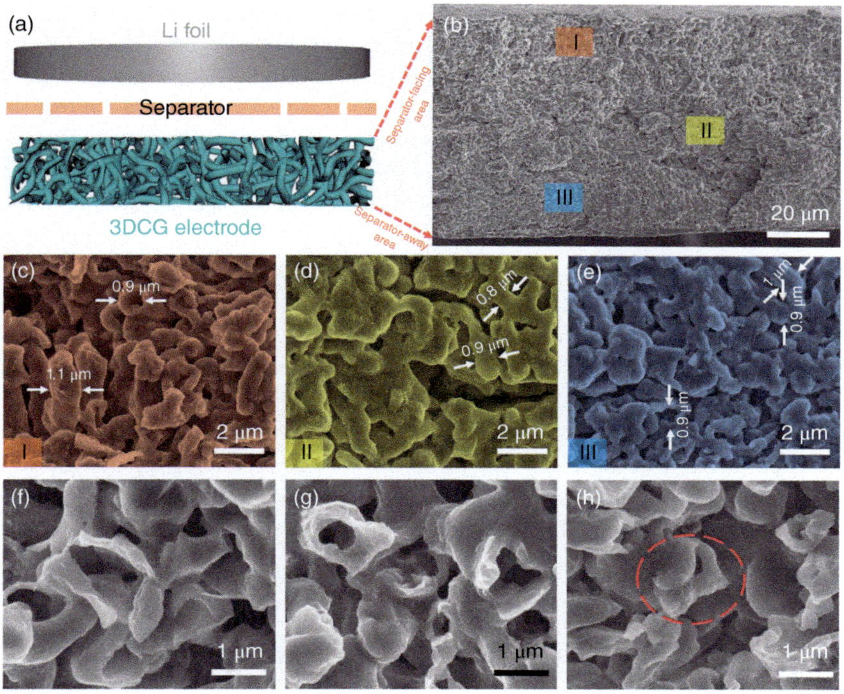

Figure 7.9 (a) Li plating process on 3DCG. (b–e) The cross-section SEM images of 3DCG with 4.0 mAh cm^{-2} Li deposition. SEM images of (f) 2.0 mAh cm^{-2}, (g) 4.0 mAh cm^{-2}, and (h) 10 mAh cm^{-2} Li plated on 3DCG at high magnification (the fracture of the duct-like graphene is marked with a red circle in (h)). Source: Zhang et al. [55]. Reproduced with permission of Elsevier.

capacity of 2588 mAh g^{-1} (based on the weight of the whole electrode) could be realized due to the lightweight of G-CNF film. In another study, hollow graphene foam was fabricated for metallic Li storage [54]. While, the composite LMAs were prepared by molten Li infusion into G-CNF film, and the Li amount could be controlled.

In addition, Zhang et al. proved that 3D graphene-based hosts with continuous duct-like structure (3DCG) could reduce the relative diffusion distance of Li ions, accelerating the Li-ions diffusion in the host. As a result, homogeneous metallic Li deposition throughout the 3D framework could be realized (Figure 7.9). Besides, in situ techniques were carried out to demonstrate homogeneous Li deposition. Spatially uniform Li deposition could improve the utilization of porous structures. As a result, 3DCG electrodes can keep high CE of nearly 98.7% for more than 1000 hours. Meanwhile, a high specific capacity of 3236 mAh g^{-1} (20 mAh cm^{-2}) could be realized by using 3DCG electrode, due to the low density and continuous porous structure of 3DCG. The Li@3DCG |S full cells with excellent electrochemical performance were assembled to demonstrate the potential practical application [55]. Furthermore, Park et al. report an electrical conductivity-controlled

3D host consisting of a glass fiber (GF) framework, size-/conductivity-controlled partially reduced graphene oxide (PrGO), and a Cu substrate (PrGO–GF/Cu). The PrGO–GF/Cu host could mitigate the Li top deposition and guides preferential Li platting at the bottom of the 3D structure [56].

From the above studies, it is can be seen that graphene-based materials with a high SSA can effectively reduce the local current density, thus suppressing the dendrites formation during the Li plating process. On the other hand, graphene with good conductivity can construct 3D hosts for Li deposition, adjusting the volume change during the cycling. Besides, an ultrahigh-specific capacity (based on the weight of the whole electrode) could be realized by using a free-standing 3D graphene-based host with a lightweight. However, it remains challenging to achieve sufficiently high CE by using 3D hosts, because it is difficult for protecting the SEI film, resulting in uneven Li nucleation and even microsized Li dendrites growth on the inner surface of 3D hosts. Besides, the Li deposition may happen on the upper surface of 3D host especially at high current density, reducing the utilization of porous structure and increasing the risk of short circuits. Therefore, the surface engineering of graphene-based hosts, contributing to homogeneous Li nucleation and plating/stripping could be more promising.

7.3 Heteroatom-Doped Graphene for Uniform Lithium Nucleation

Li nucleation and deposition are irregular in the anode, leading to dendritic Li growth that may cause short circuits. Although significant progress has been made by using 3D hosts to suppress the dendrite's growth, it is still a challenge to control the metallic Li nucleation and deposition spatially, thus reducing the utilization of porous structure. An ideal 3D host for Li deposition must regulate uniform Li nucleation and deposition. Recent works have shown that 3D host with a large electroactive surface area, can promote uniform Li nucleation and microscale homogeneous Li deposition [57, 58]. Therefore, the nucleation and deposition behavior of Li in the host need more attention for homogeneous Li deposition and improving cycling performance.

Based on the above discussion, graphene-based materials with high chemical stability, high SSA, and good conductivity are an ideal host for metallic Li storage, which can adjust the volume change and lower the local current density for high-performance LMAs. Furthermore, recent studies have shown that heteroatom-doped graphene can promote uniform Li nucleation and microscale homogeneous Li deposition, thus suppressing the dendritic Li growth. Zhang's group probed the lithiophilicity chemistry of heteroatom-doped graphene for promoting uniform Li nucleation and deposition by theoretical calculations [59]. During the nucleation and deposition process, Li ions in the electrolyte are driven to the electrode surface by the electric force, and Li ions will be reduced to metallic Li as long as the Li-ions meet electrons. Therefore, the uniform distribution of lithiophilic sites leads to uniform Li nucleation, and the binding energy between Li

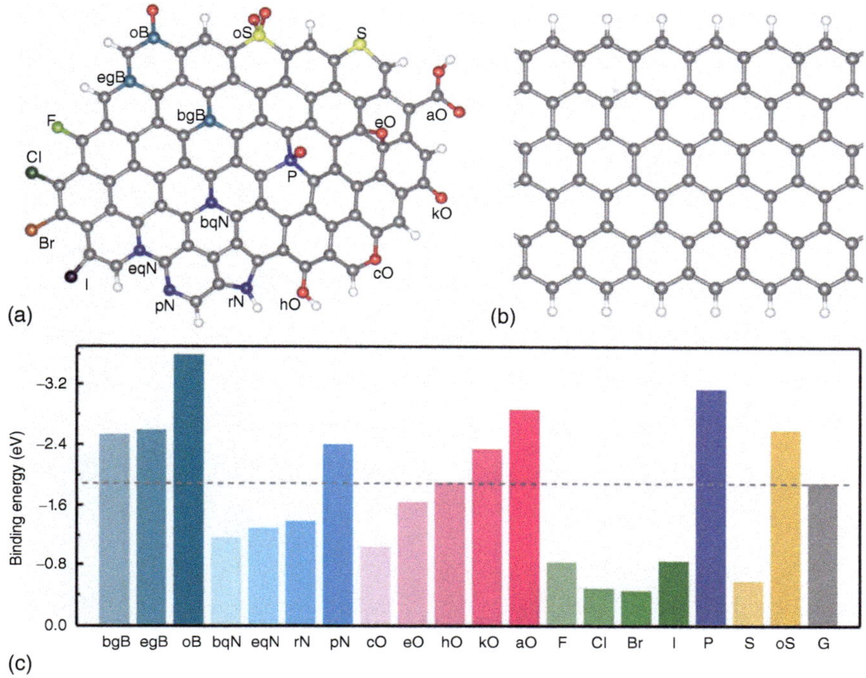

Figure 7.10 (a) Modeling of different heteroatom-doped carbons. (b) Modeling of pristine graphene nanoribbons (GNRs). (c) Summary of binding energy between heteroatom-doped graphene and a Li atom. The hydrogen, boron, carbon, nitrogen, oxygen, fluorine, phosphorus, sulfur, chlorine, bromine, and iodine atoms are marked as white, green, gray, blue, red, cyan, wathet blue, yellow, bottle green, orange, and black, respectively. Source: Chen et al. [59]. Reproduced with permission of American Association for the Advancement of Science.

atom and the graphene substrate could be suggested as the quantitative descriptor of lithiophilicity. A series of heteroatom-doped graphene nanoribbons (GNRs) were constructed (Figure 7.10a), and the binding energy based on the first-principles calculations between a Li atom and different heteroatom-doped graphene is given in Figure 7.10c. Among various single-doping GNRs, O-doping exhibits the best performance due to the higher electronegativity of the O atom (3.44) against the Li atom (0.98). Besides, graphene with N and B functional groups is also expected to reduce the Li nucleation overpotential. Meanwhile, the author also demonstrated that the O/B co-doped graphene is predicted to produce better lithiophilicity among co-doped graphene.

In addition, Zhai et al. reported the single-atom (SA) Ni doping on nitrogen-doped graphene frameworks (SANi-NG) as hosts for LMAs [60]. Isolated Li nucleation and subsequent dendritic Li growth on the surface of pristine graphene (PG) will happen due to the lack of lithiophilic site. By contrast, uniform Li nucleation and deposition can be achieved by using SANi-NG electrode. To further demonstrate the conclusion of uniform Li deposition on SANi-NG electrode, the first-principles calculations were carried out. The binding energy between Li atom and PG, NG

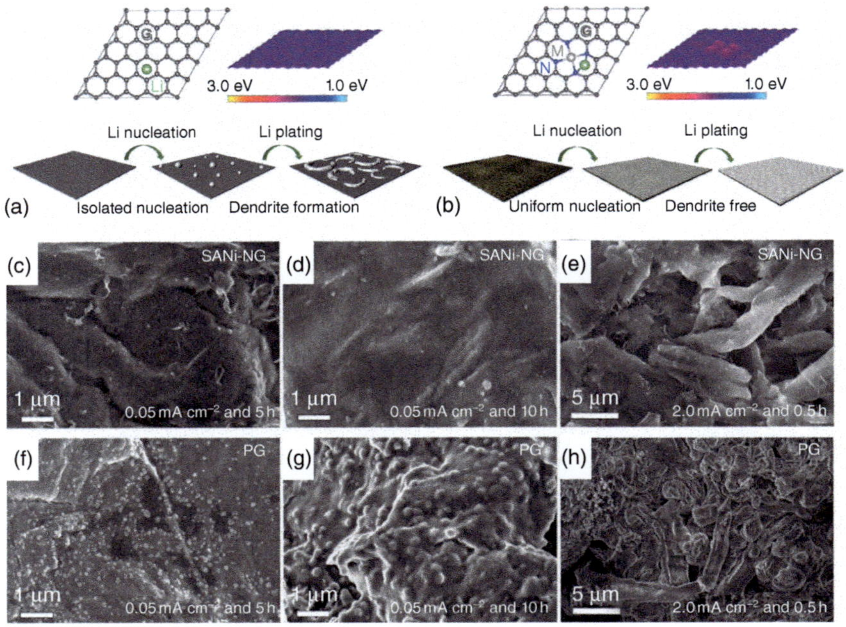

Figure 7.11 (a, b) Li adsorption energy distribution mapping and illustrations of Li nucleation and deposition process on (a) PG and (b) SANi-NG electrode. SEM images of (c–e) SANi-NG electrode and (f–h) PG electrode with metallic Li deposition (the current density and the time of Li plating are given in the picture). Source: Zhai et al. [60]. Reproduced with permission of Wiley.

(N-doped graphene), and SANi-NG are 1.23, 3.02, and 1.92 eV, which indicate that NG and SANi-NG framework can provide sufficient lithiophilic site for uniform Li nucleation (Figure 7.11a,b). According to the ex situ images of SANi-NG and PG electrodes with Li plating, homogeneous Li nucleation and deposition without Li dendrites could be achieved by using SANi-NG framework, which could be attributed to the homogenous and dense distribution of single Ni atoms (Figure 7.11c–h). As a result, the LMA with SANi-NG electrode exhibits a high CE of 98.45% over 250 cycles and an ultralow voltage hysteresis of 19 mV. In another study, Wang et al. prepared sulfur-doped graphene (SG) host to guide Li nucleation and deposition for LMA [61]. Based on first-principles simulation, S atom doping can improve the Li adsorption ability of graphene substrate. Thus, uniform Li deposition and high CE (99%) could be realized by using SG electrode. Besides, Liu et al. prepared B-doping graphene for Li-metal storage. The authors have demonstrated that the B-doping effect can trigger lithiophilic transition of graphene to guide its Li nucleation and deposition behaviors, contributing to high electrochemical performance [62].

Based on the above discussion, graphene-based materials with a high SSA can effectively reduce the local current density, thus suppressing the dendrites formation during the Li plating process. However, it remains challenging to achieve sufficiently high CE for practical applications, because Li nucleation and deposition on the surface of graphene substrate are uneven. Based on these considerations, 3D heteroatom-doped graphene can be used as host for metallic Li storage. On the one hand, 3D host can reduce the local current density and adjust the volume change, thus suppressing the Li dendrite growth and improving the cycling performance. On the other hand, heteroatom-doping site can guide Li nucleation and promote homogeneous Li deposition during the cycling, thus improving the CE and cycling performance of LMA.

To realize homogeneous Li deposition on 3D framework, Zhang et al. reported a nitrogen-doped graphene-modified 3D porous Cu (3DCu@NG) host for LMA [63]. Based on first-principles simulation, the 3D Cu framework with N-doped graphene coating has a strong interaction with Li-ion, contributing to a uniform Li nucleation and deposition. Moreover, when one Li atom is adsorbed on the NG substrate, the electrons transfer from Li atom to graphene and remain scattered throughout the graphene, resulting in homogeneous distribution of electrons on the inner surface of Cu skeleton. Therefore, uniform Li nucleation and deposition and improved cycling performance could be achieved by using 3DCu@NG electrode. According to the ex situ images of 3DCu@NG electrode with Li plating, the metallic Li coating on the inner surface of 3DCu@NG is smooth, which could be attributed to the uniform Li nucleation and deposition caused by N-doped graphene coating (Figure 7.12 a-g). As a result, with a cycling capacity of 1 $mAh\,cm^{-2}$ at 0.5 $mA\,cm^{-2}$, the CE of 3DCu@NG could reach 97.5% and remain stable for nearly 150 cycles, which was better than that of 3DCu and Cu foil electrode (Figure 7.12h). Notably, the

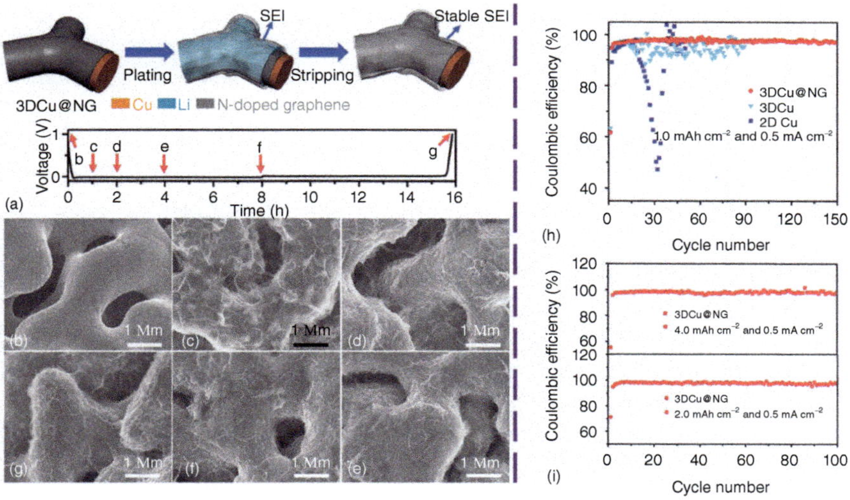

Figure 7.12 (a) Schematic illustration of Li plating/stripping on 3DCu@NG. The cross-section SEM images of (b) pristine 3DCu@NG electrode, after plating (c) 0.5 $mAh\,cm^{-2}$, (d) 1.0 $mAh\,cm^{-2}$, (e) 2.0 $mAh\,cm^{-2}$, and (f) 4.0 $mAh\,cm^{-2}$ of Li on the inner surface 3DCu@NG and (g) after stripping 4.0 $mAh\,cm^{-2}$ of Li. (h) Comparison of the CEs of 3DCu@NG, 3DCu and Cu foil electrode with a cycling capacity of 1 $mAh\,cm^{-2}$ at a current density of 0.5 $mA\,cm^{-2}$. (i) CEs of 3DCu@NG electrode with cycling capacity of 2 and 4 $mAh\,cm^{-2}$ at a current density of 0.5 $mA\,cm^{-2}$. Source: Zhang et al. [63]. Reproduced with permission of Wiley.

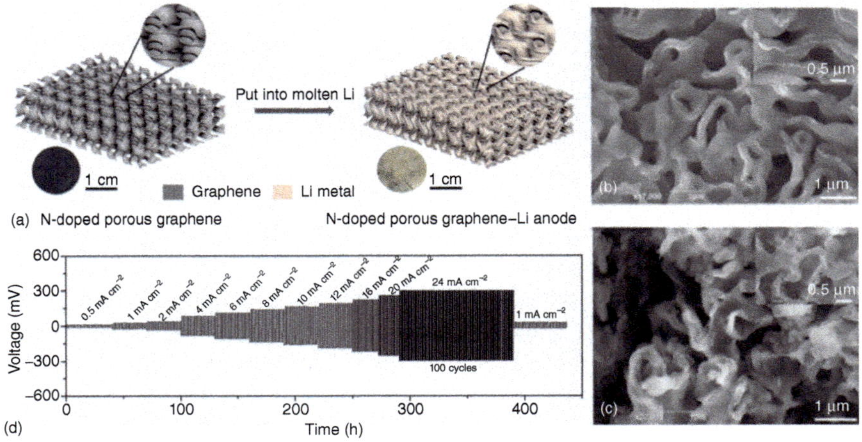

Figure 7.13 Fabrication process of the N-doped porous graphene–Li anode. SEM images of (b) N-doped porous graphene–Li anode and (c) pure graphene–Li anode (Insets are zoom-in SEM images). (d) Rate performance of N-doped porous graphene–Li anode, the current densities are given in the picture. Source: Huang et al. [64]: Reproduced with permission of Wiley.

3DCu@NG electrode could remain high CE of 97.6% or 97.8% for nearly 100 cycles even increasing the cycling capacity to 2 or 4 mAh cm^{-2} (Figure 7.12i).

In addition, heteroatom-doped graphene with lithiophilic site could promote melt Li uniformly into the 3D framework. Chen's group reported N-doped porous graphene–Li anode prepared by molten-infused lithium method [64]. 3D nanoporous graphene framework with 1.1 at.% nitrogen doping could significantly improve the lithiophilic property of graphene substrate. Thus, when N-doped porous graphene was put into contact with molten Li, Li melt readily infused into the porous channels (Figure 7.13a). Benefiting from the lithiophilic sites, Li coating on the inner surfaces of graphene substrate is homogeneous (Figure 7.13b). On the contrary, it is difficult to load the metallic Li into the porous pure graphene frameworks by molten-infused lithium method homogeneously, because of the lack of lithiophilic groups (Figure 7.13c). Besides, the large lithiophilic surfaces could well suppress uncontrolled dendrite growth and the porous structure could inhibit the volume changes of LMAs during the cycling. As a result, the Li@N-doped porous graphene composite anode showed remarkable battery performance with stabilized and low voltage hysteresis for 100 cycles even at a high current density of 24 mA cm^{-2} and a large areal capacity of 12 mAh cm^{-2} (Figure 7.13d). Besides, a large specific capacity of 3487 mAh g^{-1} can be realized due to the ultralow density of the porous N-doped graphene framework.

From the above studies, uniform Li nucleation and deposition on the electrode could be realized by using heteroatom-doped graphene hosts, resulting in improving cycling performance and ultrahigh CE. In addition, some "lithiophilic" materials also can guide the Li deposition by alloying with Li, and graphene-based materials combined with these "lithiophilic" materials as hosts can improve the electrochemical performance of LMAs.

7.4 Graphene Combined with Other "lithiophilic" Materials

Based on the above discussion, heteroatom-doped graphene hosts can promote homogeneous Li nucleation and deposition. In addition, combining graphene with uniformly distributed functional materials, such as Au, Ag, and Zn, can help to realize stable and homogeneous Li deposition by alloying with Li.

Recently, Cui's group reported an experimental discovery that the nucleation barrier is observably different among different substrates [65]. The author studied a list of 11 elemental substrate materials, they found that there was a significant voltage dip at the beginning of Li nucleation on some materials, such as Cu, Ni, and C (Figure 7.14a). The nucleation overpotential, which is defined as the difference value between the voltage dip and the later voltage platform in the voltage-capacity profile, is associated with the "lithiophilicity" of the electrode. For Cu substrate, on account of the thermodynamic mismatch between Li and Cu, the Li deposition needs to overcome the heterogeneous nucleation barrier, leading to a high nucleation overpotential. In contrast, the voltage profile of Li deposition on Au substrate showed a sharp right-angle turn at the onset (red circle,

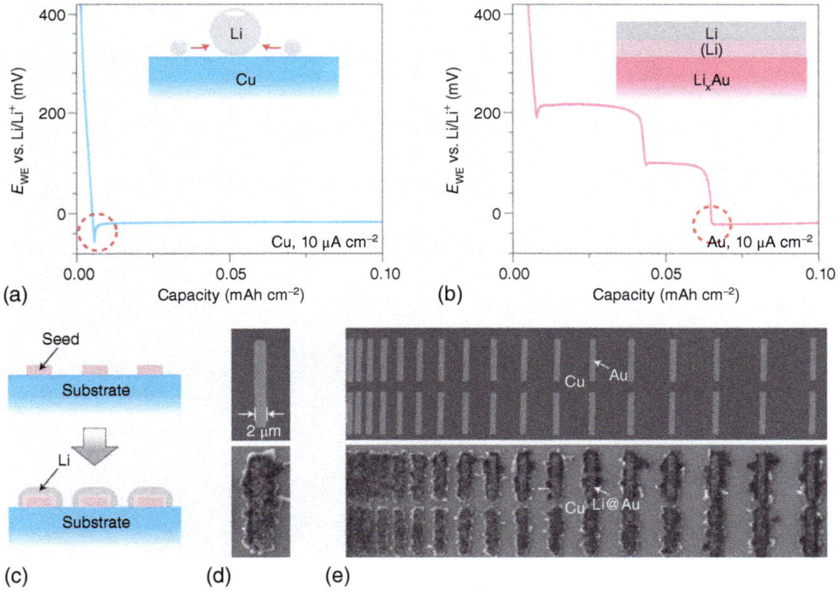

Figure 7.14 (a) Voltage-capacity profiles of Li deposition on Cu substrate at 10 μA cm^{-2}. The inset is the schematic of the mechanism of Li nucleation. (b) Voltage-capacity profiles of Li deposition on Cu substrate at 10 μA cm^{-2}. The inset shows a schematic of dissolution of the solid-solution buffer layer of Au in Li reduces the nucleation energy. (c) Schematic illustration of Li plating on the substrate with Au seeds. (d) SEM images of substrate with Au seed before and after Li deposition. (e) SEM images of Au seeds with various separations before and after Li deposition. Source: Yan et al. [65]. Reproduced with permission of Springer Nature.

Figure 7.14b). It can be seen that the nucleation overpotential of Li deposition on lithiated Au is zero. The difference in nucleation overpotential, therefore, guides Li-metal deposition. To demonstrate this opinion, Li deposition on copper substrate patterned with 2 μm wide Au strips was carried out (Figure 7.14c–e). After Li deposition on substrate, Li was deposited only on Au stripes, demonstrating that selective nucleation and deposition of Li metal could be realized by using Au seeds. Based on this consideration, hollow carbon spheres cage with nanoparticle Au seeds inside was designed and fabricated for high-performance LMA. The lithium metal is found to predominantly grow inside the hollow carbon spheres during the deposition process, leading to improved cycling performance and high CE. Besides, the author further demonstrated that Ag, Zn, Mg, Al, and Pt substrate, once fully lithiated, also exhibited ultralow nucleation overpotential during the Li deposition process.

To improve the conductivity, mechanical stability, and Li-ion conductivity of the cage for Li metal, Cui's group further prepared a new host of wrinkled graphene cage (WGC) with nanoparticle Au seeds inside (Figure 7.15a,b) [66]. The WGC with high-quality graphitic structure offers excellent mechanical strength and high surface area, and enables Li-ion intercalation and transportation, leading to excellent cycling performance. With low cycling capacities, metallic Li is preferentially nucleated and deposited inside the graphene cages under the force

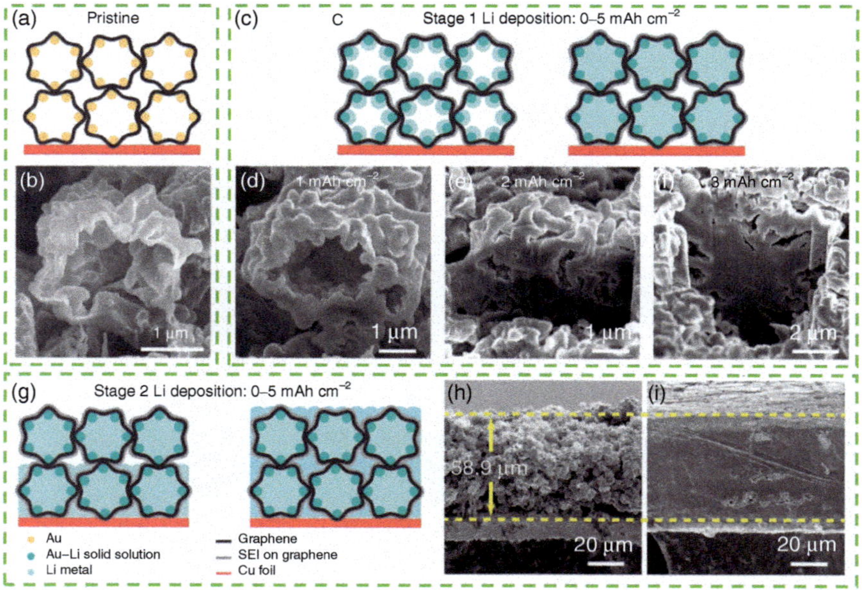

Figure 7.15 (a) Schematic and (b) SEM image of WGC. (c) Schematic and SEM image of WGC electrode after (d) 1 mAh cm^{-2}, (e) 2 mAh cm^{-2}, and (f) 3 mAh cm^{-2} of Li deposition. (g) Schematic of WGC electrode under high areal capacity Li deposition of 5–10 mAh cm^{-2}. The cross-section SEM image of (h) pristine WGC electrode and (i) electrode with 7 mAh cm^{-2} of Li deposition. Source: Wang et al. [66]. Reproduced with permission of American Chemical Society.

of Au particles (Figure 7.15c–f). Besides, uniform SEI formed outside the cage and protected the Li deposition inside during cycling. As the deposition capacity increased, Li deposition started to fill outer pore spaces of the WGC electrode. After 7 mAh cm^{-2} Li-metal deposition, the WGC electrode was densely filled with Li metal without dendrites formation and maintained the same thickness (Figure 7.15g–i). Therefore, the LMA with WGC electrode achieved a high CE of ~98.0% at 0.5 mAh cm^{-2} with 1–10 mAh cm^{-2}. Meanwhile, an ultrahigh CE of 99.1% was realized with high-concentration electrolytes at 0.5 mA cm^{-2} with an areal capacity of 3 mAh cm^{-2}. Due to the lightweight of graphene, the WGC/Li composite LMA was able to reach a high specific capacity of ~2785 mAh g^{-1}. In another study, a sandwich composite LMA of pillared rGO with gold nanoparticles inside was fabricated [67]. The Au particles could induce lithium deposition between rGO sheets, which reduce the morphological changes of SEI films and reduce the nonuniformity of lithium plating during cycling. Meanwhile, the interlay Li plating relatively suppresses surface deposition and dendrites formation. As a result, the Au–rGO composite electrode could reach an ultrahigh CE of ~98% for nearly 200 cycles with a high cycling capacity of 2 mAh cm^{-2}.

ZnO nanoparticles also can serve as "lithiophilic" seeds to guide Li deposition, because Li can react with ZnO to form Zn particles. Although lithiophilic seeds could ensure uniform Li nucleation and help achieve dendrite-free LMAs, the lithiophilic materials may detach from the substrate during cycling due to the strong interaction between the seeds and Li-ions. Thus, inhomogeneous Li deposition could happen due to the uneven distribution and loss of seeds. To improve the stability of lithiophilic seeds, Tang et al. prepared novel ZnO nanoparticle confined 3D porous carbon composite microspheres (3D ZnO@PCCMs) as 3D hosts for LMAs [68]. The ZnO nanoparticle confined within graphene could effectively inhibit the ZnO from detaching from the substrate and agglomerating during the plating/stripping process, thus improving the cycling stability (Figure 7.16a,b). The uniform Li deposition uniformly along the nanobox walls could be observed from the SEM images, owing to the strong guiding effect of homogenous ZnO NPs (Figure 7.16c,d). Notably, after 50 plating/stripping cycles with areal capacity of 2.0 mAh cm^{-2}, the 3D ZnO@PCCM electrode still had a smooth inner surface morphology without cracks or dendrites, indicating the superb structural stability of the composite anode (Figure 7.16e,f). Meanwhile, ZnO lithiophilic seeds are still evenly distributed inside the microspheres. As for the ZnO/3DPCS electrode, although lithium dendrites were inhibited with plenty of Li deposition, lithium deposition agglomeration and pulverization of 3D scaffolds could be observed after 50 cycles from the SEM images. The homogeneous Li deposition and improved structural stability of the 3D ZnO@PCCM electrode can be ascribed to the confined ZnO nanoparticles, which can effectively inhibit the ZnO from detaching from the substrate and agglomerating during the cycling. As a result, the CE of 3D ZnO@PCCM could reach 98.5%, which was better than that of ZnO/3DPCS electrode (Figure 7.16k,l). Besides, lithiophilic MgZnO nanoparticles also can serve as seeds to induce Li deposition. Le et al. prepared carbon matrix with CNF and rGO anchored with uniformly-dispersed lithiophilic MgZnO nanoparticles as framework

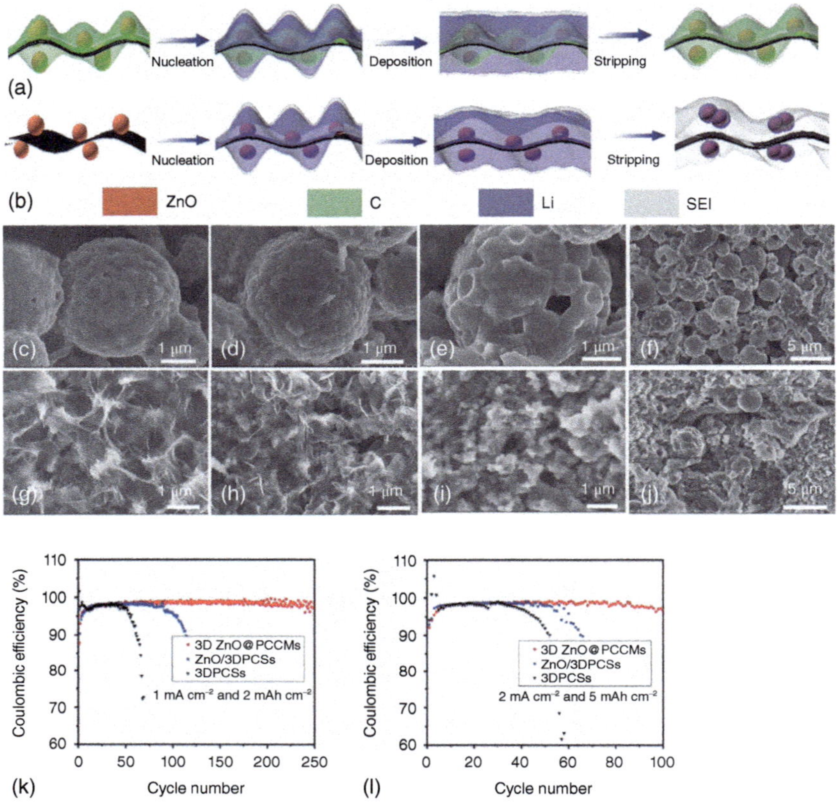

Figure 7.16 Schematic of Li deposition on the (a) 3D ZnO@PCCM and (b) ZnO/3DPCS electrodes. SEM images of the 3D ZnO@PCCM electrode with (c) 0.5 mAh cm^{-2} and (d) 2 mAh cm^{-2} of Li deposition. (e, f) SEM images of 2 mAh cm^{-2} of Li plating on ZnO@PCCM electrode after 50 cycles. SEM images of the ZnO/3DPCS electrode with (g) 0.5 mAh cm^{-2} and (h) 2 mAh cm^{-2} of Li deposition. (i, j) SEM images of 2 mAh cm^{-2} of Li plating on ZnO/3DPCS electrode after 50 cycles. Comparison of the CEs of the 3D ZnO@PCCM, ZnO/3DPCS, and 3DPCS electrodes with a cycling capacity of (k) 2 mAh cm^{-2} at a current density of 1 mA cm^{-2}, (l) 5 mAh cm^{-2} at a current density of 2 mA cm^{-2}. Source: Tang et al. [68]. Reproduced with permission of The Royal Society of Chemistry.

for Li-metal storage. The uniformly MgZnO particles could effectively guide the Li-ions to be deposited in the whole framework, and uniform Li deposition could be achieved [69]. Ma et al. fabricated scallion-like graphene-based microrods with lithiophilic Ag NWs as hosts for LMAs, Li metal was successfully siphoned and stored inside the 3D hosts. Ag NWs served as heterogeneous nucleation seeds to guide Li deposition [70].

The combination of graphene-based substrates with "lithiophilic" materials is an effective way to achieve uniform Li nucleation and deposition. However, many lithiophilic materials undergo large volume changes during the cycling, which need further consideration and investigation. Besides, the rational design of the structure and composition is still needed to be investigated for further practical application.

7.5 Outlook

Li metal has been considered as a "Holy Grail" anode for high-energy-density rechargeable batteries. A lot of studies have been done to understand the mechanism of metallic Li nucleation and dendritic Li growth. Based on the theoretical understanding of Li nucleation and deposition, various strategies have been proposed to restrain the Li dendrites and improve the cycling performance from different perspectives. Graphene-based materials with high SSA, controllable microstructure, and tunable electronic/ion conductivities are emerging as important candidates for LMAs protection. Graphene-based materials can play multifunctional roles in suppressing the dendrite growth, such as protecting the SEI layers, configuring Li^+ flux, constructing 3D hosts for Li deposition, lowering the local current density, and promoting uniform Li nucleation and deposition. The advantages of the graphene-based materials are summarized as follows: (i) high chemical stability, mechanical strength, and flexibility to maintain structural integrity during the cycling; (ii) high SSA reducing local current density; (iii) being easily decorated with lithiophilic functional groups or materials to guide Li nucleation and deposition. Based on the discussion above, graphene-based materials are preferable hosts for LMAs, which can well accommodate Li via constraining the plating behavior.

Although great advances in graphene-based materials have been achieved for LMAs, there are still some concerns to be addressed: (i) The high SSA may lead to the consumption of electrolyte and inhomogeneous distribution of the SEI, causing low Li utilization; (ii) The electrochemical behavior of LMA by using 3D host under a practical working state is poorly understood; (iii) Because of the complexity of the synthesis of graphene, the scale-up graphene-based LMA for real application is still a challenge.

References

1 Tarascon, J.-M. and Armand, M. (2001). Issues and challenges facing rechargeable lithium batteries. *Nature* 414: 359–367.

2 Goodenough, J.B. and Kim, Y. (2010). Challenges for rechargeable Li batteries. *Chemistry of Materials* 22 (3): 587–603.

3 Li, H., Wang, Z., Chen, L. et al. (2009). Research on advanced materials for Li-ion batteries. *Advanced Materials* 21 (45): 4593–4607.

4 Schmuch, R., Wagner, R., Hörpel, G. et al. (2018). Performance and cost of materials for lithium-based rechargeable automotive batteries. *Nature Energy* 3 (4): 267–278.

5 Bruce, P.G., Scrosati, B., and Tarascon, J.M. (2008). Nanomaterials for rechargeable lithium batteries. *Angewandte Chemie International Edition* 47 (16): 2930–2946.

6 Zhao, Y., Li, X., Yan, B. et al. (2016). Recent developments and understanding of novel mixed transition-metal oxides as anodes in lithium ion batteries. *Advanced Energy Materials* 6 (8): 1502175.

7 Cheng, X.B., Zhang, R., Zhao, C.Z. et al. (2017). Toward safe lithium metal anode in rechargeable batteries: a review. *Chemical Reviews* 117 (15): 10403–10473.

8 Li, S., Jiang, M., Xie, Y. et al. (2018). Developing high-performance lithium metal anode in liquid electrolytes: challenges and progress. *Advanced Materials* 30 (17): 1706375.

9 Liu, B., Zhang, J.-G., and Xu, W. (2018). Advancing lithium metal batteries. *Joule* 2 (5): 833–845.

10 Ghazi, Z.A., Sun, Z., Sun, C. et al. (2019). Key aspects of lithium metal anodes for lithium metal batteries. *Small* 15 (32): 1900687.

11 Zhang, C., Huang, Z., Lv, W. et al. (2017). Carbon enables the practical use of lithium metal in a battery. *Carbon* 123: 744–755.

12 Lin, D., Liu, Y., and Cui, Y. (2017). Reviving the lithium metal anode for high-energy batteries. *Nature Nanotechnology* 12 (3): 194–206.

13 Cohen, Y.S., Cohen, Y., and Aurbach, D. (2000). Micromorphological studies of lithium electrodes in alkyl carbonate solutions using in situ atomic force microscopy. *The Journal of Physical Chemistry B* 104 (51): 12282–12291.

14 Guo, W., Liu, S., Guan, X. et al. (2019). Mixed ion and electron–conducting scaffolds for high–rate lithium metal anodes. *Advanced Energy Materials* 9 (20): 1900193.

15 Ding, F., Xu, W., Graff, G.L. et al. (2013). Dendrite-free lithium deposition via self-healing electrostatic shield mechanism. *Journal of the American Chemical Society* 135 (11): 4450–4456.

16 Li, W., Yao, H., Yan, K. et al. (2015). The synergetic effect of lithium polysulfide and lithium nitrate to prevent lithium dendrite growth. *Nature Communications* 6: 7436.

17 Yoo, D.-J., Kim, K.J., and Choi, J.W. (2018). The synergistic effect of cation and anion of an ionic liquid additive for lithium metal anodes. *Advanced Energy Materials* 8 (11): 1702744.

18 Cheng, X.B., Zhang, R., Zhao, C.Z. et al. (2015). A review of solid electrolyte interphases on lithium metal anode. *Advanced Science* 3 (3): 1500213.

19 Zheng, G., Lee, S., Wang, H. et al. (2014). Interconnected hollow carbon nanospheres for stable lithium metal anodes. *Nature Nanotechnology* 9: 618–623.

20 Shi, L., Xu, A., and Zhao, T. (2017). First-principles investigations of the working mechanism of 2D h-BN as an interfacial layer for the anode of lithium metal batteries. *ACS Applied Materials & Interfaces* 9 (2): 1987–1994.

21 Yan, K., Lee, H.W., Gao, T. et al. (2014). Ultrathin two-dimensional atomic crystals as stable interfacial layer for improvement of lithium metal anode. *Nano Letters* 14 (10): 6016–6022.

22 Cheng, X.B., Hou, T.Z., Zhang, R. et al. (2016). Dendrite-free lithium deposition induced by uniformly distributed lithium ions for efficient lithium metal batteries. *Advanced Materials* 28 (15): 2888–2895.

23 Gao, Y., Yan, Z., Gray, J.L. et al. (2019). Polymer-inorganic solid-electrolyte interphase for stable lithium metal batteries under lean electrolyte conditions. *Nature Materials* 18: 384–389.

24 Liu, X., Li, X., Li, H. et al. (2018). Recent progress of hybrid solid-state electrolytes for lithium batteries. *Chemistry* 24 (69): 18293–18306.

25 Zhou, F., Li, Z., Lu, Y.Y. et al. (2019). Diatomite derived hierarchical hybrid anode for high performance all-solid-state lithium metal batteries. *Nature Communications* 10 (1): 2482.

26 Zhou, J., Qian, T., Liu, J. et al. (2019). High-safety all-solid-state lithium-metal battery with high-ionic-conductivity thermoresponsive solid polymer electrolyte. *Nano Letters* 19 (5): 3066–3073.

27 Wang, C., Gong, Y., Liu, B. et al. (2017). Conformal, nanoscale ZnO surface modification of garnet-based solid-state electrolyte for lithium metal anodes. *Nano Letters* 17 (1): 565–571.

28 Wang, C., Xie, H., Zhang, L. et al. (2018). Universal soldering of lithium and sodium alloys on various substrates for batteries. *Advanced Energy Materials* 8 (6): 1701963.

29 Yang, C., Xie, H., Ping, W. et al. (2018). An electron/ion dual-conductive alloy framework for high-rate and high-capacity solid-state lithium-metal batteries. *Advanced Materials* 31 (3): 1804815.

30 Chi, S.-S., Liu, Y., Song, W.-L. et al. (2017). Prestoring lithium into stable 3D nickel foam host as dendrite-free lithium metal anode. *Advanced Functional Materials* 27 (24): 1700348.

31 Ye, H., Xin, S., Yin, Y.X. et al. (2017). Stable Li plating/stripping electrochemistry realized by a hybrid Li reservoir in spherical carbon granules with 3D conducting skeletons. *Journal of the American Chemical Society* 139 (16): 5916–5922.

32 Yun, Q., He, Y.B., Lv, W. et al. (2016). Chemical dealloying derived 3D porous current collector for Li metal anodes. *Advanced Materials* 28 (32): 6932–6939.

33 Li, Q., Zhu, S., and Lu, Y. (2017). 3D porous Cu current collector/Li-metal composite anode for stable lithium-metal batteries. *Advanced Functional Materials* 27 (18): 1606422.

34 Lu, L.L., Ge, J., Yang, J.N. et al. (2016). Free-standing copper nanowire network current collector for improving lithium anode performance. *Nano Letters* 16 (7): 4431–4437.

35 Pei, F., Fu, A., Ye, W. et al. (2019). Robust lithium metal anodes realized by lithiophilic 3D porous current collectors for constructing high-energy lithium-sulfur batteries. *ACS Nano* 13 (7): 8337–8346.

36 Liu, L., Yin, Y.-X., Li, J.-Y. et al. (2017). Free-standing hollow carbon fibers as high-capacity containers for stable lithium metal anodes. *Joule* 1 (3): 563–575.

37 Xu, Y., Li, T., Wang, L. et al. (2019). Interlayered dendrite-free lithium plating for high-performance lithium-metal batteries. *Advanced Materials* 31 (29): 1901662.

38 Zuo, T.T., Wu, X.W., Yang, C.P. et al. (2017). Graphitized carbon fibers as multifunctional 3D current collectors for high areal capacity Li anodes. *Advanced Materials* 29 (29): 1700389.

39 Pu, J., Li, J., Zhang, K. et al. (2019). Conductivity and lithiophilicity gradients guide lithium deposition to mitigate short circuits. *Nature Communications* 10 (1): 1896.

40 Guo, Y., Niu, P., Liu, Y. et al. (2019). An autotransferable g-C_3N_4 Li^+-modulating layer toward stable lithium anodes. *Advanced Materials* 31 (27): 1900342.

41 Sun, Y., Zhao, Y., Wang, J. et al. (2018). A novel organic "Polyurea" thin film for ultralong-life lithium-metal anodes via molecular-layer deposition. *Advanced Materials* 31 (4): 1806541.

42 Zhang, Z., Peng, Z., Zheng, J. et al. (2017). The long life-span of a Li-metal anode enabled by a protective layer based on the pyrolyzed N-doped binder network. *Journal of Materials Chemistry A* 5 (19): 9339–9349.

43 Ren, F., Peng, Z., Wang, M. et al. (2019). Over-potential induced Li/Na filtrated depositions using stacked graphene coating on copper scaffold. *Energy Storage Materials* 16: 364–373.

44 Yao, Y., Zhao, X., Razzaq, A.A. et al. (2019). Mosaic rGO layers on lithium metal anodes for the effective mediation of lithium plating and stripping. *Journal of Materials Chemistry A* 7 (19): 12214–12224.

45 Li, C., Liu, S., Shi, C. et al. (2019). Two-dimensional molecular brush-functionalized porous bilayer composite separators toward ultrastable high-current density lithium metal anodes. *Nature Communications* 10 (1): 1363.

46 Shin, W.K., Kannan, A.G., and Kim, D.W. (2015). Effective suppression of dendritic lithium growth using an ultrathin coating of nitrogen and sulfur codoped graphene nanosheets on polymer separator for lithium metal batteries. *ACS Applied Materials & Interfaces* 7 (42): 23700–23707.

47 Foroozan, T., Soto, F.A., Yurkiv, V. et al. (2018). Synergistic effect of graphene oxide for impeding the dendritic plating of Li. *Advanced Functional Materials* 28 (15): 1705917.

48 Li, S., Wang, X.-S., Li, Q.-D. et al. (2021). A multifunctional artificial protective layer for producing an ultra-stable lithium metal anode in a commercial carbonate electrolyte. *Journal of Materials Chemistry A* 9 (12): 7667–7674.

49 Cheng, X.-B., Peng, H.-J., Huang, J.-Q. et al. (2015). Dual-phase lithium metal anode containing a polysulfide-induced solid electrolyte interphase and nanostructured graphene framework for lithium sulfur batteries. *ACS Nano* 9 (6): 6373–6382.

50 Zhang, R., Cheng, X.B., Zhao, C.Z. et al. (2016). Conductive nanostructured scaffolds render low local current density to inhibit lithium dendrite growth. *Advanced Materials* 28 (11): 2155–2162.

51 Liu, S., Wang, A., Li, Q. et al. (2018). Crumpled graphene balls stabilized dendrite-free lithium metal anodes. *Joule* 2 (1): 184–193.

52 Lin, D., Liu, Y., Liang, Z. et al. (2016). Layered reduced graphene oxide with nanoscale interlayer gaps as a stable host for lithium metal anodes. *Nature Nanotechnology* 11 (7): 626–632.

53 Zhao, C., Yu, C., Li, S. et al. (2018). Ultrahigh-capacity and long-life lithium-metal batteries enabled by engineering carbon nanofiber-stabilized graphene aerogel film host. *Small* 14 (42): 1803310.

54 Hafez, A.M., Jiao, Y., Shi, J. et al. (2018). Stable metal anode enabled by porous lithium foam with superior ion accessibility. *Advanced Materials* 30 (30): 1802156.

55 Zhang, R., Wang, N., Shi, C. et al. (2020). Spatially uniform Li deposition realized by 3D continuous duct-like graphene host for high energy density Li metal anode. *Carbon* 161: 198–205.

56 Park, K.H., Kang, D.W., Park, J.-W. et al. (2021). Modulating the electrical conductivity of a graphene oxide-coated 3D framework for guiding bottom-up lithium growth. *Journal of Materials Chemistry A* 9 (3): 1822–1834.

57 Zhang, R., Chen, X.R., Chen, X. et al. (2017). Lithiophilic sites in doped graphene guide uniform lithium nucleation for dendrite-free lithium metal anodes. *Angewandte Chemie International Edition* 129 (27): 7872–7876.

58 Li, K., Hu, Z., Ma, J. et al. (2019). A 3D and stable lithium anode for high-performance lithium-iodine batteries. *Advanced Materials* 31 (33): 1902399.

59 Chen, X., Chen, X.-R., Hou, T.-Z. et al. (2019). Lithiophilicity chemistry of heteroatom-doped carbon to guide uniform lithium nucleation in lithium metal anodes. *Science Advances* 5 (2): eaau7728.

60 Zhai, P., Wang, T., Yang, W. et al. (2019). Uniform lithium deposition assisted by single-atom doping toward high-performance lithium metal anodes. *Advanced Energy Materials* 9 (18): 1804019.

61 Wang, T., Zhai, P., Legut, D. et al. (2019). S-doped graphene-regional nucleation mechanism for dendrite-free lithium metal anodes. *Advanced Energy Materials* 9 (24): 1804000.

62 Liu, W., Zhai, P., Qin, S. et al. (2021). Boron-doping induced lithophilic transition of graphene for dendrite-free lithium growth. *Journal of Energy Chemistry* 56: 463–469.

63 Zhang, R., Wen, S., Wang, N. et al. (2018). N-doped graphene modified 3D porous Cu current collector toward microscale homogeneous Li deposition for Li metal anodes. *Advanced Energy Materials* 8 (23): 1800914.

64 Huang, G., Han, J., Zhang, F. et al. (2018). Lithiophilic 3D nanoporous nitrogen-doped graphene for dendrite-free and ultrahigh-rate lithium-metal anodes. *Advanced Materials* 31 (2): 1805334.

65 Yan, K., Lu, Z., Lee, H.-W. et al. (2016). Selective deposition and stable encapsulation of lithium through heterogeneous seeded growth. *Nature Energy* 1: 16010.

66 Wang, H., Li, Y., Li, Y. et al. (2019). Wrinkled graphene cages as hosts for high-capacity Li metal anodes shown by cryogenic electron microscopy. *Nano Letters* 19 (2): 1326–1335.

67 Pu, J., Li, J., Shen, Z. et al. (2018). Interlayer lithium plating in Au nanoparticles pillared reduced graphene oxide for lithium metal anodes. *Advanced Functional Materials* 28 (41): 1804133.

68 Tang, L., Zhang, R., Zhang, X. et al. (2019). ZnO nanoconfined 3D porous carbon composite microspheres to stabilize lithium nucleation/growth for high-performance lithium metal anodes. *Journal of Materials Chemistry A* 7 (33): 19442–19452.

69 Le, T., Yang, C., Liang, Q. et al. (2021). A fishing-net-like 3D host for robust and ultrahigh-rate lithium metal anodes. *Small* 17 (11): 2007231.

70 Ma, T., Su, T.Y., Zhang, L. et al. (2021). Scallion-inspired graphene scaffold enabled high rate lithium metal battery. *Nano Letters* 21 (6): 2347–2355.

8

Graphene-Based Materials for Li–S Batteries

Ning Wang

Hebei University of Technology, School of Electrical Engineering, State Key Laboratory of Reliability and Intelligence of Electrical Equipment, 5340 Xiping Road, Beichen District, Tianjin, 300401, P.R. China

8.1 Development History of Li–S Batteries

With the energy crisis and environmental pollution, the efficient use of clean energy (such as solar energy, wind energy, geothermal energy, and tidal energy) has attracted more and more attention, which has put forward higher requirements for the corresponding energy storage system. Chemical energy storage has become an important means of energy storage because of its good stability and movable characteristics. Among them, lithium secondary batteries represented by lithium-ion (Li-ion) batteries have made great progress since they were commercialized in 1991, and have been widely used in consumer electronics and electric vehicles. However, with the development of science and technology, high energy density and low-cost secondary batteries are required. It is clear that by 2025, the new system power battery technology will make a breakthrough, with monomer-specific energy of 500 Wh kg^{-1}. Li-ion batteries based on "state-of-the-art" mechanism are limited by their lower theoretical specific capacity, which cannot meet this demand for high-energy density. Therefore, it is crucial to find a cathode material with high specific capacity and low cost [1]. In recent years, researchers have turned their attention to lithium–sulfur batteries. The element sulfur has the highest theoretical specific capacity (1675 mAh g^{-1}) and theoretical specific energy (2600 Wh kg^{-1}) in solid cathode materials for lithium batteries. Li–S batteries are one of the best choices for energy density secondary batteries [2], and sulfur reserves are abundant, environmental-friendly, low in price, and have a good application prospect. The history of Li–S batteries can be traced back to the 1960s. In 1962, Herbet and Ulam [3] firstly took sulfur as positive electrode material in dry batteries. In 1966, Rao [4] designed a metal–sulfur battery with an organic electrolyte and calculated the theoretical energy density of the battery. However, because the actual capacity of Li–S batteries is low and the decay of the capacity is severe, it has not received much attention. Subsequent research has focused on electrolytes. After 2000, the

Templated Fabrication of Graphene-Based Materials for Energy Applications, First Edition.
Edited by Chunnian He, Naiqin Zhao, and Junwei Sha.

appearance of sulfur/carbon nanocomposite cathode materials made breakthroughs in the specific capacitance and cycle stability of Li–S batteries, which led to the research boom of Li–S batteries. But the key parameters, such as life expectancy cycle stability, energy density, and circulation, still need to be further improved to meet the requirements of commercial applications.

8.2 Working Mechanism of Li–S Battery

Li–S batteries refer to a battery system in which elemental sulfur and its composite materials or sulfur-containing compounds are used as positive electrodes, and metallic lithium is used as a negative electrode to convert electrical energy and chemical energy through the fracture and formation of sulfur–sulfur bonds. Its expression is as follows: $S + 2Li^+ + 2e^- \leftrightarrow Li_2S$. Its working process involves a complex multistep gain and loss electron reaction. In the discharging process, one sulfur atom loses two electrons, and its theoretical specific capacity is as high as 1675 mAh g^{-1}.

Mikhaylik and coworker [5] and Shim and coworker [6] studied the charging–discharging process of Li–S batteries in detail and found that during the discharging process sulfur was reduced to polysulfides, and two discharging platforms were formed. The high potential platform at 2.3 V corresponds to the conversion of S_8 to Li_2S_4 and the low potential platform at 2.1 V corresponds to the conversion of Li_2S_4 to Li_2S. Recently, Barchasz et al. [7] observed Li–S batteries in more detail. It is believed that elemental sulfur is produced by three-step of electroreduction to produce Li_2S. This process is accompanied by ten disproportionation reactions. Reaction, the first discharging platform, is very small and can only be observed during very slow discharging process, while eight polysulfide ions are detected in the electrolyte ($S_8{}^{2-}$, $S_7{}^{2-}$, $S_6{}^{2-}$, $S_5{}^{2-}$, $S_4{}^{2-}$, $S_3{}^{2-}$, $S_2{}^{2-}$, S^{2-}). These $S_n{}^{2-}$ ($n = 3$–8) are easily dissolved in the electrolyte and undergo diffusion to cause the loss of active materials. Nelson et al. [6] directly observed the dissolution and diffusion process of sulfur during discharge by transmission X-ray microscopy. The polysulfide ions entering the electrolyte could not return to the positive electrode during charging process. The working mechanism of Li–S batteries is different from the "state-of-the-art" of lithium-ion batteries. As displayed in Figure 8.1, it is a chemical reaction between elemental sulfur and metal lithium. The charging–discharging reaction of S and Li involves a multistep redox reaction and a complex lithium sulfide phase transfer process. During the discharging process, the S_8 molecule undergoes a ring-opening reaction to obtain two electrons firstly, and two lithium ions are combined to produce the long-chain polysulfide (Li_2S_8). The Li_2S_8 is further reduced to Li_2S_6 and Li_2S_4. The part of the long-chain lithium polysulfide cannot be dissolved in the ether electrolyte used in Li–S batteries. This is a liquid phase reaction with very fast kinetics, showing the high potential platform of 2.3 V. Li_2S_4 is then reduced to short-chain Li_2S_2 and Li_2S. Li_2S_2 and Li_2S are not dissolved in the electrolyte. What's more, the electronic conductivity and Li-ion conductivity of them are very low, so the kinetics of this solid phase is very slow, showing a low potential platform of 2.1 V. The charging process is just the opposite reaction process.

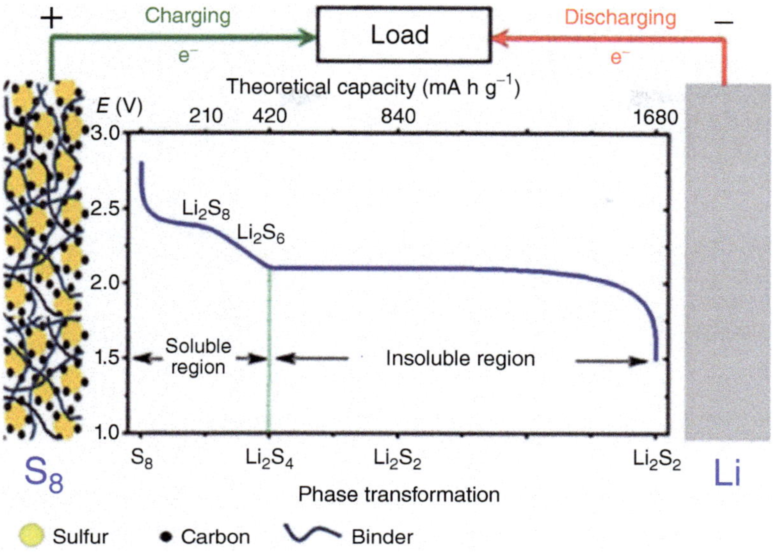

Figure 8.1 Illustration of the working mechanism of Li–S batteries. Source: Chen et al. [8]. Reproduced with permission of American Chemical Society.

8.3 Challenges of Li–S Batteries

However, elemental sulfur has some shortcomings as a positive electrode material for Li–S batteries. There are three main problems: (i) Both of the elemental S (5×10^{-30} S cm^{-1}) and its discharge products Li_2S (3.4×10^{-7} S cm^{-1}) have low conductivity, resulting in the uncomplete reaction and the reduction of the utilization of active material; (ii) The huge volume change during charging and discharging process, which can reach 80%; (iii) The long-chain lithium polysulfide is dissolved in the electrolyte, resulting in a "shuttle effect." The "shuttle effect" is a peculiar phenomenon of Li–S batteries, which means that long-chain lithium polysulfide diffuses to the negative electrode. The nonconductive lithium sulfide and short-chain lithium polysulfide are directly formed on the surface of lithium metal. This part of the short-chain lithium polysulfide returns to the positive electrode region under the action of the electric field force. The short-chain lithium polysulfide that is oxidized and converted to long-chain lithium polysulfide will run to the negative electrode, thus forming a "shuttle effect," resulting in a decrease in the coulombic efficiency (CE) of the battery and a decrease of the cycling performance. At present, most of the research work on Li–S batteries revolves around solving this problem.

Therefore, the protection of lithium metal anodes is also a huge challenge for Li–S batteries. For Li–S batteries, the deterioration of the lithium negative electrode is mainly due to the high reactivity and the infinite volume change during charging and discharging process. These two inherent characteristics lead to the appearance

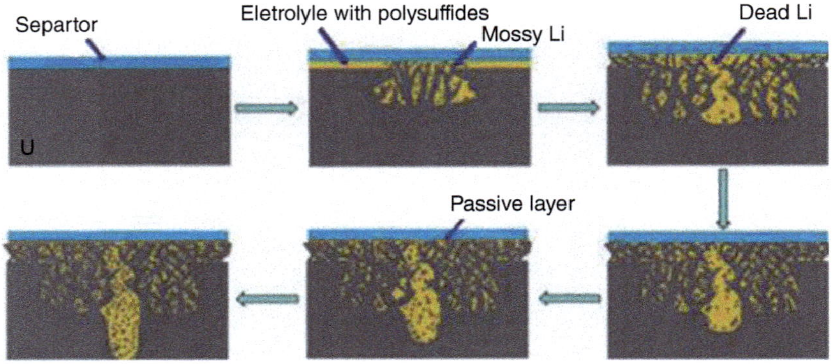

Figure 8.2 The anode corrosion and structure collapse in Li–S batteries. Schematic illustration of the evolution mechanism of Li anode. Source: Zhang et al. [11]. Reproduced with permission of American Chemical Society.

and growth of lithium dendrites, which has been confirmed in a large number of experimental and theoretical studies [9]. The formation of lithium dendrite is often affected by many factors, such as local current density, characteristics of the solid electrolyte interface (SEI), and the structure of the negative electrode. In Li–S batteries, the high activity of polysulfide in the electrolyte increases the complexity and instability of the lithium negative reaction, resulting in deterioration of the negative electrode and formation of "moss"-like lithium.

The in situ coupled spatially resolved X-ray diffraction (XRD) and absorption tomography, chemical and morphological studies of the lithium negative electrode of the actual working Li–S batteries [10] (Figure 8.2). At the end of the first charge, mossy metallic lithium was observed in the region where the lithium negative electrode was porous, which was caused by electrodeposition. Unfortunately, porous and mossy lithium with a high surface area is more susceptible to contact with polysulfides and electrolytes, regenerating SEI and forming "dead lithium," which will lose activity and do not participate in the subsequent electrochemical reactions. Li and coworkers [11] analyzed the mechanism of lithium-induced negative electrode failure caused by mossy lithium in detail. They found that the formation of mossy lithium was associated with an uneven Li surface. Mossy lithium preferentially grows from the bottom of the irregular pit on the surface of the lithium sheet, rather than at the protrusion. During the cycling process, mossy lithium is continuously oxidized to form a new SEI and "dead lithium" on the surface of lithium. The formed dense passivation layer blocks the charge transfer path, while the region not covered with the passivation layer is etched to form holes, resulting in the final destruction of metallic lithium.

The mossy lithium and an unstable SEI allow a large amount of polysulfide to penetrate the interior of a lithium metal electrode in Li–S batteries [11, 12]. Manthiram and coworkers [13] observed the Li–S batteries after 150 cycles that showed the presence of a large amount of sulfur-containing compounds in the lithium anodes indicating that lithium metal had a serious side reaction. Chen and coworkers [14] reported the effect of sulfur loading on the corrosion depth of

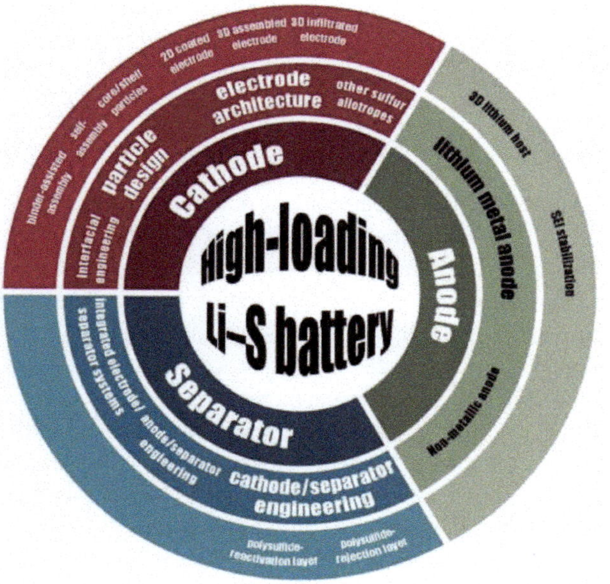

Figure 8.3 Schematic of the hierarchical design principle for Li–S batteries. Source: Peng et al. [16]. Reproduced with permission of John Wiley and Sons.

lithium anodes. The thickness of the corroded area increases with the increase of sulfur loading and cycle number, indicating that the corrosion behavior of lithium polysulfides has a great influence. In the case of high sulfur loading, more than half of the metallic lithium is corroded by the polysulfide [15]. It is worth noting that the sulfur positive electrode in the broken battery is taken out to assemble the battery with the fresh lithium anode and the fresh electrolyte. The new battery can still exhibit stable cycling performance. Therefore, severe lithium corrosion and electrolyte decomposition and depletion are the main reasons for the failure of high sulfur-loaded Li–S batteries. A similar failure mechanism is more obvious in large Li–S batteries.

To obtain high-energy-density Li–S batteries, the main contents of the research are shown in Figure 8.3. The rational material selection and structure optimization of Li–S batteries mainly contain three parts – cathode electrode, anode electrode, and separator [16]. Graphene is widely used in all three parts.

8.4 Overview of the Graphene as Host for S

In recent years, research on the materials used as the carriers for cathode material in Li–S batteries has focused on two types – (i) various carbon materials (such as porous carbon, graphene, carbon nanotubes (CNTs), and their composite materials); (ii) various metal compounds (including metal oxides, sulfides, nitrides, phosphides, carbides, selenides, etc.). The metal compounds are generally used in combination with carbon materials. Both of these materials inhibit the dissolution and migration of polysulfide ions to different extents, improving the utilization of sulfur and the cycling performance of Li–S batteries [17].

8.4.1 High-Quality Graphene

The ideal graphene is a single-layer and two-dimensional material which is connected by carbon atoms in an sp^2 hybrid manner. Graphene has excellent physical properties (high electrical conductivity, high mechanical strength, large specific surface area, and low density) with widespread use. The application is in Li–S batteries. Ma et al. [18] used plasma discharge assisted high-energy ball milling to load nanosulfur on exfoliated graphite. The plasma discharging process opens the sheet structure of graphite and performs hetero atom doping, increasing the active site for adsorption of polysulfide. The nanosulfur is not only more closely contacted with the graphene carrier, but also has a significantly reduced volume expansion when discharged compared to the bulk sulfur. The electrode has a sulfur content of 70% and a capacity with a fade rate of 0.07% per cycle over 500 cycles at 0.5 C.

Cui and coworkers [19] took the lead in using graphene-coated sulfur particles as a cathode material for Li–S batteries. The encapsulation by graphene will be used to capture lithium polysulfide species and improve electronic conductivity and mitigate the volume change of sulfur during charging and discharging process. As a result, the prepared graphene-coated sulfur composite exhibits enhanced cycling stability with a specific capacity of 750 mAh g^{-1} compared to the uncoated counterpart, and the capacity retention is 70% after 100 cycles at a current density of 0.2 C.

Zhang and coworkers [20] used calcined mesoporous layered bimetallic hydroxide (MGallDH) as a template to deposit graphene layers on its surface by chemical vapor deposition (CVD) method. (Figure 8.4a) Some of the carbon was deposited in the mesoporous LDH to form processes. After removing the template, double-layer templated graphene (DTG) separated by a large number of mesosized protuberances were obtained. The specific surface area of the DTG up to 1628 m^2 g^{-1}. The mesoporous protrusions between these lamellas can encapsulate the LiPSs and inhibit the shuttle effect. Therefore, DTG–sulfur composites (DTG/S) exhibit excellent rate performance, with initial specific capacities as high as 1034 and 734 mAh g^{-1} at high current densities of 5 and 10 C, respectively, and remaining at 530 and 380 mAh g^{-1} after 1000 cycles.

They [21] also used porous calcium oxide as a template to grow porous graphene by CVD (Figure 8.4b). The pore size and pore volume of graphene can be adjusted by adjusting carrier gas and catalyst. The prepared graphene has both vacancies in tiny planes, medium-sized pleated pores, and large cavities. The hierarchical structure facilitates the capture of polysulfides during the reaction, shortens diffusion path of the Li-ion, accelerates mass transfer, and reduces interfacial impedance. The pore size distribution of the hierarchical structure shows a clear graded pore structure. The capacity retention is 83.5% after 150 cycles, showing high cycle stability. The defects and the mesopores on graphene increase its specific

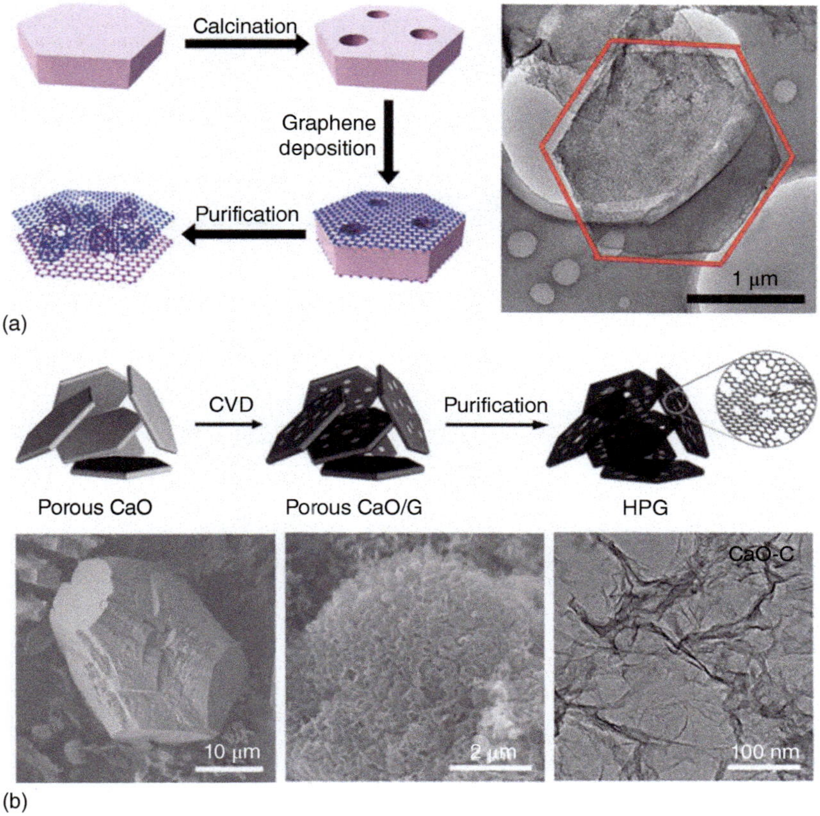

Figure 8.4 (a) Schematic for the synthesis of unstacked DTG and the TEM image of the cast DTG flakes. Source: Zhao et al. [20]. Reproduced with permission of Springer Nature. (b) Schematic illustration of hierarchical porous graphene obtained by CVD growth on CaO templates and the SEM images. Source: Tang et al. [21]. Reproduced with permission of John Wiley and Sons.

surface area, which improves the electrochemical performance with high sulfur loading. The micropores of graphene help to capture polysulfides, which reduce the "shuttle effect" and increase the CE of Li–S batteries. The interconnected mesopores and macropores shorten the Li-ion transmission distance, improve the electrolyte permeability, and have good rate performance.

8.4.2 Heteroatom-Doped Graphene

However, the effect of pure graphene materials on lithium polysulfide is physical adsorption, and it is difficult to obtain long-cycling stability. At present, most of the methods for preparing graphene introduce many defects, which makes it difficult for graphene to achieve rational performance. To improve the quality of graphene and improve the adsorption of lithium polysulfide, doping treatment of graphene materials is a common strategy. The substitution of atoms such as N, S, B, O, and

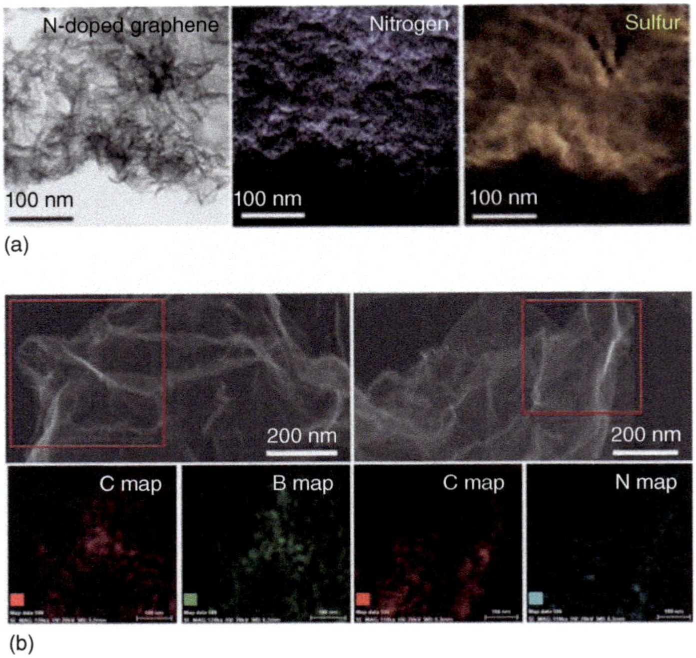

Figure 8.5 (a) The EF-TEM image and corresponding electron energy loss spectroscopy (EELS) elemental mapping of the wrinkled N-doped graphene. Source: Song et al. [22]. Reproduced with permission of American Chemical Society. (b) The SEM images of the B-doped graphene and N-doped graphene and the corresponding carbon, boron, and nitrogen elemental mapping result from the area squared. Source: Zhou et al. [23]. Reproduced with permission of John Wiley and Sons.

P to a polar state enhances the adsorption of lithium polysulfide. What's more, the heteroatom-doping improves the conductivity of the defective graphene, which is beneficial to the improvement of electrochemical performance. For example, Wang Donghai and coworkers [22] prepared a highly wrinkled N-doped graphene through a simple thermal-induced expansion strategy (Figure 8.5a). Benefiting from its abundant pore structure and chemical affinity of nitrogen-containing functional groups, the battery exhibits a high capacity of 1227 mAh g^{-1} and long cycle life of 75% retention after 300 cycles. Zhang and coworkers [24] used nickel-doped graphene to achieve a long-cycling performance of 2000 cycles. Spectral analysis and calculations show that the excellent performance can be attributed to the fact that the graphene has a good recovery of the C–C lattice after nitrogen-doping. There is unique binding ability of the nitrogen-containing functional group to lithium polysulfide.

Yuan et al. [25] centrifuged graphene in an ionic liquid and calcined the obtained product under Ar atmosphere to prepare graphene-based N-doped and B-doped porous carbon for the storage of sulfur (Figure 8.5b). The transmission electron microscopy test can clearly show that the prepared graphene has a small layer structure and a large number of wrinkles, which indicates that graphene has a large specific surface area. There is no obvious sulfur particle aggregation outside the graphene. It is indicated that the sulfur nanoparticles enter the interior of the graphene fold and are densely bonded to the sheet. The electrode retains a capacity of 301 mAh g^{-1} with a fade rate of 0.035% per cycle over 1500 cycles at 2 C. The excellent performance results from the rich pleats of the graphene sheets that provide a larger pore volume to accommodate the deformation of the active material after reaction. Zhou et al. [23] used three-dimensional nitrogen–boron-doped graphene aerogel as a self-supporting electrode for a Li–S battery. The graphene aerogel has a large pore diameter of 5–10 μm. After the nitrogen–boron doping, the diameter of graphene aerogel pore is slightly reduced, and the sheet is thinned. After 100 cycles at 0.3 C, the capacity retention of nitrogen-doped graphene composite electrode is as high as 79%, and the capacity retention of boron-doped graphene composite electrode is only 73%, which is much higher than that of undoped graphene composite. The doped graphene composite electrode was also far superior to the undoped control group in the rate performance test. Nitrogen is more electronegative than boron and is more easily doped into the graphene lattice to increase graphene conductivity. The reason for the high performance of the electrode is the strong interaction between the doped graphene and the lithium polysulfide in the reaction process. The interconnected pore structure ensures rapid electron transfer of the material that the active material can be in close contact with the surface of the graphene. What's more, this structure allows for large volume changes of sulfur during the reaction and reduces the reaction resistance.

8.4.3 Functionalized Graphene

The introduction of functional groups on graphene to produce functionalized graphene is beneficial to the improvement of the electrochemical performance of Li–S batteries. Graphene oxide is a poor electrical conductor due to a large amount of oxygen-containing functional groups on its surface. After partly reduction treatment, the conductivity is improved, and the functional group remains partly. Zhang and coworkers [26] used the reduced graphene oxide (rGO) as the carrier for S (Figure 8.6). The oxygen-containing functional groups on the graphene can be used to improve the interaction between sulfur and graphene, and the sulfur is anchored on the graphene network by the C—S bonding, thereby improving the cycling performance of the Li–S battery. GO–S composites are exhibiting excellent electrochemical performance with a reversible capacity of 1000 mAh g^{-1} at 2 C in ionic liquid electrolytes.

8.4.4 Structure-Designed Graphene

Graphene has attracted more and more attention because of its high electrical conductivity, high theoretical specific surface area, and good mechanical flexibility. However, graphene cannot effectively limit polysulfide in the matrix material to achieve stable cycling performance due to its open structure. Because of the

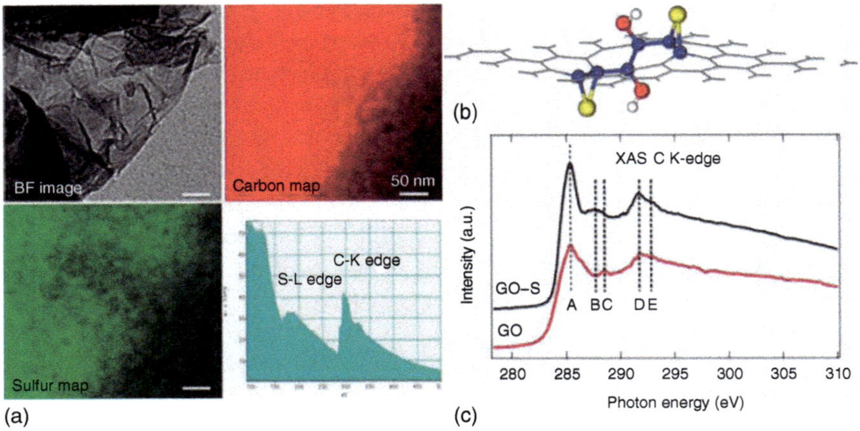

Figure 8.6 TEM bright-field (BF) image (a) and the corresponding elemental mapping for carbon and S and the EELS spectrum. The scale bars are 50 nm. (b) Representative pattern of GO immobilizing S. The hydroxyl enhances the binding of S to the C—C bond due to the induced ripples by epoxy or hydroxyl group. Yellow, red, and white balls denote S, O, and H atoms, respectively, while the others are C atoms. Note that the C atoms bonding to S or O are highlighted as blue balls. (c) C K-edge XAS spectra of GO and GO–S nanocomposites. Source: Ji et al. [26]. Reproduced with permission of American Chemical Society.

electrostatic attraction, the single graphene layer is easily agglomerated so that the theoretical specific surface area is not obtained. Porous carbon materials and CNTs are added as "spacers" to the interior of the graphene layer to prevent self-aggregation to achieve a large surface area [27].

A flexible rGO/S composite self-supporting film was prepared by simultaneously reducing and assembling GO on the surface of zinc metal to form rGO/S [28] (Figure 8.7a–e). The self-supporting rGO/S film can be frequently folded into the structure of "folding fan." The flexible Li–S battery based on the rGO/S film can maintain stable electrochemical performance under different bending conditions. In addition to the direct synthesis of flexible graphene/sulfur positive electrodes, there is another concept by designing flexible graphenes as flexible positive current collectors [30]. This design allows the Li–S battery to deliver a high discharge capacity of 1345 mAh g^{-1} and the CE is 98%.

Therefore, it is a good method to make two-dimensional graphene into three-dimensional graphene. As shown in Figure 8.7f, Cheng and coworkers [29] used a hydrothermal method to produce the composite of graphene aerogel and sulfur. The composite material exhibits the porous network structure composed of fibrous graphene, has a good electron transport network. It is easy to adjust the sulfur content and can be directly cut into an electrode for use as a positive electrode of Li–S battery without using a metal current collector, binder, and conductive additive. The oxygen-containing groups (mainly composed of hydroxyl groups/epoxides) exhibit strong binding to polysulfides to prevent them from dissolving into the electrolyte. Therefore, G/S composites have high capacity, excellent rate performance, and long cycling life.

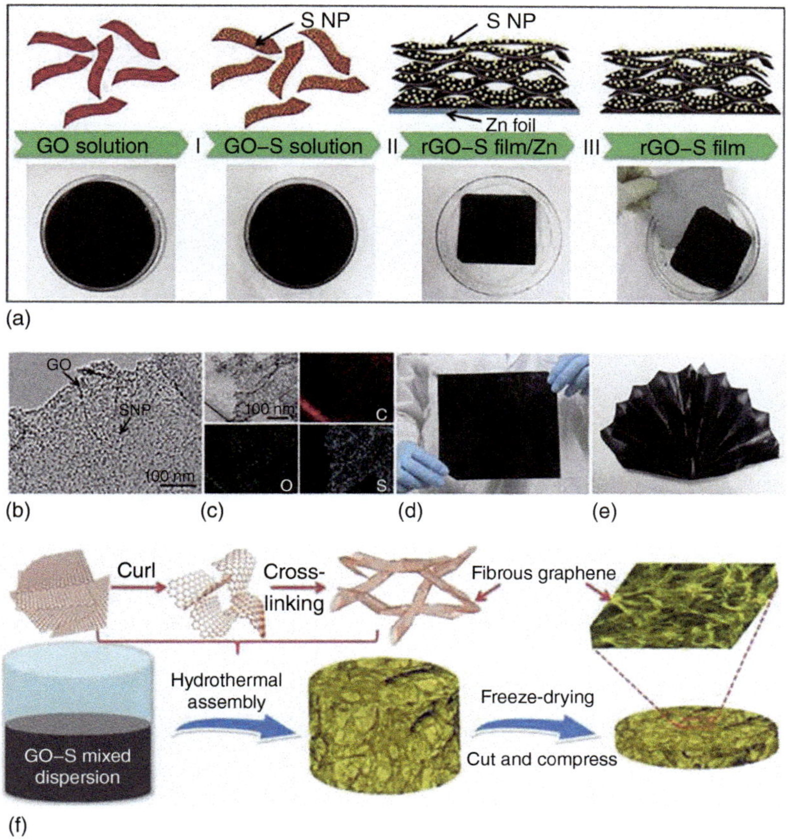

Figure 8.7 (a) Schematic process of fabricating free-standing rGO–S composite films by: (I) depositing S NPs on the surface of the GO sheets, (II) self-assembling the GO sheets with S NPs on the Zn foil surface, (III) peeling off the rGO–S composite film from the Zn foil. (b) TEM image of a GO sheet with S NPs. (c) TEM elemental mappings of elemental C, O, and S on the surface of an rGO sheet in an rGO–S composite film. (d, e) Optical images of an rGO–S composite film and the film is folded into a "folding fan" structure. Source: (a–e) Cao et al. [28]. Reproduced with permission of John Wiley and Sons. (f) Illustration of the formation process of the G/S hybrid. Source: Zhou et al. [29]. Reproduced with permission of American Chemical Society.

8.4.5 Graphene-Based Composites

CNTs are a kind of one-dimensional carbon materials that are made of graphene. They have a large aspect ratio and tend to have excellent electrical conductivity. They are well used in Li–S batteries. Chen and coworkers [31] designed and synthesized a graphene-based layered porous carbon material to support sulfur as a cathode material for Li–S batteries (Figure 8.8a). A thin layer of porous carbon uniformly covers both surfaces of the graphene, and sulfur is highly dispersed in its pores. The porous carbon material itself can also provide greater surface area and larger pore volume for higher sulfur loading. In addition, porous carbon can also adsorb polysulfides

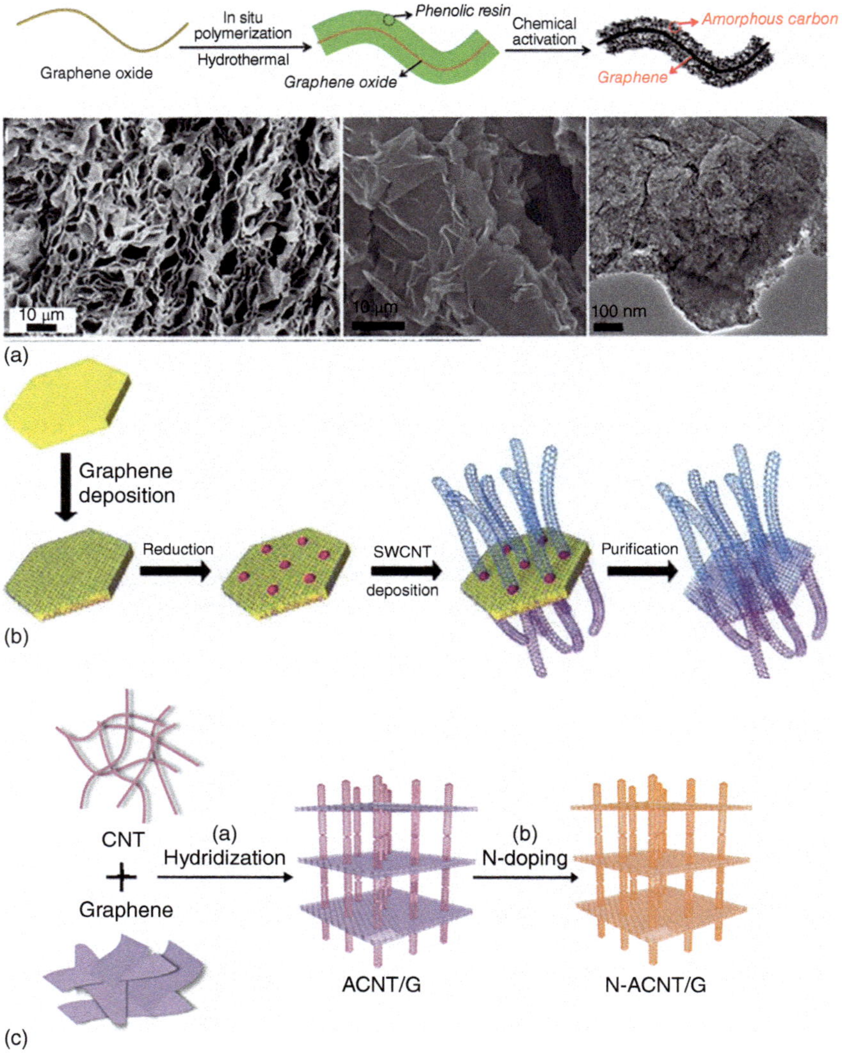

Figure 8.8 (a) A schematic showing the preparation of L-GPC material. Source: Yang et al. [31]. Reproduced with permission of American Chemical Society. (b) Schematic illustration of catalytic CVD of G/SWCNT hybrids on LDH flakes. Source: Zhao et al. [32]. Reproduced with permission of John Wiley and Sons. (c) Conceptual scheme of the design of N-ACNT/G hybrids with graphene and aligned CNTs as building blocks. Source: Tang et al. [33]. Reproduced with permission of American Chemical Society.

and accelerate electron/ion transfer to achieve fast kinetic processes. The composites have better electrochemical performance and cycling stability. Liu et al. [34] prepared a novel coaxial structure of graphene–sulfur–carbon nanofiber (G–S–CNF) self-supporting cathode. Conductive carbon nanofibers act as the current collector and deliver good mechanical support. Graphene is beneficial for relieving volume changes and inhibiting the dissolution of polysulfides in the electrolyte. The Li–S battery with G–S–CNF as the cathode electrode delivers a high capacity with a fade rate of 0.043% per cycle over 1500 cycles.

Zhang and coworkers [32] prepared graphene/single-walled carbon nanotubes (G/SWCNTs) hybrid materials by means of a double hydroxide (LDH) template, and fixed the SWCNTs in the graphene plane by covalent bonding (Figure 8.8b). The strong connection promotes the construction of a high conductive path between the internal space layer of the graphene and the SWCNT to provide a space for storing sulfur. Therefore, the obtained G/SWCNT-S cathode electrode exhibits excellent electrochemical performance in a Li–S battery and delivers a capacity of 650 mAh g^{-1} after 100 cycles even at 5 C. In the subsequent work, Zhang and coworkers [33] further prepared a composite carbon material with N-doped CNT arrays on the surface of graphene by dual-mode method (Figure 8.8c). The sandwich-structured composite provides excellent electron and ion transport paths, and nitrogen doping produces more active sites on the composite while increasing the adsorption of lithium polysulfide. The cell delivers an initial capacity of 1152 mAh g^{-1} at a high current of 1 C, and the capacity maintains 880 mAh g^{-1} after 80 cycles.

8.4.6 Metal Compound Anchored on Graphene

Graphene with good conductivity is one of the ideal carriers for metal compounds. Since the metal compound is denser than the carbon material, and the conductivity of most metal compounds is inferior to the carbon material, it is difficult to obtain a high-loading sulfur cathode electrode. Therefore, it is a better choice to load the metal compound on the carbon material. Most of the current research work on metal compounds is also based on this choice.

Yu et al. [35] use atomic layer deposition to deposit titanium dioxide on the surface of nitrogen-doped graphene as a positive carrier for Li–S batteries (Figure 8.9a). The high-energy ball milling uniformly distributes the sulfur nanoparticles on the surface of the N-doped graphene, which improves the electron conductivity of the graphene skeleton. After the deposition of the titanium dioxide layer, the rate performance of the composite is significantly improved (Figure 8.9b). The thin layer of titanium dioxide plays a certain slowing effect on the volume expansion of the active material during the cycles. At the same time, it further enhances the electronegativity of graphene and adsorption capacity of polysulfides. It has been found that an excessively thick layer of titanium dioxide reduces the capacity of the electrode due to the lower electronic conductivity of titanium dioxide. The electrode has an optimized electrode retention capacity of 918 mAh g^{-1} and capacity retention of 86% after 500 cycles at 1 C. The capacity decay is only 0.024% per cycle before the 5th and 500th cycles. The reversible capacity of the electrode is still as high as 833 mAh g^{-1} even at 4 C (Figure 8.9c).

Pyrite CoS_2 has been shown to have high catalytic activity and is highly conductive, with an electron conductivity of 6.7×10^3 S cm^{-1} at 300 K. Zhang and coworkers [36] used CoS_2 as a catalyst to promote the reaction kinetics of Li–S batteries by promoting polysulfide redox reactions (Figure 8.10a,b). Graphene/CoS_2 (15% or

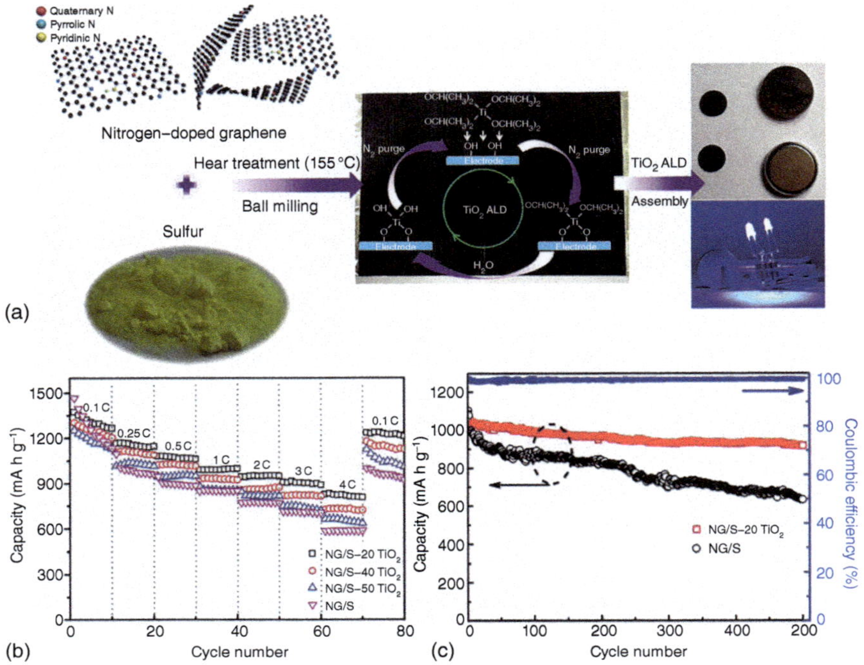

Figure 8.9 (a) Schematic illustration of the NG/S–TiO_2 preparation process and the experimental demonstration showing that the battery can light up two blue LEDs. (b) The rate performances of the electrodes at various current densities. (c) Cycling performances of NG/S and NG/S–20TiO_2 electrodes at 1 C. Source: Yu et al. [35]. Reproduced with permission of Royal Society of Chemistry.

30%) was prepared by mixing CoS_2 and graphene materials. CoS_2 particles with a diameter of about 1 μm adhere to the highly wrinkled graphene substrate, achieving an electron conduction path from the conductive skeleton to the surface of the polar CoS_2 (Figure 8.10c,d). The initial discharge capacity of the G/CoS_2 (15%) sample was 1368 mAh g^{-1}, and the initial discharge capacity of G/CoS_2 (30%) was 1174 mAh g^{-1}, indicating that the addition of polar CoS_2 increased capacity due to its adsorption and conversion capacity of lithium polysulfide. But content of CoS_2 is not higher than better, the low specific surface area of CoS_2 will decrease the sulfur loading of the composite carrier that it is more important to choose a suitable ratio.

To increase the Mo edge sites of MoS_2, Lin et al. [37] heat treated the MoS_2 under H_2/Ar (Figure 8.10e). The S atom in MoS_2 reacts with H_2 to form H_2S gas, which causes sulfur vacancies on the surface of MoS_2. Thus, more Mo atoms are exposed to improve the catalytic activity of MoS_2. Adding a small amount of sulfur-vacuum MoS_{2-x}/rGO composite to the positive electrode of Li–S batteries can significantly improve the rate performance of the battery. The battery delivers a specific capacity of 826.5 mAh g^{-1} at 8 C.

Sun et al. [38] reported that a highly conductive porous vanadium nitride nanobelt/graphene composite used as a carrier olefinic composite of a sulfur

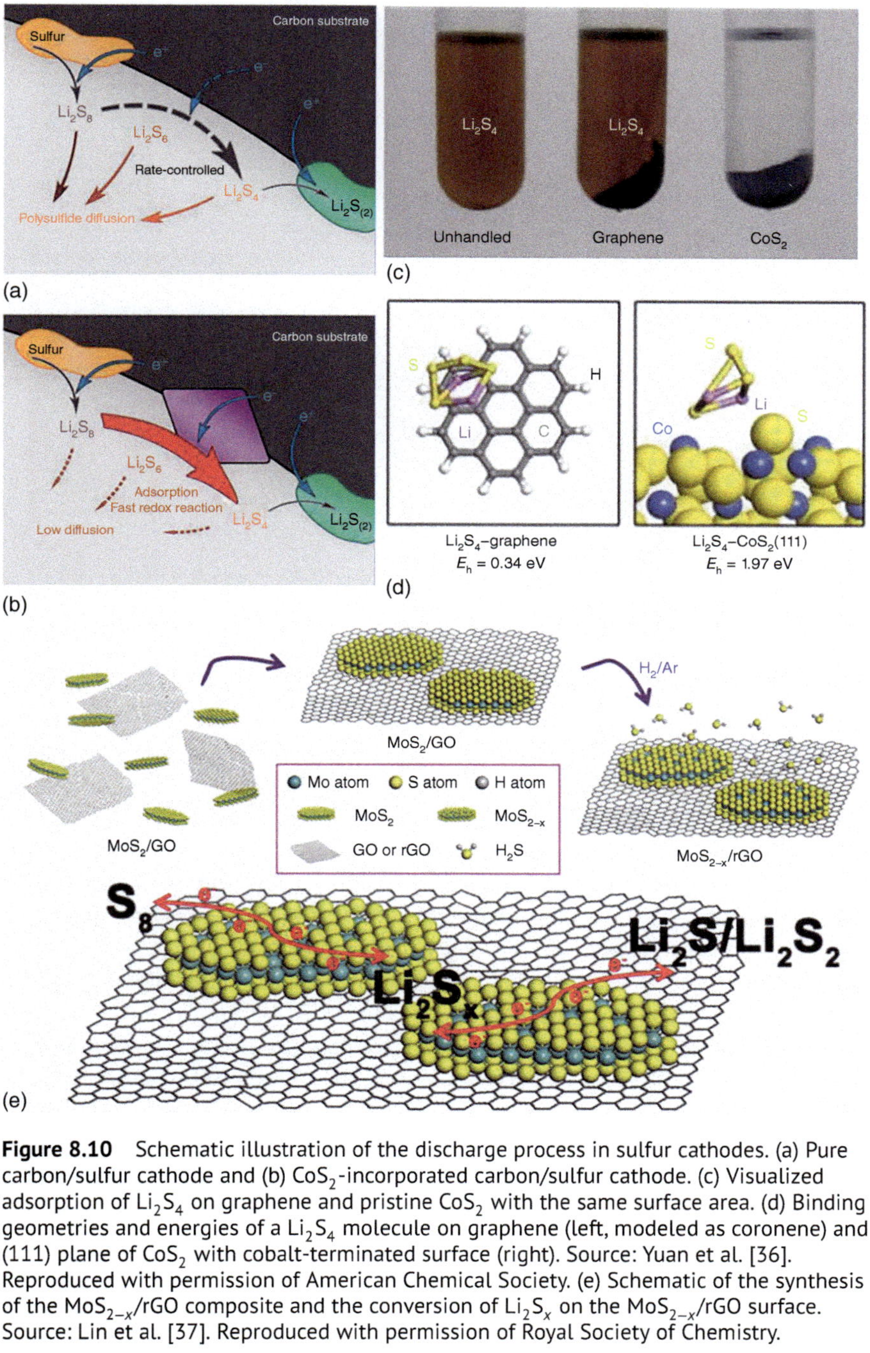

Figure 8.10 Schematic illustration of the discharge process in sulfur cathodes. (a) Pure carbon/sulfur cathode and (b) CoS_2-incorporated carbon/sulfur cathode. (c) Visualized adsorption of Li_2S_4 on graphene and pristine CoS_2 with the same surface area. (d) Binding geometries and energies of a Li_2S_4 molecule on graphene (left, modeled as coronene) and (111) plane of CoS_2 with cobalt-terminated surface (right). Source: Yuan et al. [36]. Reproduced with permission of American Chemical Society. (e) Schematic of the synthesis of the MoS_{2-x}/rGO composite and the conversion of Li_2S_x on the MoS_{2-x}/rGO surface. Source: Lin et al. [37]. Reproduced with permission of Royal Society of Chemistry.

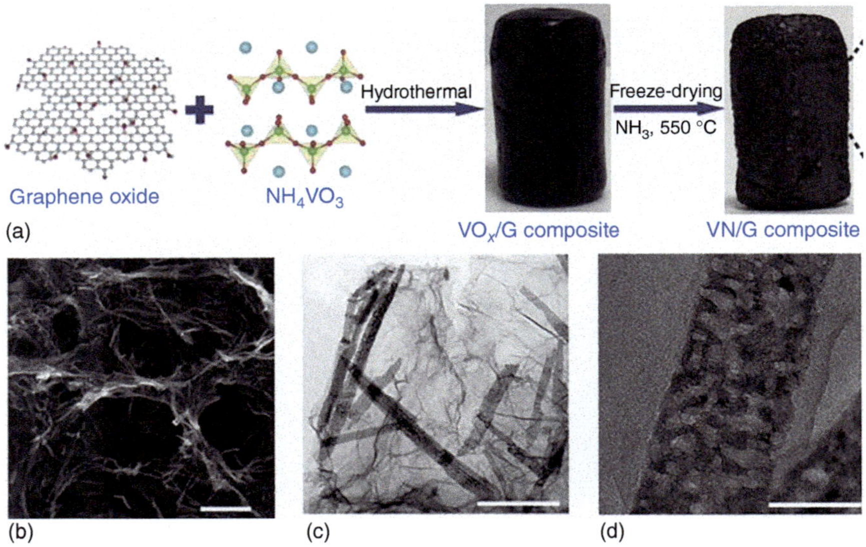

Figure 8.11 (a) Schematic of the fabrication of a porous VN/G composite and the cell assembly with corresponding optical images of the material obtained. (b) High-magnification SEM image, (c, d) TEM images of the as-prepared porous VN/G composite. Scale bars, (b) 2 mm; (c) 500 nm; (d) 50 nm. Source: Sun et al. [38]. Reproduced with permission of Springer Nature.

positive electrode has an efficient anchoring and rapid conversion of polysulfide (Figure 8.11). The anchoring effect of vanadium nitride was further confirmed by experimental and theoretical results. Due to the high conductivity of vanadium nitride, the composite positive electrode exhibits less polarization and faster redox reaction kinetics than the rGO positive electrode, showing good rate performance and cycle stability. At 0.2 C, the initial capacity reached 1471 mAh g^{-1}, and after 100 cycles, the capacity was 1252 mAh g^{-1}, and the capacity loss was only 15%. Li et al. [39] prepared mesoporous cerium nitride microspheres/niobium-doped graphene nanosheet composites as carrier materials for sulfur cathodes. The mesoporous NbN microspheres chemically immobilize lithium polysulfide by Nb–S chemical bonding and catalytically convert soluble polysulfide into insoluble cerium U_2S, thereby accelerating the oxidative redox reaction kinetics. Moreover, highly conductive NbN and N-doped graphene nanosheets can also provide rapid electron transport for the electrochemical reaction of Li–S batteries. Therefore, the S/NbN@NG cathode electrode exhibits a specific capacity of 948 mAh g^{-1} at a current density of 1 C, and the capacity is still as high as 739 mAh g^{-1} at 5 C.

8.4.7 Metal Compounds Anchored on Carbon Composite Material

Due to the high theoretical specific capacity (1166 mAh g^{-1}), especially can be paired with lithium-free anodes, lithium sulfide (Li_2S) is considered the next

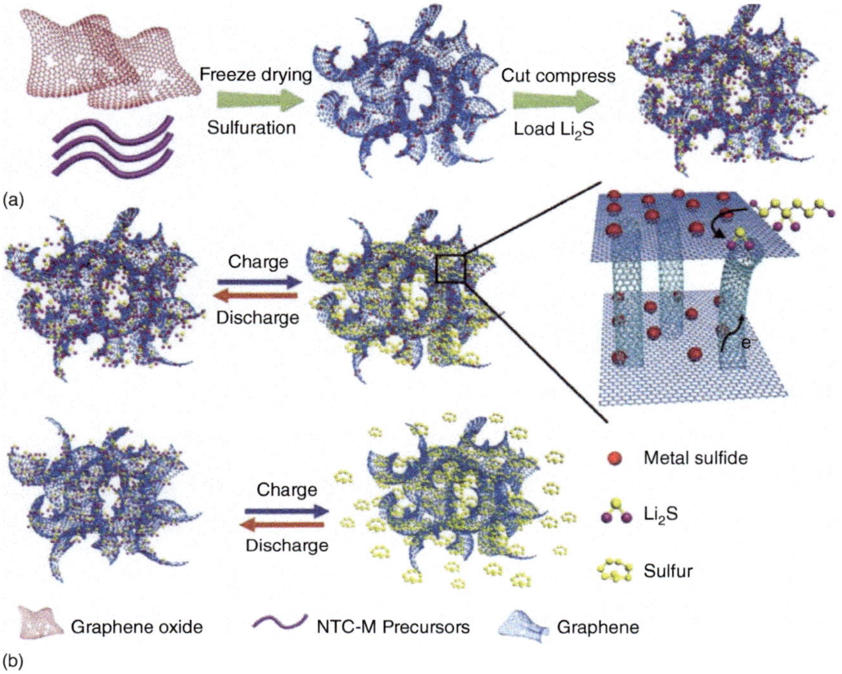

Figure 8.12 (a) Synthesis of the 3DTSC composite. (b) Advantages of the 3DTSC composite with respect to 3DC. Source: He et al. [40]. Reproduced with permission of John Wiley and Sons.

generation of advanced Li–S batteries. However, the low conductivity of Li_2S and the severe "polysulfide shuttle" of lithium polysulfide, have seriously hampered their practical application in Li–S batteries. To solve these problems, a bifunctional 3D metal sulfide decorative carbon sponge (3DTSC) was used, which consists of 1D carbon nanowires, 2D graphene nanosheets with high conductivity, and polar 0D metal sulfide nanoparticles (Figure 8.12). It is made of cross-linking and has high electrocatalytic activity and strong chemical bonding to LiPS. Benefits from well-designed multiscale, multidimensional 3D porous nanostructures with high electrical conductivity, effective electrocatalytic and absorption capabilities, 3DTSC can significantly inhibit LiPS shuttle and improve the utilization of Li_2S. Li_2S promotes the transport of electrons and ions. As a result, the free-standing 3DTSC-Li_2S cathode without polymer binder and metal current collector provides excellent electrochemical performance even at a high Li_2S loading of 8 mg cm^{-2}, with an area capacity of 8.44 mAh cm^{-2}.

The development of high-performance S cathode with high sulfur load is an important challenge for the practical application of lithium–sulfur batteries. Our group [41] successfully prepared 3DC–CNTS–Ni composite with soluble salt NaCl as template by spray drying and growth CNTs by CVD process (Figure 8.13). 1D CNTs were grown in situ on the surface of the 3D carbon network, and 0D Ni nanoparticles were located on the top of 1D CNTs. The 3DC–CNTS–NiS_2

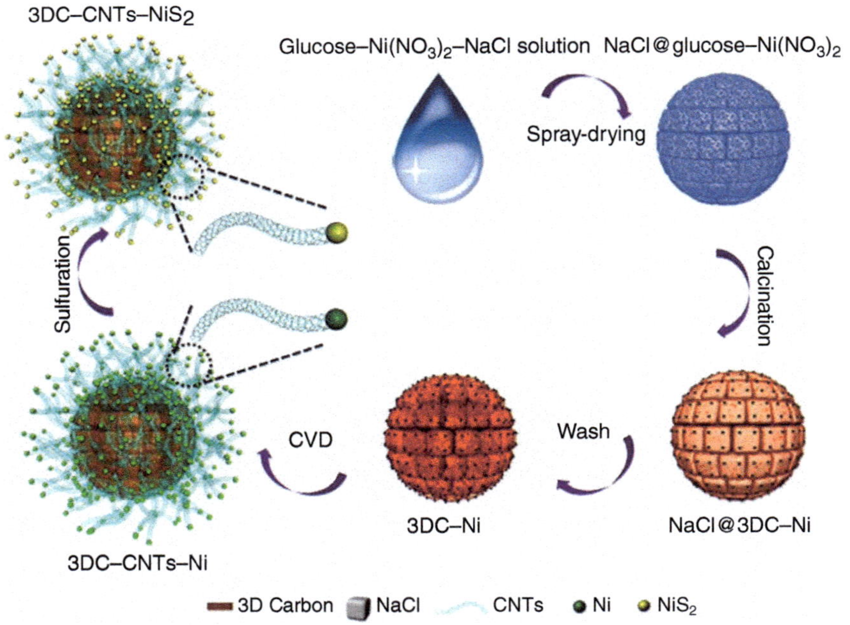

Figure 8.13 Schematic of the fabrication process of the hierarchical architecture. Source: Wang et al. [41]. Reproduced with permission of American Chemical Society.

hierarchical structure was further prepared by curing, and the Ni at the top of CNT was in situ vulcanized into NiS_2. The 3DC–CNTS–NiS_2 fractional composite material has high surface area, good electrical conductivity, and low NiS_2 content, while the top NiS_2 nanoparticles are capable of highly efficient LiPS adsorption and catalysis. As the carrier of S positive electrode, it shows high capacity, excellent magnification performance, and stable cycling performance. The initial specific capacity is 1053 mAh g^{-1}, and the attenuation rate of each cycle is 0.043% at 1C. The high area capacity of 4.8 mAh cm^{-2} can be maintained even under high S load (6.5 mg cm^{-2}, overall S content of the electrode is 76.5 wt%). The soft pack full battery with S@3DC–CNTs–NiS_2 positive and a Li negative electrode has a high initial capacity of 1200 mAh g^{-1} and a capacity retention rate of 87% after 50 stable cycles.

8.4.8 Graphene Used in Separator

The role of the separator for lithium batteries is to isolate electrons but allow Li^+ to pass. Commercially available separators used in Li batteries are typically made from high strength, thinned polyethylene (polyethylene), and polypropylene (PP) porous membranes. These commercial separators have high porosity and high ionic conductivity for Li^+ with good permeability, at the same time soluble lithium polysulfide will also pass, forming a so-called "shuttle effect," causing the loss of active substances, corrosion of the lithium negative electrode, and battery capacity attenuation. Based on this, it is a very effective method to modify the traditional separator and apply a coating on the side of the sulfur positive electrode to effectively limit the lithium sulfide shuttle to the anode region.

8.4.8.1 Carbon Material as a Coating Layer

Early researchers used a variety of carbon materials as coatings on the separators and achieved certain results. Carbon materials are electronically conductive, and conductive coatings help to disperse the current through the battery and reduce the internal resistance of the battery. If the carbon material is porous, this can greatly increase the surface area of the carbon and increase the ability to adsorb lithium polysulfide. Therefore, the electron-conducting porous carbon layer can not only capture lithium polysulfide but also the captured lithium polysulfide is reactivated in the subsequent charging process, thereby improving the utilization rate of the active material.

Zhou et al. [42] used graphene as a coating layer for the separator of Li–S battery. The battery delivers a capacity of 1006 mAh g^{-1} with a fade rate of 0.10% per cycle over 300 cycles under a high sulfur load of 4 mg cm^{-2}. (Figure 8.14a). Zhou et al. [43] further optimized the structural design in the subsequent work. Using the integrated structural design, the sulfur–carbon composite positive electrode was directly coated on the separator covered with the graphene coating layer to prepare

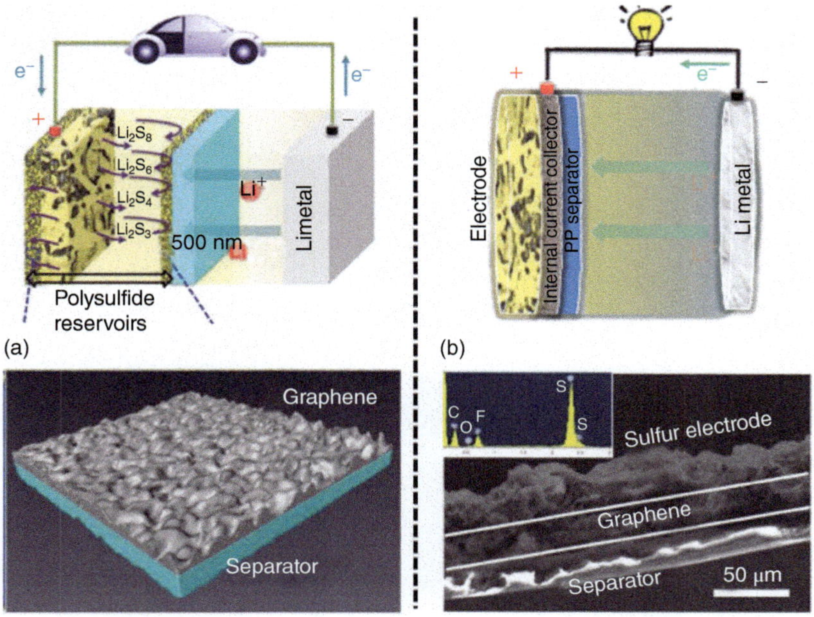

Figure 8.14 (a) Schematic of a Li–S battery with electrode configuration and photograph of a large-area G-separator. Source: Zhou et al. [42]. Reproduced with permission of John Wiley and Sons. (b) Schematic of the electrode configuration using an integrated structure of sulfur and G@PP separator and cross section of an integrated sulfur electrode. Source: Zhou et al. [43]. Reproduced with permission of John Wiley and Sons.

an integrated flexible battery. The battery with a fade rate of 0.05% per cycle over 500 cycles at 0.9 C (Figure 8.14b). Considering the oxygen-containing functional groups and structural defects of the rGO basal and edge. It provides a strong fixed point, and its two-dimensional sheet structure effectively promotes the formation of autonomous film. Wang et al. [44] prepared rGO film as an interlayer for Li–S battery system. The enriched epoxy and carboxyl groups on the surface of rGO can effectively adsorb sulfur and polysulfides, and promote the utilization of active substances. Further, the conductive carbon black KetjenBlack (CB) was added to the rGO preparation process to obtain the rGO–CB composite interlayer to shorten the activation time and improve the cycle performance. The rGO–CB composite interlayer is more porous, and is an electrolyte and polysulfide. The permeation provides a channel. When using the 2rGO-CB interlayer (rGO:CB = 2 : 1), the battery delivers a capacity of 1260 and 895 mAh g^{-1} after 100 cycles.

8.4.8.2 Carbon Material/Inorganic Metal Compound Composite as a Coating Layer

The inorganic material enhances the ability to capture lithium polysulfide by Lewis acid–base interaction and also has a catalytic effect, effectively suppressing the shuttle effect and increasing the utilization of the active material. Due to the "pro-lithium" of N-doped graphene (NG), the two-dimensional structure of the double metal hydroxide LDH has excellent "sulfophilicity," and the combination of NG and LDH adsorbs lithium polysulfide (Figure 8.15). Using two materials as a barrier layer for polysulfide shuttle, synergistic adsorption of lithium polysulfide, showing better sulfur fixation effect. By modifying the membrane, using

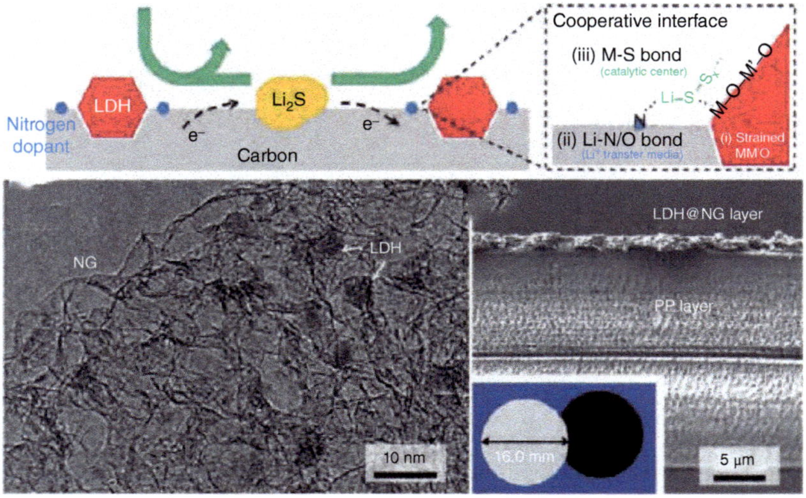

Figure 8.15 Schematic diagram of synergistic adsorption reduction of polysulfide ions at LDH@NG interface, TEM image of LDH@ NG composite, and SEM image of LDH@ NG modified diaphragm and its digital photos. Source: Peng et al. [45]. Reproduced with permission of John Wiley and Sons.

the synergistic sulfur fixation and the excellent conductivity and physical barrier of NG/LDH, the battery exhibits excellent electrochemical performance under extremely low modification loading.

The heterostructure of the surface interface design of new inorganic materials has attracted much attention as a sulfur cathode carrier in the field of Li–S batteries. A new and versatile LiPS binding mechanism is realized in the heterostructure carrier to improve the kinetics of the LiPS redox reaction by improving the chemical interaction and enhanced interaction with LiPS, thereby improving the performance of Li–S batteries. Yang and coworkers [46] designed a twin TiO_2–TiN heterostructure as a high-efficiency catalyst for LiPSs. The highly adsorptive TiO_2 shows a strong interaction with LiPS, while the highly conductive TiN accelerates the conversion kinetics of LiPS during the cycle. Using such an efficient LiPS trapping and redox-converted TiO_2–TiN heterostructure, the Li–S battery exhibited 2000 cycles at 1 C, and the capacity retention was 73%.

Our group [47] designed a heterostructure of Co_9S_8 modified by nano-CoO particles. CoO has a strong adsorption function, and Co_9S_8 has a remarkable electrocatalytic effect (Figure 8.16). The Co_9S_8/CoO heterostructure with reasonable structural design makes the two synergistic effects, which can not only improve the chemical capture ability of CoO to LiPS but also promote the electrocatalytic activity of Co_9S_8, especially the Li_2S_2 to Li_2S, which is difficult to react. The heterojunction is compounded with graphene, and a multifunctional modified layer is formed on the separator by filtration. The conductive graphene enhances the adsorption and catalysis of the polysulfide lithium, which effectively inhibits lithium polysulfide shuttles to the lithium negative electrode region. A Li–S battery using a graphene/sulfur composite as cathode electrode and a separator with a coating layer delivers a specific capacity of 925 mAh g^{-1} with a fade rate of 0.049% per cycle over 1000 cycles under a high load of 2.5 mg cm^{-2}.

Table 8.1 shows the electrochemical properties of some materials used for membrane modification and Li–S batteries using them as separators. By comparison, it

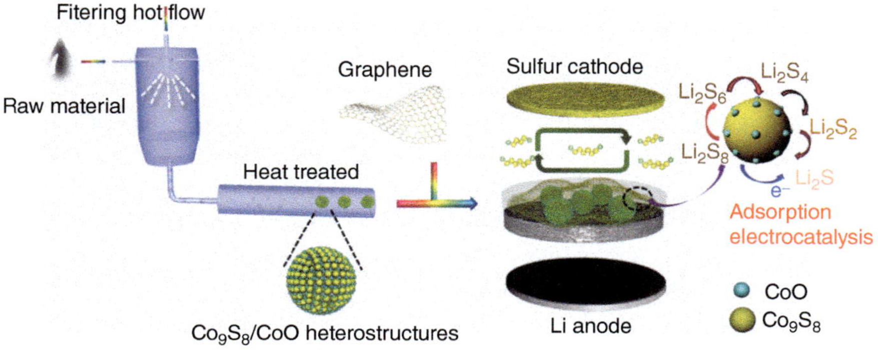

Figure 8.16 Schematic of the preparation of the Co_9S_8/CoO heterostructures and the Li–S cell configuration employing the Co_9S_8/CoO–G coated separator. Source: Wang et al. [47]. Reproduced with permission of Elsevier.

Table 8.1 Overview of the materials used to modify the separator.

Materials	Electrochemical performance	S content	References
Graphene	1006 mAh g^{-1} at 0.9 C 0.1% for 300 cycles	Carbon black/S 70 wt%/3 mg cm^{-2}	[42]
Graphene	932 mAh g^{-1} at 0.9 C 0.064% for 500 cycles	Carbon black/S 70 wt%/1.5 mg cm^{-2}	[43]
TiO_2–TiN/G	684 mAh g^{-1} at 1 C 0.013% for 2000 cycles	Graphene/S 80 wt%/3.1 mg cm^{-2}	[46]

can be found that the Li–S battery with metal compound added as a modification material has better cycling stability than pure carbon material, especially the use of multifunctional metal compound heterostructures.

References

1 Ji, X., Lee, K.T., and Nazar, L.F. (2009). A highly ordered nanostructured carbon-sulphur cathode for lithium-sulphur batteries. *Nature Materials* 8 (6): 500–506.

2 Bruce, P.G., Freunberger, S.A., Hardwick, L.J. et al. (2011). Li-O_2 and Li-S batteries with high energy storage. *Nature Materials* 11 (1): 19–29.

3 Danuta H, Juliusz U. (1962) Electric dry cells and storage batteries. U. S. Patent 3,043,896.

4 Rao, M.L.B. (1966). Organic electrolyte cells. US Patent 3, 154.

5 Mikhaylik, Y.V. and Akridge, J.R. (2004). Polysulfide shuttle study in the Li/S battery system. *Journal of the Electrochemical Society* 151 (11): A1969–A1976.

6 Shim, J., Striebel, K.A., and Cairns, E.J. (2002). The lithium/sulfur rechargeable cell - Effects of electrode composition and solvent on cell performance. *Journal of the Electrochemical Society* 149 (10): A1321–A1325.

7 Barchasz, C., Molton, F., Duboc, C. et al. (2012). Lithium/sulfur cell discharge mechanism: an original approach for intermediate species identification. *Analytical Chemistry* 84 (9): 3973–3980.

8 Chen, J.J., Yuan, R.M., Feng, J.M. et al. (2015). Conductive lewis base matrix to recover the missing link of Li_2S_8 during the sulfur redox cycle in Li-S battery. *Chemistry of Materials* 27 (6): 2048–2055.

9 Cheng, X.B., Zhang, R., Zhao, C.Z. et al. (2017). Toward safe lithium metal anode in rechargeable batteries: a review. *Chemical Reviews* 117 (15): 10403–10473.

10 Tonin, G., Vaughan, G., Bouchet, R. et al. (2017). Multiscale characterization of a lithium/sulfur battery by coupling operando X-ray tomography and spatially-resolved diffraction. *Scientific Reports* 7 (1): 2755–2762.

11 Zhang, Y., Heim, F.M., Song, N. et al. (2017). New insights into mossy Li induced anode degradation and its formation mechanism in Li–S batteries. *ACS Energy Letters* 2 (12): 2696–2705.

12 Xiao, J., Hu, J.Z., Chen, H. et al. (2015). Following the transient reactions in lithium-sulfur batteries using an in situ nuclear magnetic resonance technique. *Nano Letters* 15 (5): 3309–3316.

13 Zu, C.X., Dolocan, A., Xiao, P.H. et al. (2016). Breaking down the crystallinity: the path for advanced lithium batteries. *Advanced Energy Materials* 6 (5): 1501933.

14 Han, Y., Duan, X., Li, Y. et al. (2015). Effects of sulfur loading on the corrosion behaviors of metal lithium anode in lithium–sulfur batteries. *Materials Research Bulletin* 68: 160–165.

15 Qie, L., Zu, C., and Manthiram, A. (2016). A high energy lithium-sulfur battery with ultrahigh-loading lithium polysulfide cathode and its failure mechanism. *Advanced Energy Materials* 6 (7): 1502459.

16 Peng, H.-J., Huang, J.-Q., Cheng, X.-B. et al. (2017). Review on high-loading and high-energy lithium-sulfur batteries. *Advanced Energy Materials* 7: 1700260.

17 Wang, J.L., Yao, Z.D., Monroe, C.W. et al. (2013). Carbonyl-beta-cyclodextrin as a novel binder for sulfur composite cathodes in rechargeable lithium batteries. *Advanced Functional Materials* 23 (9): 1194–1201.

18 Ma, Z.L., Tao, L., Liu, D.D. et al. (2017). Ultrafine nano-sulfur particles anchored on in situ exfoliated graphene for lithium-sulfur batteries. *Journal of Materials Chemistry A* 5 (19): 9412–9417.

19 Wang, H.L., Yang, Y., Liang, Y.Y. et al. (2011). Graphene-wrapped sulfur particles as a rechargeable lithium-sulfur battery cathode material with high capacity and cycling stability. *Nano Letters* 11 (7): 2644–2647.

20 Zhao, M.Q., Zhang, Q., Huang, J.Q. et al. (2014). Unstacked double-layer templated graphene for high-rate lithium-sulphur batteries. *Nature Communications* 5: 3410.

21 Tang, C., Li, B.-Q., Zhang, Q. et al. (2016). CaO-templated growth of hierarchical porous graphene for high-power lithium-sulfur battery applications. *Advanced Functional Materials* 26 (4): 577–585.

22 Song, J., Yu, Z., Gordin, M.L. et al. (2016). Advanced sulfur cathode enabled by highly crumpled nitrogen-doped graphene sheets for high-energy-density lithium-sulfur batteries. *Nano Letters* 16 (2): 864–870.

23 Zhou, G., Paek, E., Hwang, G.S. et al. (2016). High-performance lithium-sulfur batteries with a self-supported, 3D Li_2S-doped graphene aerogel cathodes. *Advanced Energy Materials* 6 (2): 1501355.

24 Qiu, Y., Li, W., Zhao, W. et al. (2014). High-rate, ultra long cycle-life lithium/sulfur batteries enabled by nitrogen-doped graphene. *Nano Letters* 14 (8): 4821–4827.

25 Yuan, S., Bao, J.L., Wang, L. et al. (2016). Graphene-supported nitrogen and boron rich carbon layer for improved performance of lithium-sulfur batteries due to enhanced chemisorption of lithium polysulfides. *Advanced Energy Materials* 6 (5): 1501733.

26 Ji, L., Rao, M., Zheng, H. et al. (2011). Graphene oxide as a sulfur immobilizer in high performance lithium/sulfur cells. *Journal of the American Chemical Society* 133 (46): 18522–18525.

27 Tian, P., Tang, L., Teng, K.S. et al. (2018). Graphene quantum dots from chemistry to applications. *Materials Today Chemistry* 10: 221–258.

28 Cao, J., Chen, C., Zhao, Q. et al. (2016). A flexible nanostructured paper of a reduced graphene oxide-sulfur composite for high-performance lithium-sulfur batteries with unconventional configurations. *Advanced Materials* 28 (43): 9629–9636.

29 Zhou, G., Yin, L.C., Wang, D.W. et al. (2013). Fibrous hybrid of graphene and sulfur nanocrystals for high-performance lithium-sulfur batteries. *ACS Nano* 7 (6): 5367–5375.

30 Zhou, G., Li, L., Wang, D.W. et al. (2015). A flexible sulfur-graphene-polypropylene separator integrated electrode for advanced Li-S batteries. *Advanced Materials* 27 (4): 641–647.

31 Yang, X., Zhang, L., Zhang, F. et al. (2014). Sulfur-infiltrated graphene-based layered porous carbon cathodes for high-performance lithium-sulfur batteries. *ACS Nano* 8 (5): 5208–5215.

32 Zhao, M.-Q., Liu, X.-F., Zhang, Q. et al. (2012). Graphene/single-walled carbon nanotube hybrids: one-step catalytic growth and applications for high-rate Li-S batteries. *ACS Nano* 6 (12): 10759–10769.

33 Tang, C., Zhang, Q., Zhao, M.Q. et al. (2014). Nitrogen-doped aligned carbon nanotube/graphene sandwiches: facile catalytic growth on bifunctional natural catalysts and their applications as scaffolds for high-rate lithium-sulfur batteries. *Advanced Materials* 26 (35): 6100–6105.

34 Lu, S., Cheng, Y., Wu, X. et al. (2013). Significantly improved long-cycle stability in high-rate Li-S batteries enabled by coaxial graphene wrapping over sulfur-coated carbon nanofibers. *Nano Letters* 13 (6): 2485.

35 Yu, M.P., Ma, J.S., Song, H.Q. et al. (2016). Atomic layer deposited TiO_2 on a nitrogen-doped graphene/sulfur electrode for high performance lithium-sulfur batteries. *Energy & Environmental Science* 9 (4): 1495–1503.

36 Yuan, Z., Peng, H.J., Hou, T.Z. et al. (2016). Powering lithium-sulfur battery performance by propelling polysulfide redox at sulfiphilic hosts. *Nano Letters* 16 (1): 519–527.

37 Lin, H., Yang, L., Jiang, X. et al. (2017). Electrocatalysis of polysulfide conversion by sulfur-deficient MoS_2 nanoflakes for lithium–sulfur batteries. *Energy & Environmental Science* 10 (6): 1476–1486.

38 Sun, Z., Zhang, J., Yin, L. et al. (2017). Conductive porous vanadium nitride/graphene composite as chemical anchor of polysulfides for lithium-sulfur batteries. *Nature Communications* 8: 14627.

39 Li, X.X., Gao, B., Huang, X. et al. (2019). Conductive mesoporous niobium nitride microspheres/nitrogen-doped graphene hybrid with efficient polysulfide anchoring and catalytic conversion for high-performance-lithium sulfur batteries. *ACS Applied Materials & Interfaces* 11 (3): 2961–2969.

40 He, J., Chen, Y., and Manthiram, A. (2019). Metal sulfide-decorated carbon sponge as a highly efficient electrocatalyst and absorbant for polysulfide in high-loading Li_2S batteries. *Advanced Energy Materials* 9 (20): 1900584.

41 Wang, N., Chen, B., Qin, K. et al. (2020). Octopus-inspired design of apical NiS_2 nanoparticles supported on hierarchical carbon composites as an efficient host for lithium sulfur batteries with high sulfur loading. *ACS Applied Materials & Interfaces* 12 (15): 17528–17537.

42 Zhou, G.M., Pei, S.F., Li, L. et al. (2014). A graphene-pure-sulfur sandwich structure for ultrafast, long-life lithium-sulfur batteries. *Advanced Materials* 26 (4): 625–631.

43 Zhou, G., Li, L., Wang, D.W. et al. (2015). A flexible sulfur-graphene-polypropylene separator integrated electrode for advanced Li-S batteries. *Advanced Materials* 27 (4): 641–647.

44 Wang, X.F., Wang, Z.X., and Chen, L.Q. (2013). Reduced graphene oxide film as a shuttle-inhibiting interlayer in a lithium-sulfur battery. *Journal of Power Sources* 242: 65–69.

45 Peng, H.J., Zhang, Z.W., Huang, J.Q. et al. (2016). A cooperative interface for highly efficient lithium-sulfur batteries. *Advanced Materials* 28 (43): 9551–9558.

46 Zhou, T., Lv, W., Li, J. et al. (2017). Twinborn TiO_2–TiN heterostructures enabling smooth trapping–diffusion–conversion of polysulfides towards ultralong life lithium–sulfur batteries. *Energy & Environmental Science* 10 (7): 1694–1703.

47 Wang, N., Chen, B., Qin, K.Q. et al. (2019). Rational design of Co_9S_8/CoO heterostructures with well-defined interfaces for lithium sulfur batteries: A study of synergistic adsorption-electrocatalysis function. *Nano Energy* 60: 332–339.

9

Graphene-Based Materials for Supercapacitors

Shan Zhu[1,2], Chunnian He[1,2], and Naiqin Zhao[1]

[1] *Tianjin University, School of Materials Science and Engineering, 135 Yaguan Rd., Jinnan Dist., Tianjin, 300350, P.R. China*
[2] *Joint School of National University of Singapore and Tianjin University, International Campus of Tianjin University, Binhai New City, Fuzhou, 350207, P.R. China*

9.1 Supercapacitor

9.1.1 Fundamentals

The supercapacitor is a kind of high-capacity capacitor, its capacitance value is much higher than that of the traditional capacitor [1]. Compared with lithium-ion batteries, supercapacitors have higher power density, better stability, and a wider working temperature range [2, 3]. In 1879, the concept of "double layer" was put forward by Helmholtz [4], which provided a theoretical basis for the construction of supercapacitors. In 1957, Beckers used activated carbon as an electrode material to assemble capacitors and measured high device capacitance. In 1970, Donald L. Boos applied for the patent of activated carbon double-layer capacitor, and the supercapacitor officially entered the stage of practical production and application [5]. With the continuous expansion of the supercapacitor market, several excellent supercapacitor enterprises have been born, such as Maxwell [6]. Governments around the world have also put forward their policy plans for the development of supercapacitors, and the new energy vehicle represented by supercapacitor electric bus has been popularized step by step [7].

Compared with batteries and traditional physical capacitors, the characteristics of supercapacitors are mainly reflected in the following aspect. Firstly, the power density of supercapacitors can reach $10^2 \sim 10^4$ W kg^{-1}, which is much higher than the power density of the battery (Figure 9.1) [1]. Secondly, supercapacitors enjoy long-cycle life, which can conduct the high-speed deep charge–discharge cycles up to 1 million cycles. Thirdly, the working temperature limit of supercapacitors is wide. Because the adsorption and desorption rate of ions in supercapacitors change little at low temperature, which is much smaller than that of batteries. The working temperature range of commercial supercapacitors can reach −40 to 80 °C. Moreover, the supercapacitor is a kind of environmental-friendly device, which does not use heavy

Templated Fabrication of Graphene-Based Materials for Energy Applications, First Edition.
Edited by Chunnian He, Naiqin Zhao, and Junwei Sha.

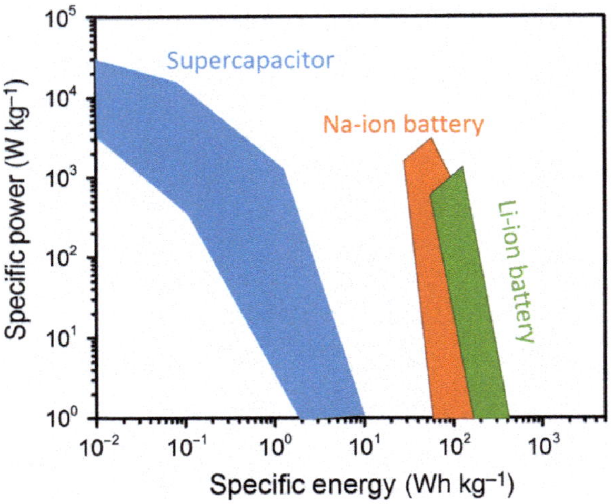

Figure 9.1 Ragone plot supercapacitors and other energy-storage devices.

metals and other harmful chemicals in the production process, thus it is a new type of green environmental protection power supply.

The research system involved in a supercapacitor is complex. For example, to improve the performance of a supercapacitor, we need to consider many factors, including material selection, device assembly, testing system, and so on (Table 9.1).

9.1.2 Mechanism

The supercapacitor mainly consists of active material, electrolyte, separator, and current collector. Its working principle can be divided into electrical double-layer mechanism and pseudocapacitance mechanism.

Electrical double-layer capacitors (EDLCs) mainly depend on the active material to adsorb charge at the electrode/electrolyte interface and to realize electrochemical energy storage. When charging, under the action of the external electric field, the positive charge will be adsorbed on the surface of the positive electrode and the negative charge will be adsorbed on the surface of the negative electrode. To maintain the balance of the electric field, a corresponding reverse charge layer is formed at the interface between the positive/negative electrode and the electrolyte. During discharge, through the external circuit, the electron moves from the negative electrode to the positive electrode, thus releasing the charge originally stored in the positive and negative electrode, the double electric layer disappears, and the charge and discharge process is completed. The charge–discharge mechanism based on the double layer depends on the adsorption and desorption of charged ions in electrolytes. This physical process not only occurs quickly but also does not cause loss to the electrode or electrolyte. Generally, EDLCs have the advantages of high-power density, good rate performance, long service life, and so on. At present, the materials suitable for EDLCs should have a large specific surface area (SSA) and good electrical conductivity at the same time. Therefore, carbon materials represented by activated carbon, carbon nanotubes, and graphene are the first choice. The electrode materials used in

Table 9.1 Common considerations for optimizing the performance of supercapacitors.

Material level	Active material	Activated carbon	Template carbon	Carbide-derived carbon	Carbon nanotube	Graphene	
	Microstructure	Surface area	Pore size	Ionic pathway			
	Pseudocapacitance	Heteroatom-doping	Redox-active electrolyte	Metal carbides and nitrides	Intercalation pseudocapacitance	Metal oxides	Conducting polymers
Device level	Structure	Parallel plate	Fiber	3D structures			
	Shape	Coin	Laminate	Cylindrical			
Testing system	Voltage	Aqueous cells	Ionic liquid cells	Li-ion capacitors			
	Power density	Current collector	Electrode materials with high conductivity	Electrolytes with high ionic conductivity	Contact resistance	Separators with fast ion transport	

commercial supercapacitors are often activated carbon, and the SSA of these materials is often up to 2000 $m^2 g^{-1}$, which can greatly enhance the energy-storage effect of double layers.

The energy-storage mechanism of pseudocapacitors can be divided into three kinds – (i) surface Faraday pseudocapacitance, (ii) intercalation pseudocapacitance, and (iii) underpotential deposition [8]. The third one mainly occurs on the surface of inert metals such as Au, which is not common and is not discussed here. The surface Faraday pseudocapacitance occurs on the surface of the electrode material and relies on the reversible redox reaction between the active material and the electrolyte to store energy. Common materials with surface Faraday pseudocapacitance include transition metal inorganic materials (metal oxides, metal sulfides, etc.) and conductive polymers (polyaniline [PANi], polypyrrole [PPy], etc.). Another pseudocapacitance mechanism is the intercalation pseudocapacitor, which mostly occurs in layered materials (Nb_2O_5) and relies on the process of electrolyte ions insertion and extraction in the active material to store energy. In fact, for Faraday pseudocapacitors, charge storage often has both pseudocapacitance and EDLC. Generally speaking, compared with the EDLC supercapacitor, the pseudocapacitance mechanism often has a higher specific capacity and energy density, but its rate performance and cyclic stability are poor.

9.1.3 Comparison Between Supercapacitor and Li-Ion Battery

Lithium-ion batteries and supercapacitors have different performances in electrochemical measurements. The performance and energy-storage mechanism of various electrode materials can be deeply studied based on correlation characterization. The main difference between supercapacitors and lithium-ion batteries is their different physical/chemical mechanisms. The battery stores energy chemically and has a much higher energy density; the supercapacitor stores charge physically, and it has no change in the microstructure during charge/discharge processes.

In the electrochemical characterization of supercapacitors, for typical EDLC energy-storage materials, the ideal cyclic voltampere (CV) curve is rectangular (Figure 9.2a), indicating that there are only capacitive elements in the circuit. However, as typical pseudocapacitance materials, the CV curves of RuO_2 and MnO_2 in aqueous electrolytes are also approximate rectangles (Figure 9.2b). Not only in the curve of CV, but also in the galvanostatic charge/discharge test, the pseudocapacitors behave similarly to EDLCs (Figure 9.2c). Because the redox reaction in this kind of pseudocapacitor electrode mainly occurs on the surface of the material, and the reaction occurs rapidly, the performance of this kind of pseudocapacitor material is similar to that of EDLC mechanism material in electrochemical measurements. The CV curve for intercalation pseudocapacitance materials shows a wide reversible redox peak (Figure 9.2), e.g. Ti_3C_2 in acidic aqueous electrolytes and Nb_2O_5 in organic electrolytes containing lithium ions. The materials with redox peaks in this part of the CV curve are also classified as pseudocapacitors, provided that the difference between the peak positions of the oxidation peak and the reduction peak is small (Figure 9.2e,f). In Figure 9.2g,h,

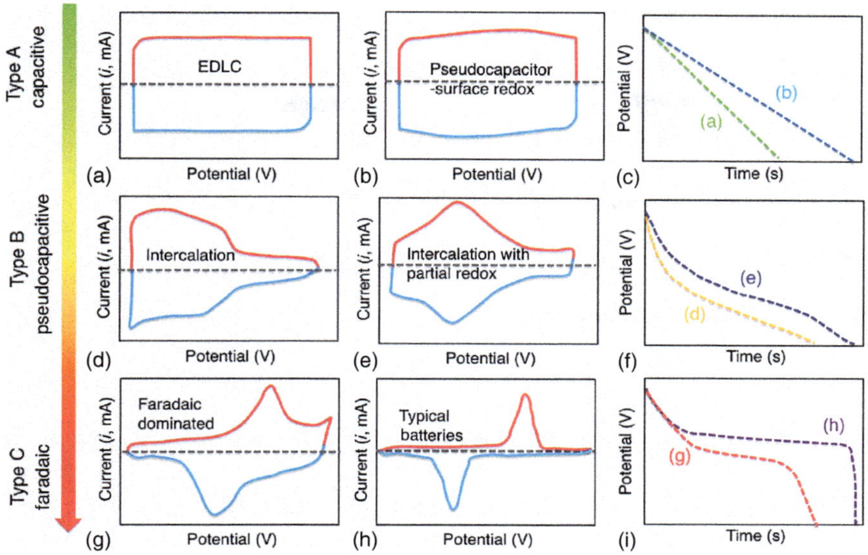

Figure 9.2 Comparison between double-layer capacitor, pseudocapacitor, and lithium-ion battery: (a, b, d, e, g, h) Schematic cyclic voltammograms and (c, f, i) corresponding galvanostatic discharge curves for various kinds of energy-storage materials. Source: Gogotsi and Penner [9]. Reproduced with permission of American Chemical Society.

the electrochemical response corresponds to the battery material, and the sharp peak of Li-ion intercalation/deintercalation appears in the CV test. There is a large difference in the value of the oxidation/reduction peak corresponding to the insertion and extraction peak. Battery-type of material will show obvious discharge platform characteristics in CP testing.

9.1.4 Influencing Factors of Carbon-Based Supercapacitor

Before introducing graphene-based supercapacitors, it is necessary to review the research status of carbon-based supercapacitors.

Carbon materials are widely used in supercapacitors. Activated carbon is the most successful supercapacitor electrode material. According to the different raw materials and activation methods, the SSA of these activated carbons can be obtained from 900 to 3500 $m^2\ g^{-1}$. These materials can show a specific capacity of 200–550 $F\ g^{-1}$ in aqueous electrolytes, and 130–230 $F\ g^{-1}$ in nonaqueous electrolytes. The EDLC mechanism of carbon materials depends on the ion adsorption on the surface of the materials to complete the energy-storage process, so the SSA is positively correlated with the amount of charge stored. At the same time, the characteristics of the pore structure are also the key factors affecting the capacitance of carbon materials. Many kinds of carbon materials are gradually used in supercapacitors, such as carbon nanotube, carbon nanofiber, and carbon nanosheets. With the deepening of the research, scholars have explored several factors that affect the carbon-based supercapacitors (Figure 9.3).

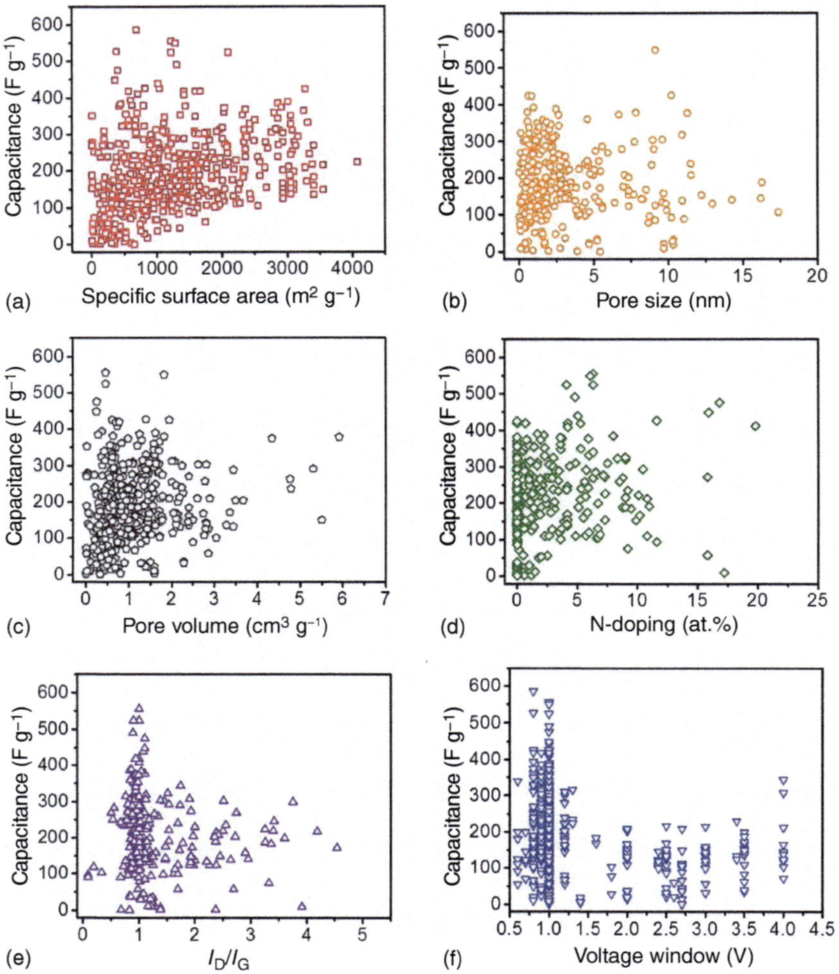

Figure 9.3 The relation between capacitance and (a) specific surface area, (b) pore size (calculated), (c) pore volume, (d) N-doping amount, (e) the ratio of I_D/I_G, and (f) voltage window of carbon samples reported in relative researches Source: Zhu et al. [10]. Reproduced with permission of Elsevier.

The key feature of carbon-based materials is their SSA (Figure 9.3a), whose relation with capacitance (C) is

$$C = \frac{\varepsilon\varepsilon_0 A}{d}$$

where ε is the electrolyte dielectric constant, ε_0 is the vacuum permittivity, and d is the distance between ion and carbon surface.

Then, the porous structure of the electrode should be considered because it can influence the accessibility of electrolyte ions (Figure 9.3b,c). According to IUPAC regulations, the pore size of materials can be divided into three types – pore size less than 2 nm is microporous, pore size greater than 50 nm is macroporous, and pore size between micropores and macropores is mesoporous. Since there are various strategies to define the pore distribution of porous carbon, several researchers prefer to use the mathematical calculation approximation to reflect the pore size [10]:

$$\text{Pore size (calculated)} = \frac{4000 * \text{Pore volume}}{\text{SSA}}$$

In addition to the design of the structure, the components greatly affect the capacitive properties of carbon-related materials. Taking into account the nature of various elements, nitrogen is the most popular doping element in the carbon electrodes (Figure 9.3d). The N-doping effect can enhance the capacitance by adding pseudocapacitance.

Supercapacitors need electrode materials with excellent conductivity during operation. The crystallinity of the carbon materials affects their conductivity. To express the crystallinity of carbon materials, we can use the ratio of the D-band and G-band (I_D/I_G) in the Raman spectroscopy (Figure 9.3e) [11].

Besides, the capacitor test conditions are needed to study. Among various test features, the electrolyte plays an important role. Due to the characteristics of various electrolytes, the corresponding voltage window is different. Accordingly, the energy density (E) of the supercapacitor is proportional to the square of the voltage window (V). In this regard, the voltage window should be taken into consideration when exploring the influencing factors of supercapacitors (Figure 9.3f).

9.2 Graphene-Based Supercapacitor

Under the current situation of the gradual depletion of global nonrenewable resources and the trend of low-carbon environmental protection, the research of new energy materials is of great significance, and the study of high-performance supercapacitors has a broad application prospect and strategic significance. For example, in the field of automobile manufacturing, the electric vehicle has become the most attractive alternative to traditional cars. If it is combined with lithium-ion power cells, fuel cells, etc., as auxiliary power, a supercapacitor can meet the requirements of instantaneous high-power density and high-energy density. The extremely high-power density of graphene-based supercapacitors can meet the needs of starting, accelerating, and climbing electric vehicles. Therefore, electric vehicle companies are committed to the research and development of energy-storage equipment with short charging time and strong endurance.

At present, the world is in the upsurge of graphene research, graphene applications in various fields emerge endlessly. In the field of application, graphene-based supercapacitors have made breakthroughs and are of great strategic significance.

9.2.1 Advantages of Graphene Used in Supercapacitors

Because of its unique structure, graphene has a series of outstanding physical and chemical properties. These properties are as follows:

(1) the theoretical specific surface area of graphene is as high as 2675 $m^2\ g^{-1}$.
(2) the conductivity of graphene can reach 10^4 to 10^6 $S\ m^{-1}$, and the carrier mobility can exceed 20 000 cm^2 V s.

(3) graphene is the material with the highest known strength, which is 100 times higher than that of iron and steel, and Young's modulus is as high as 1 TPa.
(4) the chemical stability of graphene is very high, the surface is in a state of inertia, and will not react with acid and base.

Based on these excellent properties, graphene is considered to be an ideal electrode material for supercapacitors.

9.2.2 Improving the Performance of Graphene-Based Supercapacitors

Regarding the previous exploration of the influencing factors of carbon-based supercapacitors, to improve the performance of graphene in the application of supercapacitors, we can first consider improving the design of graphene materials themselves.

9.2.2.1 Design of Graphene Electrode

To increase the capacitance, two main factors are considered in graphene-based materials, which are improving the structures (SSA and pore size distribution) and introducing heteroatoms in the carbon matrix.

Since the EDLC is a surface phenomenon, a larger surface area is required to achieve high capacitance. Zhu et al. prepared a porous graphene electrode (a-MEGO) by KOH activation [12]. The SSA of the porous graphene electrode based on BET is as high as $3100\,m^2\,g^{-1}$. Tested in the organic electrolyte, the specific capacity of the active fossil graphene is as high as $165\,F\,g^{-1}$, and the corresponding bulk energy density is $60\,F\,cm^{-3}$. Chen and coworkers obtained graphene materials by reducing graphene oxide. This graphene electrode achieved the specific capacitance of $205\,F\,g^{-1}$, the power density of $10\,kW\,kg^{-1}$, and the energy density of $28.5\,Wh\,kg^{-1}$ in an aqueous electrolyte (Figure 9.4) [13].

At present, porous graphene materials have attracted great interest because of their unique porous structure and inherent properties of graphene due to their high SSA, excellent chemical stability, and rapid ion/electron transport. To synthesize these types of graphene, numerous methods are applied, such as chemical synthesis, hydrothermal method, electrochemical reduction method, and template-guided chemical vapor deposition (CVD) method. In these carbonaceous porous materials, all-carbon porous graphene shows great application potential in electrode materials of supercapacitors. Yu et al. prepared layered porous graphene nanoparticles (HPGN), with a SSA of up to $1942\,m^2\,g^{-1}$ [14]. The current density of the device is $16\,A\,g^{-1}$ and the specific capacitance is as high as $112\,F\,g^{-1}$. However, the serious problem faced by graphene-based materials is that this kind of two-dimensional materials will have a serious agglomeration phenomenon in practical use, thus greatly reducing the effective utilization area of the materials. Therefore, avoiding graphene stacking is one of the technical problems in the preparation of high-performance graphene-based supercapacitors. There are two kinds of solutions, one is to construct a graphene three-dimensional self-supporting structure, and the other is to introduce the second material between graphene

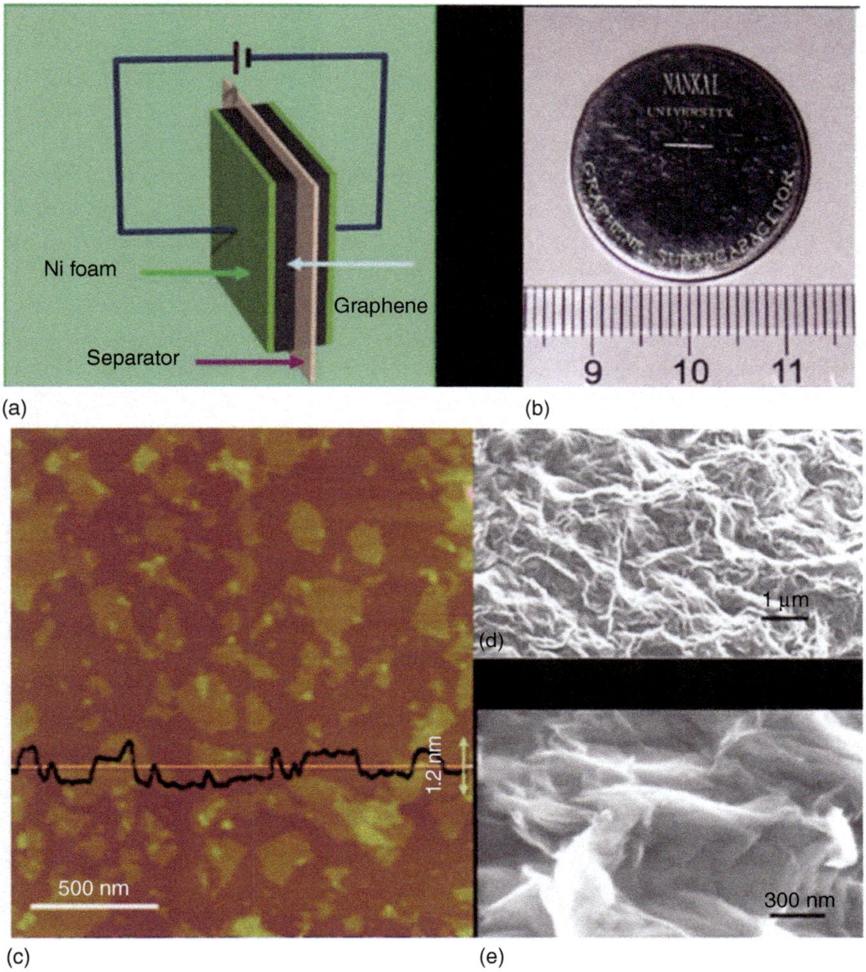

Figure 9.4 (a) Graphene-based supercapacitor device and (b) its optical image. (c) AFM image of graphene oxide, (d, e) SEM images of graphene with different scale bars. Source: Wang et al. [13]. Reproduced with permission of American Chemical Society.

sheets to prevent agglomeration. Wei and coworkers prepared the "sandwich" structure of carbon nanotube and graphene by the CVD method [15], in which carbon nanotube grew between graphenes and effectively prevented the stacking of graphene in 6 M KOH solution. It shows a high specific capacitance of 385 $F g^{-1}$.

Graphene can also be applied as an interfacial layer to enhance the performance of active materials. Wang et al. reported that they introduced 3D graphene between active materials (reduced graphene oxide [rGO]) and the Ni collector. The electrochemical analysis indicates that the graphene interfacial layer can improve the charge transport in the device. As the result, the power density of the graphene-enhanced electrode can be increased by 220% [16].

9.2.2.2 Heteroatom-Doping of Graphene

Besides the structure optimization of graphene, when the graphene is modified or doped by heteroatoms, it will show excellent electrochemical properties. For example, N-doped graphene has better electronic conductivity and wettability, more ionic active sites, and N-containing functional groups can provide additional pseudocapacitance by redox reaction on the surface of graphene. As a result, the electrochemical performance of the N-graphene electrode material is greatly improved. In the fabrication of N-doped graphene, the widely used N dopants are polyaniline, polypyrrole, ammonia, melamine, and diamine derivatives. Jeong et al. introduced N into the surface of graphene by plasma treatment (Figure 9.5), and the specific capacitance of the electrode material is as high as $280\,F\,g^{-1}$, which is four times that of pure graphene, and the capacitance decay is not obvious after 200 000 long-cycle test [17]. The authors found that N doping has a significant effect on the thinning properties of graphene. Also, it is found that P-doped and N-doped graphenes show n-type semiconductor properties at the same time, but the former is much more stable in the air. The P-doped graphene prepared by H_3PO_4 reduction of graphene oxide can reach 367 F ($5\,mV\,s^{-1}$), which is much higher than that of undoped graphene. However, this work does not give the amount of P doping and group bond components, and the related research work needs to be further studied [18]. Moreover, if graphene is doped by two kinds of heteroatoms, its conductivity and specific capacitance can be obviously improved. Li et al. prepared graphene composites with N and S doping. This research demonstrates that the co-doping of nitrogen and sulfur can greatly improve the capacitive performance of graphene with a surface area of $275.8\,m^2\,g^{-1}$. Tested in KOH-electrolyte, this co-doped graphene shows the specific capacitance of $386.5\,F\,g^{-1}$ [19]. Besides, the

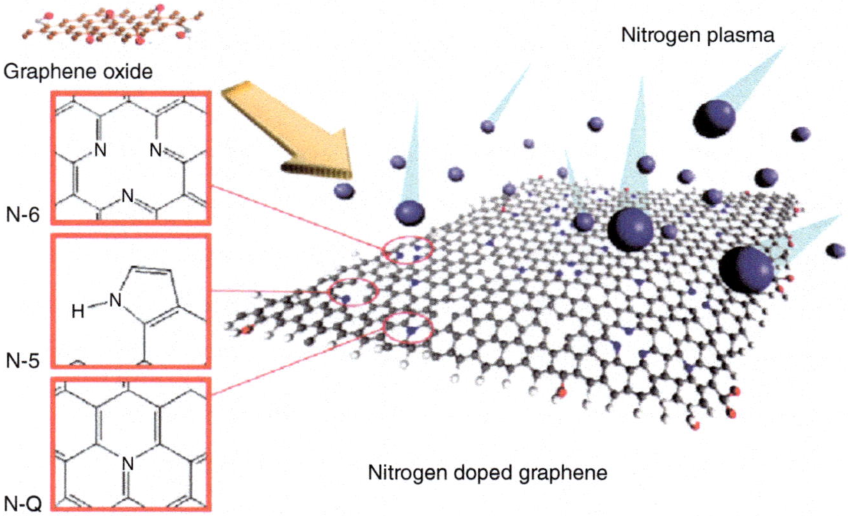

Figure 9.5 Schematic illustration of the N-doping process. Source: Jeong et al. [17]. Reproduced with permission of American Chemical Society.

heteroatom-doping can broaden the working voltage window of the carbon-based material in aqueous electrolyte, which is an effective method to obtain high-energy density [20].

9.2.2.3 Constructing 3D Graphene by Template Method

When graphene is used, agglomeration will reduce the effective utilization area of graphene, and is not conducive to the transport and diffusion of ions. One of the ways to solve agglomeration is to construct three-dimensional (3D) graphene. Moreover, compared with 0D, 1D, and 2D carbon nanomaterials, 3D graphene has the advantages of large SSA, high porosity, and efficient mass transfer.

Among various strategies, the template-based methods were the most effective and feasible to produce 3D graphene. Until now, many kinds of templates have been developed, including metal, salt, silica, and so on. Because of the different features of these templates, the as-produced 3D graphene materials have different structures and properties. Zhu et al. have summarized several types of hard templates based on the pore sizes of templates (Figure 9.6) [21].

Qin et al. prepared nitrogen and oxygen co-doped 3D nanoporous duct-like graphene@carbon nano-cage hybrid films (NO-3DG@CNC) by a CVD method (Figure 9.7) [22]. The obtained NO-3DG@CNC exhibits a high SSA of 972.2 $m^2\ g^{-1}$, and the concentrations of the N and O elements are 2.45 at.% and 5.91 at.%, respectively. This 3D graphene material shows high power density and energy density in both aqueous electrolytes and ionic electrolytes. Tested in 2 M $LiClO_4$ electrolyte, NO-3DG@CNC can achieve a specific capacitance of 122 $F\ g^{-1}$ at

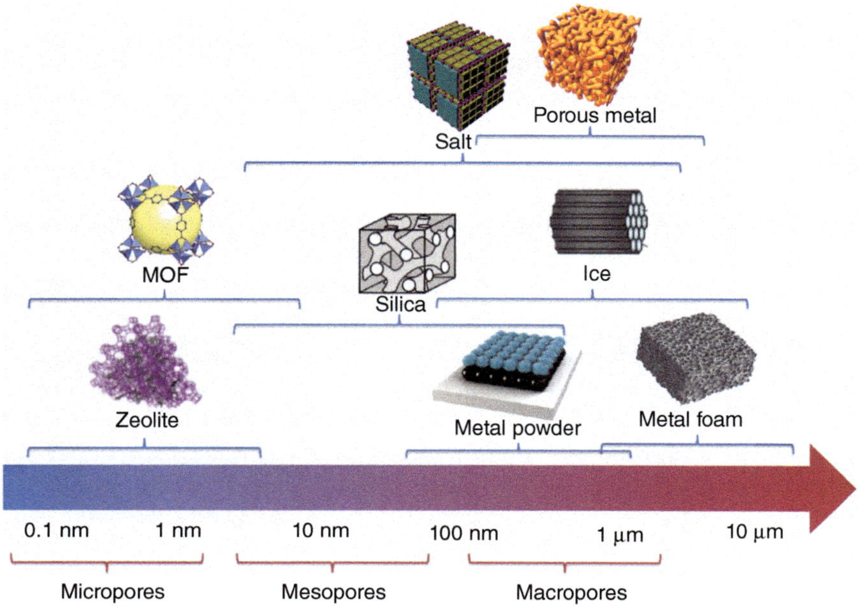

Figure 9.6 The different types of hard templates and their corresponding length scales. Source: Zhu et al. [21]. Reproduced with permission of Elsevier.

$1\,A\,g^{-1}$ and remain $75\,F\,g^{-1}$ at $500\,A\,g^{-1}$. Tested in $EMIMBF_4$ electrolyte, this 3D graphene-based electrode obtains a specific capacitance of $162\,F\,g^{-1}$ at $0.5\,A\,g^{-1}$.

A kind of N-doped ordered mesoporous few-layer carbon (OMFLC) was produced by Huang's group with the assistance of silica template [23]. Such template graphene-like carbon shows the capacitance of $855\,F\,g^{-1}$. The reason for this high capacitance is that the N-associated defects in carbon layer lead to redox reactions. Also, this carbon-based electrolyte has a specific energy of $41\,Wh\,kg^{-1}$. Wu et al. reported a three-dimensional (3D) nitrogen and boron co-doped monolithic graphene aerogels (BN-GAs) (Figure 9.7) [24]. The 3D structure provided high surface area and enhanced ion diffusion and electron transport in the bulk electrode. The nitrogen and boron doping in graphene matrix can facilitate charge transfer. Tested in polyvinyl alcohol (PVA)/H_2SO_4 gel electrolyte, BN-GAs achieved high specific capacitance ($62\,F\,g^{-1}$), high-energy density ($8.65\,Wh\,kg^{-1}$), and good power density ($1600\,W\,kg^{-1}$).

The pore size of 3D graphene can also be adjusted by the different templates. For example, Zhao's group has demonstrated a common strategy for synthesizing the 3D graphene-like carbon with the assistance of NaX ($X = Cl^-$, CO_3^{2-}, SiO_3^{2-}) salts [21]. Such salt-based method can be divided into the three steps – (i) the precursors and salts were dissolved in water and were dried to make the salt coated by carbon precursors; (ii) the precursor–salt mixture was treated at high temperature to enable the precursors to convert into graphene-like carbon nanosheets; (iii) the following washing step was conducted to remove the salts. In this strategy, the pore size of carbon networks can be tuned by choosing the different salts. For example, the NaCl particles can produce the micrometer-scale pores in 3D carbon; the pores caused by Na_2CO_3 are 50–100 nm; the nanopores (5–10 nm) can be derived from Na_2SiO_3 (Figure 9.7). By mixing these three kinds of salts, the produced 3D graphene-like carbon networks exhibit multiscale pores, which can be adjusted by varying the ratio of the salt in raw materials.

9.2.2.4 Introducing Composition on Graphene

To further increase the capacitance of graphene-based supercapacitors, researchers introduced other materials to combine with graphene.

Transition metal oxides, such as MnO_2, Mn_3O_4, RuO_2, Co_3O_4, Fe_3O_4, SnO_2, NiO, Nb_2O_5, and ZnO, are the ideal materials for supercapacitors. Their materials exert the pseudocapacitance, which mainly exists on the surface and inside of the transition metal oxide and is much higher than that of the EDLC material. However, the application of metal oxide in supercapacitors is limited because of its low conductivity and poor-cycle life. By compounding it with graphene materials, the above problems can be effectively improved. There are many kinds of literature about the combination of graphene and metal oxides. On the one hand, the composite can improve the conductivity of metal oxides, and the dispersion of metal oxides on the surface of graphene can also effectively inhibit the agglomeration of graphene and improve the effective utilization rate of graphene. Dong et al. prepared graphene by vapor deposition on nickel foam and further loading Co_3O_4 nanowires on graphene [25]. The maximum specific capacitance of the complex as an electrode

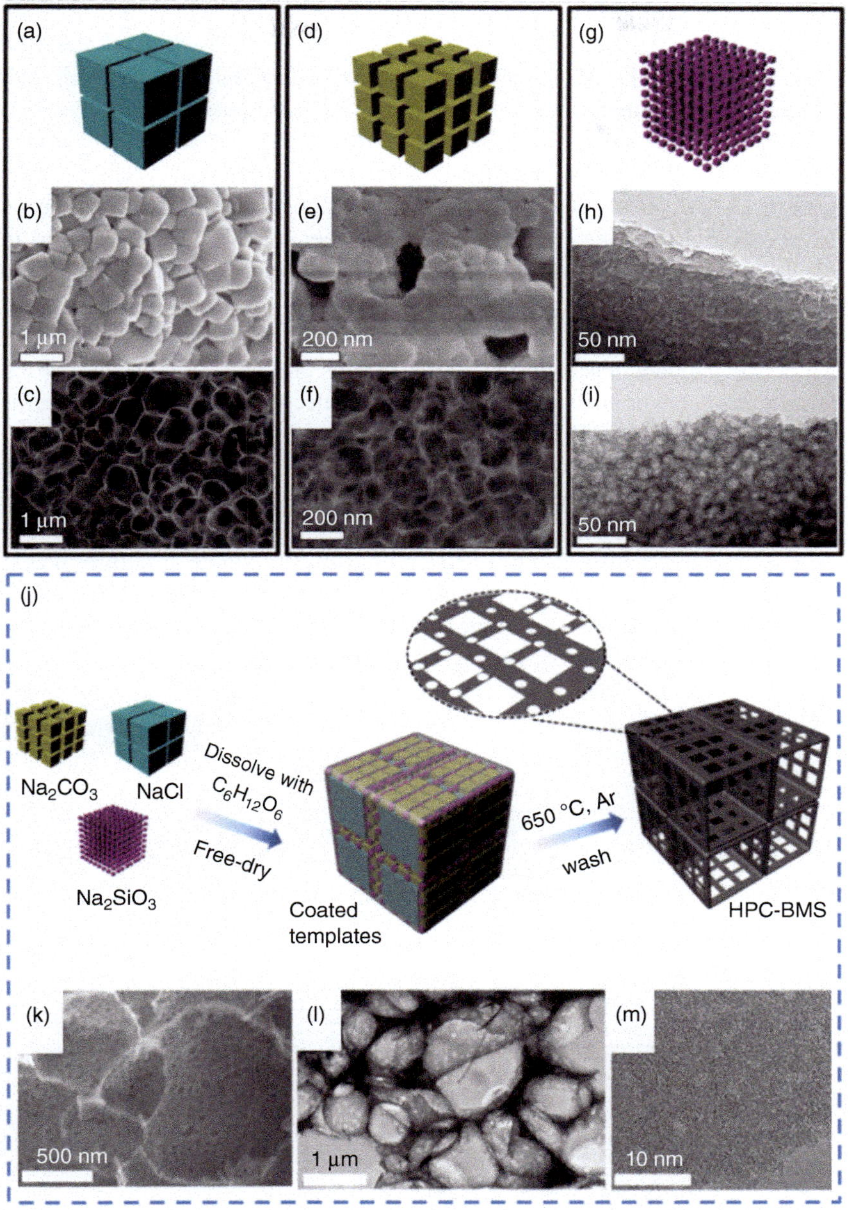

Figure 9.7 (a, d, g) Schematic of NaX template, (b, e, h) SEM image of carbon-coated NaX and (c, f, i) the obtained 3D graphene-like carbon. (j) Schematics for producing 3D porous graphene based on the combination of three types NaX, and (k–m) are the corresponding microstructure images. Source: Zhu et al. [21]. Reproduced with permission of Elsevier.

is 1100 F g^{-1}. In addition to these complexes, there are graphene–Mn_3O_4 [26], graphene–RuO_2 [27], graphene–Fe_3O_4 [28], graphene–SnO_2 [29], graphene–NiO [30], graphene–ZnO [31], and other complexes have been widely reported with high performance. For example, Choi used graphene–MnO_2 composites (MnO_2/e-CMG) [32], which showed excellent capacitance performance in 1 M Na_2SO_4 electrolyte with specific capacitance of 389 F g^{-1}. At the current density of 35 A g^{-1}, the capacity retention rate is 97.7% (Figure 9.8). Graphene-transition metal oxide electrode materials have been reported a lot, but there are still some problems in the application of graphene-transition metal oxide electrode materials, such as the initial irreversible capacity of graphene-transition metal oxide electrode materials is too large. Most of the future research will focus on the optimization of the structure.

Conductive polymer has the advantages of low price and lightweight. What is more important is that when it is used as the electrode material, the specific capacitance and working voltage are both higher than common carbon-based supercapacitors. At present, many conductive polymer materials are reported, including poly(3,4-ethylenedioxythiophene) (PEDOT), polyaniline (PANi), polythiophene (PTh), polypyrrole (PPy), and their derivatives. The advantage of conductive polymer is mainly because they have the conjugated electronic structure, whose conjugated electrons can carry on the transfer and the migration freely. But the conductivity of this polymer is poor, and the repeated insertion and removal of electrons and ions between layers of polymer materials during charge and discharge make them mechanically unstable. These shortcomings restrict the practical application of conductive polymers in supercapacitors. In recent years, graphene materials are

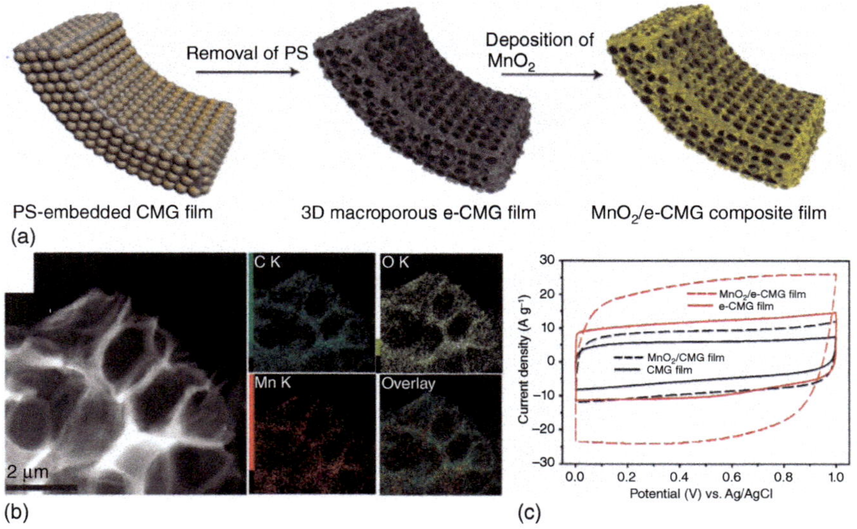

Figure 9.8 (a) Schematic illustration of producing the MnO_2/e-CMG process. (b) STEM image and EDS element mapping of C, O, and Mn on the materials. (c) CV curves of MnO_2/e-CMG composites. Source: Choi et al. [32]. Reproduced with permission of American Chemical Society.

becoming more and more prosperous so that conductive polymers also ushered in a new opportunity for development. The composite electrode material of graphene and polymer can improve these shortcomings of polymer to a great extent. Rajesh et al. prepared a polypyrrole/graphene film by the electrochemical method [33]. When the current density is 5 mV s^{-1}, the maximum specific capacitance of the film can reach 343 F g^{-1}. After 4000 charge and discharge cycles, the specific capacitance can still be kept at about 91%. Zhang et al. coated conducting polymers (PEDOT, PANi, and PPy) on the rGO sheets by the polymerization process (Figure 9.9) [35]. In the obtained conducting-polymer-rGO composites, ethanol played an important role to obtain the uniform coating. Among various samples, rGO–PANi exhibited a specific capacitance of 361 F g^{-1} at a current density of 0.3 A g^{-1}. Cong et al. synthesized graphene tablets by the one-step method and loaded the PANi on the graphene surface [36]. These graphene sheets are large in size, flexible, and light in weight. The utility model has the advantages of high density (2 g cm^{-3}) and high

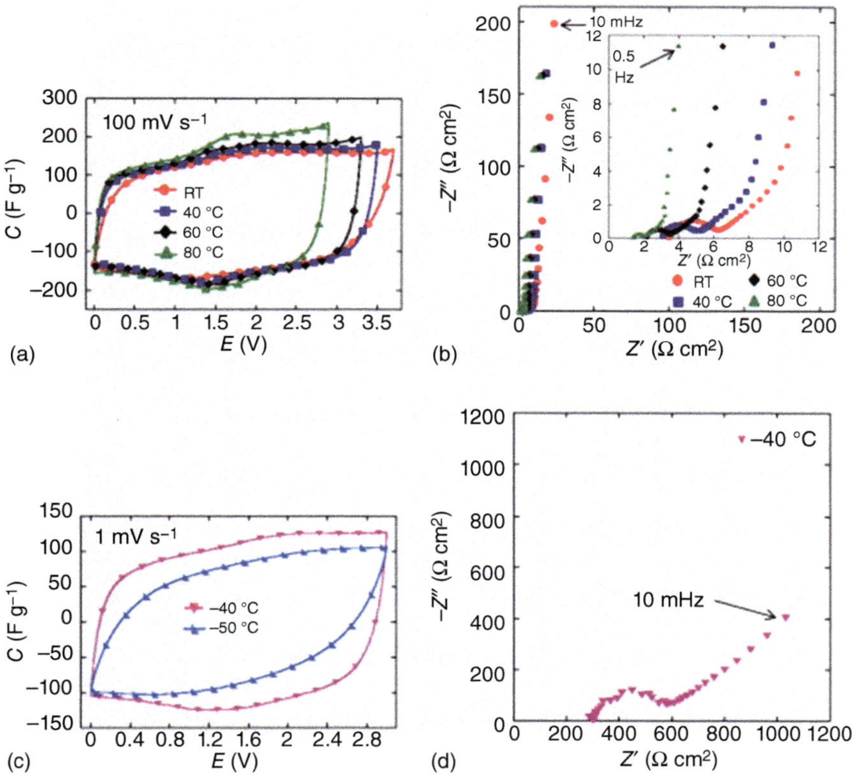

Figure 9.9 Electrochemical characterization of a-MEGO electrodes in the electrolyte of $(PIP13\text{-}FSI)_{0.5}(PYR14\text{-}FSI)_{0.5}$: (a) CVs at 100 mV s^{-1}, and (b) corresponding EIS Nyquist plots at room and elevated temperatures (RT ~80 °C). (c) CVs at 1 mV s^{-1} (−40 °C, −50 °C) and (d) the corresponding EIS Nyquist plot (−40 °C). Source: Tsai et al. [34]. Reproduced with permission of Elsevier.

conductivity (15 Ω per square). PANi/graphene composite shows great potential as the electrode for supercapacitors, whose specific capacitance is up to 763 F g^{-1}.

Qin et al. designed a method to synthesize NiCo-layered double hydroxide (NiCo-LDH) and derived sulfide on heteroatom-doped edge-enriched 3D rivet graphene films, which can be used as the electrode of asymmetric supercapacitor (ASC) (Figure 9.9). In this research, the free-standing 3D graphene film served as the substrate to grow the NiCo-LDH nanoneedles. Then, the NiCo-LDH was converted to $NiCo_2S_4$ by anion exchange. The 3D porous graphene has an interconnected structure, which is in favor of reaction kinetics. As a result, this 3D graphene and $NiCo_2S_4$ composite show high-energy densities of 37.6 Wh kg^{-1} in ASC devices [37].

9.2.3 Advanced Graphene-Based Supercapacitors

9.2.3.1 Electrolyte Design

To improve the overall performance of the supercapacitor, one feasible method is to extend the voltage of devices.

According to equation $E = CV^2/2$, the energy density (E, in Wh/kg) is proportional to specific capacitance (C, in F/g) and the square of voltage window (V, in V). Thus, the careful selection of electrolytes with a large voltage window is one effective route toward high-energy density. Currently, the widely used aqueous electrolytes, such as KOH or H_2SO_4, are limited by water's decomposition voltage (~1.23 V). Therefore, nonaqueous electrolytes have attracted great attention, including organic electrolytes and ionic liquids. Although the regular organic electrolytes used in such applications can sustain voltage windows higher than 2.5 V, they are usually flammable and toxic. The latter issues can be in part addressed using ionic liquids enabling a voltage window as high as 4.0 V, thanks to their low vapor pressure and thermal stability. Tsai et al. used KOH-activated microwave exfoliated graphite oxide ("a-MEGO") as the electrode in an ionic liquid mixture. This novel electrolyte was consisted of *N*-methyl-*N*-propylpiperidinium bis(fluorosulfonyl)imide (PIP13-FSI) and *N*-butyl-*N*-methylpyrrolidinium bis(fluorosulfonyl)imide (PYR14-FSI) with a weight of 1:1. In this ionic liquid electrode, a-MEGO shows an outstanding capacitance of 180 F g^{-1} with a wide voltage window of 3.5 V (Figure 9.9). More importantly, due to the chemical stability of ionic liquid, this supercapacitor can work in a wide temperature range from −50 to 80 °C. Such a superior electrochemical response is derived from the optimization of the carbon/electrolyte interface [34]. Also, new electrolytes should consider the adaptability of electrode materials. Zhu et al. demonstrate that the mesopores of 3D carbon can influence the ion-transfer ability of ionic liquids [38].

9.2.3.2 Asymmetric Supercapacitors

Another effective method to extend the voltage window is to assemble ASCs. To obtain high-energy density and power density at the same time, researchers began to design a new type of asymmetric electrochemical supercapacitor, that is one electrode is double-layer electrode and the other is Faraday pseudocapacitor

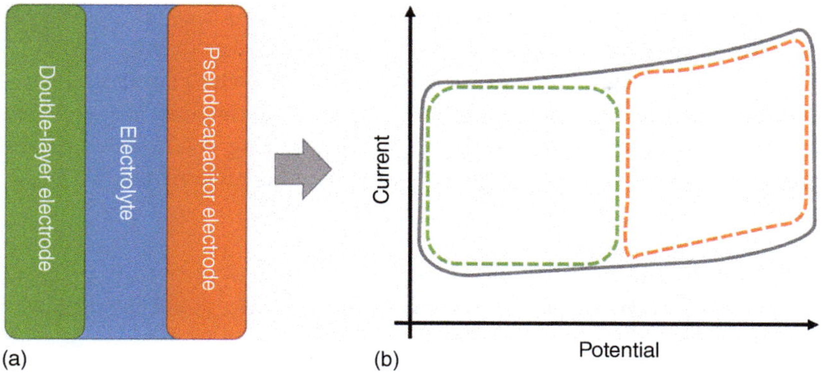

Figure 9.10 (a) Schematic diagram of asymmetric supercapacitors configuration and (b) its typical CV curve.

electrode (Figure 9.10). ASCs combine the advantages of two kinds of electrochemical capacitors and can better meet the overall requirements of load on energy density and power density of the power supply system in practical application. Qin et al. synthesized a free-standing 3D duct-like graphene (3D-DG) by porous metal template, which acts as the robustness backbone for depositing MnO_2 nanosheets (3D-DG@MnO_2) [39]. To assemble ASC, 3D-DG@MnO_2 acted as the positive electrode and 3D-DG served as the negative electrode. This device displayed ultrahigh volumetric energy density (28.2 mWh cm^{-3}) and power density (55.7 W cm^{-3}) at 2.0 V in $LiNO_3$-PVA gel electrolyte. Furthermore, a piece of this kind of ACS can illuminate 25 LED lamps at the same time, supporting its high-energy density and promising application prospect. Wang et al. coated graphene on carbon cloth as a negative electrode by drop method, and the energy density of asymmetric devices was up to 0.234 mWh cm^{-3} [40]. In addition, the preparation of graphene film by vacuum filtration using a microporous membrane is also a commonly used square method for the preparation of the graphene electrode. Yang group directly uses the graphene thin film as a negative electrode to assemble an ASC with a mass-specific capacitance of 108.5 F g^{-1} [41].

Alshareef research group adheres the graphene film to the carbon cloth as the negative electrode [42]. The ASC based on the negative electrode has an energy density of 60 Wh kg^{-1} when the output power density is 1.8 kW kg^{-1}. The device also shows excellent cyclic stability, maintaining a capacitance of 90.1% after 10 000 charge and discharge cycles. Two-dimensional graphene materials are also used to assemble three-dimensional porous graphene materials. Three-dimensional porous graphene is usually prepared by the hydrothermal method using graphene or graphite oxide solution as a precursor. This porous structure makes graphene with larger contact surface area and more active sites, which greatly enhances the electrochemical activity. Using three-dimensional nickel foam as a template, rGO can be grown on the surface of this three-dimensional template by the hydrothermal method [43]. Yu et al. assembled the rGO anode material with the activated nickel foam cathode to form an ASC [44]. The three-dimensional graphene electrode shows excellent capacitance in the voltage range of −0.8 to 0 V. The assembled ASC has a stable working voltage window of 1.5 V. The energy density is as high as 1.06 mWh cm^{-3}. In addition to nickel foam, other templates can also be used to synthesize 3D porous graphite. For example, Choi et al. reported the synthesis of three-dimensional macroporous graphene materials using polystyrene spheres as templates, which were used as negative electrode materials for ASCs [32]. The maximum energy density of the whole device is 44 Wh kg^{-1}.

9.2.3.3 Metal-Ion Capacitor

As a new type of capacitor-battery hybrid energy-storage device, a metal ion capacitor has attracted wide attention because it can not only achieve high-power density but also ensure energy density. The positive electrode material is the activated carbon material with EDLC energy storage, the negative electrode material is the intercalation material with the function of metal ion deintercalation, and the electrolyte is the corresponding metal salt electrolyte. It has the characteristics of high voltage, energy density, and power density between lithium-ion battery and supercapacitor.

For example, the lithium-ion capacitors (LICs) capture the merits of EDLC and Li-ion battery – its positive electrode utilizes the materials for EDLCs, and the negative electrode material uses the material for Li-ion battery. Tested as electrodes of a LIC, a 3D rebar graphene foam (GF) made by Ni powder template achieved the capacity of 320 mAh g^{-1} and the energy density as high as 32 Wh kg^{-1} (Figure 9.11) [45]. Theoretical and experimental studies have shown that heteroatom doping can further adjust the electronic structure and enhance ion storage performance. Therefore, heteroatom doping graphene has become a potential battery electrode material for lithium/potassium-ion capacitors. Cao used $(NH_4)_3PO_4$ as solid nitrogen and phosphorus source, 2 to 6 layers of nitrogen and phosphorus heteroatom co-doped graphene anode materials were obtained by one-step arc discharge method [46]. The material has the characteristics of good electrical conductivity and large layer spacing. When nitrogen and phosphorus co-doped graphene is used as anode materials for lithium-ion batteries and potassium ion batteries, respectively, they show good rate capacity and stable cycle performance, which lays a foundation for the construction of high-performance ion capacitors. For this reason, all-carbon LICs and potassium ion capacitors were constructed by using nitrogen and phosphorus co-doped graphene as negative electrodes and activated carbon as positive electrodes. These two kinds of ion capacitors show high-energy density and power density, as well as excellent cyclic stability. This work provides a new strategy for designing high-performance lithium/potassium ion batteries or capacitors.

Meanwhile, zinc-ion hybrid supercapacitors are developed in recent years. In this type of energy-storage application, zinc ion acts as a similar role as lithium-ion in LICs. Deng and his coworkers developed a new 3D carbon cathode for zinc-ion supercapacitors. In this composite, the graphene-like carbon layers have existed on the surfaces, and the N, O were co-doped in the carbon matrix. Tested in the aqueous ZHSs, this carbon cathode delivered a specific capacity of 138.5 mAh g^{-1} with good cycling stability [47].

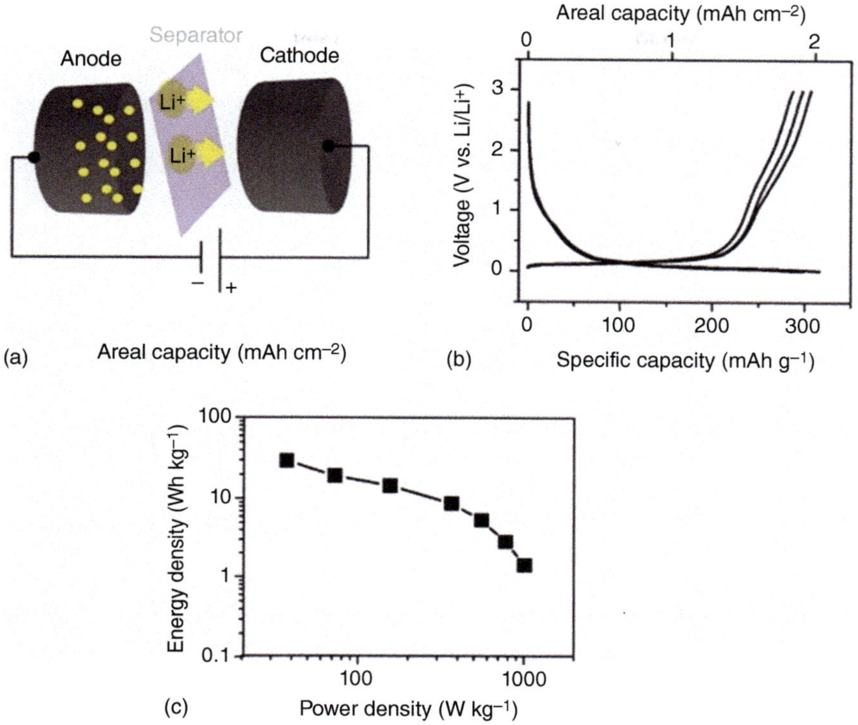

Figure 9.11 (a) Scheme of LIC, (b) Galvanostatic charge–discharge curve at 0.1 A g^{-1}, and (c) Ragone plot of the 3D rebar GF. Source: Salvatierrra et al. [45]. Reproduced with permission of American Chemical Society.

9.2.3.4 Flexible Supercapacitor

With the rapid development of portable electronic products, wearable and flexible energy-storage devices are favored by researchers. It is of great significance to develop energy-storage devices with flexibility, small size, and high electrochemical performance [48]. Among many electrochemical energy-storage devices, supercapacitors have been widely studied by researchers because of their fast charge and discharge speed, long-cycle life, and high-power density. However, the shape of the device is greatly limited because of the non-bending of the electrode, and the metal collector and binder are involved in the preparation of the electrode. This will also reduce the electrochemical performance of supercapacitors [49]. Therefore, the development of a flexible supercapacitor matching with portable electronic products has become the development direction of the next generation of flexible energy-storage devices.

Flexible supercapacitors are usually composed of flexible electrode materials, gel electrolytes, membranes, and packaging materials. The main difference between flexible supercapacitors and traditional supercapacitors is that the electrode materials and electrolytes are flexible. The electrode materials of flexible supercapacitors mainly include carbon-based materials, metal oxides, conductive polymers, and composite electrode materials. In the application of flexible supercapacitors, graphene is considered to be a good flexible electrode material because of its good mechanical and electrical properties. There are three common configurations of flexible supercapacitors, namely, fiber supercapacitor, planar supercapacitor, and in-plane supercapacitor.

Fiber flexible supercapacitor is a supercapacitor with fibrous structure and can be easily made by a single-electrode coaxial structure, double-electrode parallel structure, or winding structure [50, 51]. Because of their unique structure, fiber supercapacitors not only have good flexibility but also can be woven and integrated into other fabrics by using traditional textile technology. At present, the construction of fiber supercapacitors is mostly based on the winding structure of double electrodes, which are separated by gel electrolytes. Because of its large SSA and excellent electrical and mechanical properties, graphene fiber is widely used in the construction of fiber flexible supercapacitors [52]. Meng et al. prepared all graphene fibers with graphene fiber in the core and graphene networks in the shell [53]. The graphene fiber with a multilayer hybrid structure not only retains the high conductivity and flexibility of the core fiber but also further increases the SSA of the outer graphene network. Based on the graphene fiber, the specific capacity of the double-electrode wound fiber supercapacitor is up to $1.7\,\mathrm{mF\,cm^{-2}}$. In the bending state, the electrochemical properties of the devices remain unchanged and show good flexibility. Kou et al. prepared core–shell polymer electrolyte-coated graphene and carbon nanotube composite fibers by wet spinning technology [54]. The fiber supercapacitor based on the double-electrode winding structure of the composite fiber has ultrahigh area specific capacity ($177\,\mathrm{mF\,cm^{-2}}$) and high-energy density ($3.84\,\mu\mathrm{Wh\,cm^{-2}}$). During 2000 charge–discharge cycles, its capacity has no obvious attenuation and has a higher capacity.

Traditional planar supercapacitors usually require metal as charge collection fluid, which not only increases the quality and cost of devices but also limits the flexibility of supercapacitors [55]. Because of its excellent electrical and mechanical properties, graphene thin films can be used as charge collectors and active electrode materials for planar supercapacitors at the same time, and highly flexible devices can be realized [51]. Based on the graphene film electrode with ultrahigh mechanical strength, the area-specific capacity of the flexible supercapacitor is up to $71\,\mathrm{mF\,cm^{-2}}$, which has almost no degradation by bending to any angle [56]. Wu et al. prepared a flexible supercapacitor with high-rate performance by compounding graphene with thiophene nanoparticles [57]. The results show that the supercapacitor has an obvious redox peak and shows very high pseudocapacitance. The volume-specific capacity, energy density, and power density can reach $375\,\mathrm{F\,cm^{-3}}$, $13\,\mathrm{mWh\,cm^{-3}}$, and $776\,\mathrm{W\,cm^{-3}}$, respectively. This new type of extendable supercapacitor can maintain the stability of performance under the action of certain destructive forces and has a wide range in the field of wearable energy devices. Kaner and coworkers used a standard LightScribe DVD optical drive to conduct the reduction of graphite oxide films. This laser scribing graphene (LSG) enjoys high electrical conductivity and high SSA of $1520\,\mathrm{m^2\,g^{-1}}$. The LSG can be applied in the planer supercapacitor as the electrodes without the binders and current collectors. The LSG supercapacitor shows high-energy density and high

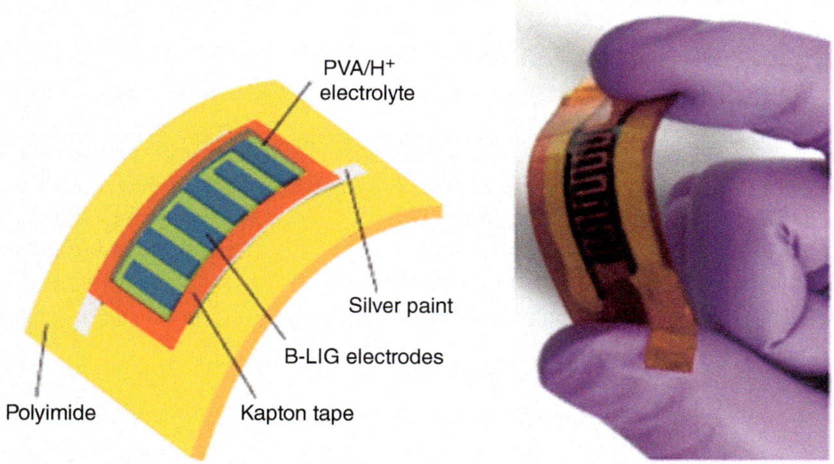

Figure 9.12 Schematic of a LIG-MSC device and the digital photograph. Source: Peng et al. [58]. Reproduced with permission of American Chemical Society.

stability in the bending station. Smalley and coworkers produced the boron-doped porous graphene by treating boric acid-containing polyimide sheets with a laser. The flexible supercapacitors (B-LIG-MSCs) based on this boron-doped graphene show an area capacitance of 16.5 mF cm^{-2} with superior stability and mechanical flexibility (Figure 9.12) [58].

The serious stacking between graphene sheets reduces the utilization rate of the surface area of graphene electrode materials, and the planar supercapacitors based on graphene electrode materials do not meet the expectations. Therefore, the in-plane structure was put forward. In the in-plane flexible supercapacitors, the charge or ion is diffused and transported directly along with the graphene lamellar without the need to transport it across the layers. It can greatly improve the performance of capacitors. In addition, the in-plane supercapacitor can also make use of the existing Wiener processing technology to realize the miniaturization and integration of the device [59, 60]. It has important application value in the field of micro and portable electronic devices [61]. The in-plane flexible supercapacitor can be prepared by a patterned design strategy [62]. During the process, a layer of GO was coated on the copper foil. After that, graphene films were obtained by reduction treatment with methane plasma. Then, this graphene film was transferred to the PET substrate and graphically designed. The in-plane flexible supercapacitor was obtained by dripping gel electrolytes on the patterned graphene electrode. The area-specific capacity of the supercapacitor is 80.7 μF cm^{-2}, and it has high-energy density (2.5 mWh cm^{-3}) and high-power density (495 W cm^{-3}). Moreover, this device can maintain its performance at 1000 V s^{-1}. Based on graphene electrode material and the introduction of pseudocapacitance material polyaniline, the flexible in-plane supercapacitor has not only a high volume-specific capacity

(75 F cm^{-3}) but its energy density can reach 6.67 mWh cm^{-3}, which is comparable to the commercial thin-film lithium battery [63].

9.2.3.5 Microsupercapacitor

Microsupercapacitor is a new type of high-power micro electrochemical energy-storage device, which can not only be directly integrated with microelectronic devices (such as sensors and radio-frequency devices) as the micro power source but also provide effective peak power in an instant [64]. Moreover, it can be easily integrated with other microsystems, such as nano-piezoelectric generators, solar cells, and thermoelectric components, to build a new flexible and miniaturized self-power supply energy system [65, 66].

Recently, microsupercapacitors based on graphene are produced with the development of microfabrication technology. The scale of these devices is tens to hundreds of micrometers, which can serve as the power source of microelectronic devices. However, due to the high resistance of the electrodes and limited capacitance, the device exhibits only modest performance. By focusing on graphene and its composites, many research efforts have focused on the development of flexible microsupercapacitors. For example, by using GO and solid electrolytes, a laser-patterned microsupercapacitor is prepared by directly laser-writing a GO film. This rGO-GO microsupercapacitor exhibits good electrochemical performance [67]. Based on the entire device, the microsupercapacitor has an area-specific capacitance of 0.51 mF cm^{-2}, which is almost twice that of a traditional supercapacitor. The main difficulty of the microsupercapacitors is to assemble two small electrodes on the flexible substrate. Moreover, in the microsupercapacitor, the electrolyte ions should have high transportation efficiency in the narrow spaces, which can ensure a high-energy density and power density. In this regard, Niu et al. apply photolithography and electrophoretic deposition to produce graphene-based microelectrodes. Then, two pieces of microelectrode are fabricated into an all-solid-state device. The specific capacitance of this microsupercapacitor is 285 F g^{-1}, which is much higher than the conventional graphene supercapacitor (86 F g^{-1}) [68]. These superior performances are derived from two aspects. Firstly, the open edges of the graphene enhance the diffusion path of electrolyte ions. Secondly, the high surface area is suitable for electrolyte ions accessibility.

Although microsupercapacitors show great potential to replace batteries in some special applications. However, the traditional microfabrication techniques are not cost-effective and hinder their further application. Kaner and coworkers reported a scalable method to fabricate graphene-based microsupercapacitors by laser writing on graphite oxide films. They approve that over 100 microsupercapacitors can be fabricated within 30 minutes. Moreover, these microsupercapacitors deliver a power density of 200 W cm^{-3} [61]. Wu's research group successfully prepared continuous and uniform sulfur-doped graphene films by bottom-up step-by-step pyrolysis based on the two-dimensional nano-limiting effect and gold catalysis of gold nanoparticles. Using the thin film as the electrode, the microsupercapacitor was constructed by dry lithography. The microsupercapacitor has the advantages of high volume-specific volume (582 F cm^{-3}), high-power density (1191 W cm^{-3}), and good

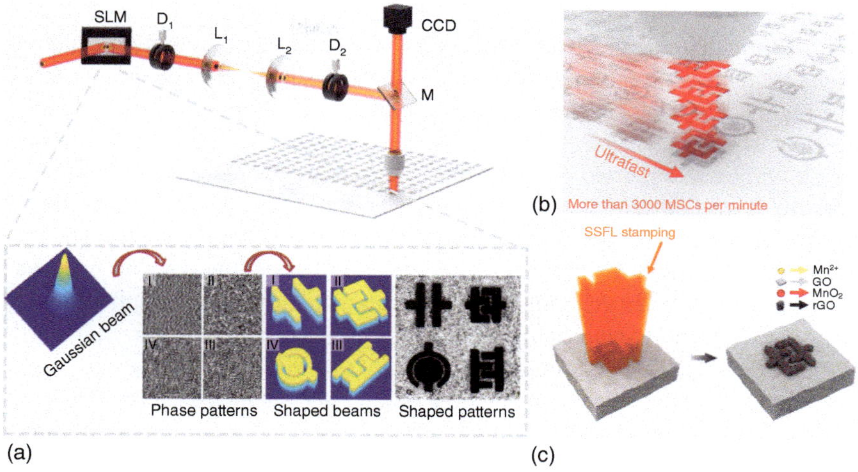

Figure 9.13 The schematics and graphene-electrode patterns of laser photonic-reduction stamping method to fabricate microsupercapacitors: (a) the illustration for program-controlled laser photonic-reduction processing; (b) the illustration for the ultrafast processing of laser-based stamping method; (c) the illustration for fabricating the graphene/ MnO_2 micro-supercapacitors. Source: Yuan et al. [70]. Reproduced with permission of Springer Nature.

cyclic stability [69]. To further improve the capacity of the device, two-dimensional thiophene nanosheets, and graphene laminated composite films were prepared by layer-by-layer alternating deposition, resulting in a high performance, flexibility, and miniaturization supercapacitors.

To enable the requirements of mass production, Yuan et al. developed a single pulse laser photonic-reduction stamping method to produce graphene-based microsupercapacitors. By applying numerous laser pulses, over 30 000 microsupercapacitors can be prepared within 10 minutes (Figure 9.13). Due to the 3D feature of these graphene-based electrodes, a single microsupercapacitor shows the energy density of 0.23 Wh cm^{-3} and the capacitance of 426.7 F cm^{-3} [70].

9.3 Future Prospects

Graphene-based supercapacitors have excellent energy-storage performance, so they are used in many fields, such as energy, automobile, information technology, and so on. From the point of view of the development trend, the future will expand to the field of transportation and new energy (wind energy, solar power intelligent grid, etc.). For the research on graphene-based supercapacitors, it is necessary to define some future development directions. The main research directions include energy-storage mechanism, preparation process optimization of graphene materials, electrode design, supercapacitor design, and packaging, etc.

The industry of graphene and materials is developing rapidly, but there are still many problems, especially the typical application is less and the market share needs to be improved. In many application fields of graphene, the graphene-based supercapacitor is the upgrade of traditional supercapacitor and has better performance. With the promotion of national policy, the graphene-based supercapacitor market will be highly prosperous in the future. Among them, the market size of electrode materials will also quadruple. The main problem in the development of electrode materials is the agglomeration of graphene layers, so graphene must be combined and modified with other materials to improve its properties. In addition, there are numerous uncontrollable factors in the preparation of graphene. For example, the problems of the area, quality, and the number of layers, the relationship between the structural properties of graphene composites and the preparation process, structural optimization and control, and the safety of supercapacitors in the actual installation process. However, with the continuous improvement of the preparation technology of graphene and graphene-based materials (such as graphene powder and thin-film materials), the problems hindering graphene supercapacitors, such as high production cost, are gradually solved.

Besides, the fabrication of electrodes plays a key role in the feasibility of energy-storage technologies. Nowadays, the electrode of the supercapacitor is mainly produced by the roll coating method, which enjoys the advantages of low cost and high-efficient. As a result, the production efficiency of the prepared electrodes is up to hundreds of meters per hour. Most of the technology used in graphene-based devices is limited to research laboratories and difficult to scale up. Future research on graphene should focus on alternative large-scale production techniques. It is believed that the graphene supercapacitor will become one of the most breakthrough applications of graphene in a short time.

With the unremitting efforts of many chemists, material scientists, physicists, and characterization experts, batteries, and supercapacitors have been developed rapidly. With the deepening of theoretical research, the development of computational simulation methods, and the application of advanced characterization techniques, especially in situ characterization technology, researchers have a more in-depth understanding of electrochemical energy-storage devices. At the same time, the development of supercapacitors has also ushered in new opportunities and challenges. In this regard, the energy-storage mechanism, double-layer capacitance, and pseudocapacitance of supercapacitors need to be further studied. Also, the electrolyte needs to be further optimized to broaden the working voltage and energy density of supercapacitors. More advanced in-situ characterization techniques are needed to study the physicochemical properties of the double-layer capacitive materials and the pseudocapacitive materials during the charge/discharge processes to help understand their energy storage mechanisms.

Although the research on graphene supercapacitors is still in its infancy, researchers have designed and prepared graphene composite electrodes. A lot of work and effort need to be done to improve the performance (energy density, power density, and cycle life) of supercapacitors. From the current situation of the downstream application of graphene, we believe that graphene supercapacitors will promote the further development of electrochemical energy-storage devices and be widely used in other fields.

References

1 Simon, P. and Gogotsi, Y. (2020). Perspectives for electrochemical capacitors and related devices. *Nature Materials* 19 (11): 1151–1163.

2 Jiang, H., Lee, P.S., and Li, C. (2013). 3D carbon based nanostructures for advanced supercapacitors. *Energy & Environmental Science* 6 (1): 41–53.

3 Lobato, B., Vretenár, V., Kotrusz, P. et al. (2015). Reduced graphite oxide in supercapacitor electrodes. *Journal of Colloid and Interface Science* 446: 203–207.

4 Ozoemena, K.I. and Chen, S. (2016). *Nanomaterials in Advanced Batteries and Supercapacitors*. Springer.

5 Liu, F. and Xue, D. (2015). Advanced graphene nanomaterials for electrochemical energy storage. *Materials Research Innovations* 19 (1): 7–19.

6 Devillers, N., Jemei, S., Péra, M.-C. et al. (2014). Review of characterization methods for supercapacitor modelling. *Journal of Power Sources* 246: 596–608.

7 Hall, P.J. and Bain, E.J. (2008). Energy-storage technologies and electricity generation. *Energy Policy* 36 (12): 4352–4355.

8 Augustyn, V., Simon, P., Dunn, B.J.E. et al. (2014). Pseudocapacitive oxide materials for high-rate electrochemical energy storage. *Energy & Environmental Science* 7 (5): 1597–1614.

9 Gogotsi, Y. and Penner, R.M. (2018). Energy storage in nanomaterials–capacitive, pseudocapacitive, or battery-like? *ACS Nano* 12 (3): 2081–2083.

10 Zhu, S., Li, J., Ma, L. et al. (2018). Artificial neural network enabled capacitance prediction for carbon-based supercapacitors. *Material Letters* 233: 294–297.

11 Zhu, S., Li, J., He, C. et al. (2015). Soluble salt self-assembly-assisted synthesis of three-dimensional hierarchical porous carbon networks for supercapacitors. *Journal of Materials Chemistry A* 3 (44): 22266–22273.

12 Zhu, Y., Murali, S., Stoller, M.D. et al. (2011). Carbon-based supercapacitors produced by activation of graphene. *Science* 332 (6037): 1537–1541.

13 Wang, Y., Shi, Z., Huang, Y. et al. (2009). Supercapacitor devices based on graphene materials. *The Journal of Physical Chemistry C* 113 (30): 13103–13107.

14 Yu, S., Liu, Y.-D., Li, Y. et al. (2016). Preparation of hierarchical porous graphene nanosheets with high specific surface area and their electrochemical behaviors in supercapacitors. *Materials Chemistry and Physics* 177: 171–178.

15 Fan, Z., Yan, J., Zhi, L. et al. (2010). A three-dimensional carbon nanotube/graphene sandwich and its application as electrode in supercapacitors. *Advanced Materials* 22 (33): 3723–3728.

16 Wang, Z., Liu, J., Hao, X. et al. (2020). Enhanced power density of a supercapacitor by introducing 3D-interfacial graphene. *New Journal of Chemistry* 44 (31): 13377–13381.

17 Jeong, H.M., Lee, J.W., Shin, W.H. et al. (2011). Nitrogen-doped graphene for high-performance ultracapacitors and the importance of nitrogen-doped sites at basal planes. *Nano Letters* 11 (6): 2472–2477.

18 Karthika, P., Rajalakshmi, N., and Dhathathreyan, K.S. (2013). Phosphorus-doped exfoliated graphene for supercapacitor electrodes. *Journal of Nanoscience and Nanotechnology* 13: 1746–1751.

19 Li, Z., He, W., Wang, X. et al. (2020). N/S dual-doped graphene with high defect density for enhanced supercapacitor properties. *International Journal of Hydrogen Energy* 45 (1): 112–122.

20 Yu, M., Lin, D., Feng, H. et al. (2017). Boosting the energy density of carbon-based aqueous supercapacitors by optimizing the surface charge. *Angewandte Chemie* 129 (20): 5546–5551.

21 Zhu, S., Zhao, N., Li, J. et al. (2019). Hard-template synthesis of three-dimensional interconnected carbon networks: Rational design, hybridization and energy-related applications. *Nano Today* 29: 100796.

22 Qin, K., Wang, L., Wang, N. et al. (2017). Nitrogen and oxygen co-doped 3D nanoporous duct-like graphene@carbon nano-cage hybrid films for high-performance multi-style supercapacitors. *Journal of Materials Chemistry A* 5 (35): 18535–18541.

23 Lin, T., Chen, I.-W., Liu, F. et al. (2015). Nitrogen-doped mesoporous carbon of extraordinary capacitance for electrochemical energy storage. *Science* 350 (6267): 1508–1513.

24 Wu, Z.S., Winter, A., Chen, L. et al. (2012). Three-dimensional nitrogen and boron co-doped graphene for high-performance all-solid-state supercapacitors. *Advanced Materials* 24 (37): 5130–5135.

25 Dong, X.-C., Xu, H., Wang, X.-W. et al. (2012). 3D graphene-cobalt oxide electrode for high-performance supercapacitor and enzymeless glucose detection. *ACS Nano* 6 (4): 3206–3213.

26 Liu, Y., He, D., Wu, H. et al. (2013). Graphene and nanostructured Mn_3O_4 composites for supercapacitors. *Integrated Ferroelectrics* 144 (1): 118–126.

27 Hu, C.-C., Wang, C.-W., Chang, K.-H. et al. (2015). Anodic composite deposition of RuO_2/reduced graphene oxide/carbon nanotube for advanced supercapacitors. *Nanotechnology* 26 (27): 274004.

28 Ullah, W., Anwar, A., Majeed, A. et al. (2015). Cost-effective and facile development of Fe_3O_4-reduced graphene oxide electrodes for supercapacitors. *Materials Technology* 30 (sup3): 144–149.

29 Wang, Y., Liu, Y., and Zhang, J. (2015). Colloid electrostatic self-assembly synthesis of SnO_2/graphene nanocomposite for supercapacitors. *Journal of Nanoparticle Research* 17 (10): 1–10.

30 Bello, C.L., Smith, E., Ruiz-Garcia, A. et al. (2013). A phase I, open-label, mass balance study of C-14 dacomitinib (PF-00299804) in healthy male volunteers. *Cancer Chemotherapy and Pharmacology* 72 (2): 379–395.

31 Li, Z., Zhou, Z., Yun, G. et al. (2013). High-performance solid-state supercapacitors based on graphene-ZnO hybrid nanocomposites. *Nanoscale Research Letters* 8 (1): 1–9.

32 Choi, B.G., Yang, M., Hong, W.H. et al. (2012). 3D macroporous graphene frameworks for supercapacitors with high energy and power densities. *ACS Nano* 6 (5): 4020–4028.

33 Rajesh, M., Raj, C.J., Kim, B.C. et al. (2016). Supercapacitive studies on electropolymerized natural organic phosphate doped polypyrrole thin films. *Electrochimica Acta* 220: 373–383.

34 Tsai, W.-Y., Lin, R., Murali, S. et al. (2013). Outstanding performance of activated graphene based supercapacitors in ionic liquid electrolyte from −50 to 80 C. *Nano Energy* 2 (3): 403–411.

35 Zhang, J. and Zhao, X. (2012). Conducting polymers directly coated on reduced graphene oxide sheets as high-performance supercapacitor electrodes. *The Journal of Physical Chemistry C* 116 (9): 5420–5426.

36 Cong, H.-P., Ren, X.-C., Wang, P. et al. (2013). Flexible graphene-polyaniline composite paper for high-performance supercapacitor. *Energy & Environmental Science* 6 (4): 1185–1191.

37 Qin, K., Wang, L., Wen, S. et al. (2018). Designed synthesis of NiCo-LDH and derived sulfide on heteroatom-doped edge-enriched 3D rivet graphene films for high-performance asymmetric supercapacitor and efficient OER. *Journal of Materials Chemistry A* 6 (17): 8109–8119.

38 Zhu, S., Taberna, P.-L., Zhao, N. et al. (2018). Salt-template synthesis of mesoporous carbon monolith for ionogel-based supercapacitors. *Electrochemistry Communications* 96: 6–10.

39 Qin, K., Liu, E., Li, J. et al. (2016). Free-Standing 3D nanoporous duct-like and hierarchical nanoporous graphene films for micron-level flexible solid-state asymmetric supercapacitors. *Advanced Energy Materials* 6 (18): 1600755.

40 Zilong, W., Zhu, Z., Qiu, J. et al. (2014). High performance flexible solid-state asymmetric supercapacitors from MnO_2/ZnO core–shell nanorods//specially reduced graphene oxide. *Journal of Materials Chemistry C* 2 (7): 1331–1336.

41 Qiu, Y., Zhao, Y., Yang, X. et al. (2014). Three-dimensional metal/oxide nanocone arrays for high-performance electrochemical pseudocapacitors. *Nanoscale* 6 (7): 3626–3631.

42 Chen, W., Xia, C., and Alshareef, H.N. (2014). One-step electrodeposited nickel cobalt sulfide nanosheet arrays for high-performance asymmetric supercapacitors. *ACS Nano* 8 (9): 9531–9541.

43 Zhai, T., Wang, F., Yu, M. et al. (2013). 3D MnO_2–graphene composites with large areal capacitance for high-performance asymmetric supercapacitors. *Nanoscale* 5 (15): 6790–6796.

44 Yu, M., Wang, W., Li, C. et al. (2014). Scalable self-growth of Ni@NiO core–shell electrode with ultrahigh capacitance and super-long cyclic stability for supercapacitors. *NPG Asia Materials* 6 (9): e129–e129.

45 Salvatierra, R.V., Zakhidov, D., Sha, J. et al. (2017). Graphene carbon nanotube carpets grown using binary catalysts for high-performance lithium-ion capacitors. *ACS Nano* 11 (3): 2724–2733.

46 Luan, Y., Hu, R., Fang, Y. et al. (2019). Nitrogen and phosphorus dual-doped multilayer graphene as universal anode for full carbon-based lithium and potassium ion capacitors. *Nano-Micro Letters* 11 (1): 30.

47 Deng, X., Li, J., Shan, Z. et al. (2020). A N,O co-doped hierarchical carbon cathode for high-performance Zn-ion hybrid supercapacitors with enhanced pseudocapacitance. *Journal of Materials Chemistry A* 8 (23): 11617–11625.

48 Rogers, J.A., Someya, T., and Huang, Y. (2010). Materials and mechanics for stretchable electronics. *Science* 327 (5973): 1603–1607.

49 Gao, Y., Ota, H., Schaler, E.W. et al. (2017). Wearable microfluidic diaphragm pressure sensor for health and tactile touch monitoring. *Advanced Materials* 29 (39): 1701985.

50 Hu, Y., Cheng, H., Zhao, F. et al. (2014). All-in-one graphene fiber supercapacitor. *Nanoscale* 6 (12): 6448–6451.

51 Nguyen, B.H. and Nguyen, V.H. (2016). Promising applications of graphene and graphene-based nanostructures. *Advances in Natural Sciences: Nanoscience and Nanotechnology* 7 (2): 023002.

52 Sheng, L., Wei, T., Liang, Y. et al. (2017). Vertically oriented graphene nanoribbon fibers for high-volumetric energy density all-solid-state asymmetric supercapacitors. *Small* 13 (22): 1700371.

53 Meng, Y., Zhao, Y., Hu, C. et al. (2013). All-graphene core-sheath microfibers for all-solid-state, stretchable fibriform supercapacitors and wearable electronic textiles. *Advanced Materials* 25 (16): 2326–2331.

54 Kou, L., Huang, T., Zheng, B. et al. (2014). Coaxial wet-spun yarn supercapacitors for high-energy density and safe wearable electronics. *Nature Communications* 5 (1): 1–10.

55 Koga, H., Tonomura, H., Nogi, M. et al. (2016). Fast, scalable, and eco-friendly fabrication of an energy storage paper electrode. *Green Chemistry* 18 (4): 1117–1124.

56 Xiong, Z., Liao, C., Han, W. et al. (2015). Mechanically tough large-area hierarchical porous graphene films for high-performance flexible supercapacitor applications. *Advanced Materials* 27 (30): 4469–4475.

57 Wu, Z.-S., Zheng, Y., Zheng, S. et al. (2017). Stacked-layer heterostructure films of 2D thiophene nanosheets and graphene for high-rate all-solid-state pseudocapacitors with enhanced volumetric capacitance. *Advanced Materials* 29 (3): 1602960.

58 Peng, Z., Ye, R., Mann, J.A. et al. (2015). Flexible boron-doped laser-induced graphene microsupercapacitors. *ACS Nano* 9 (6): 5868–5875.

59 Yuan, K., Xu, Y., Uihlein, J. et al. (2015). Straightforward generation of pillared, microporous graphene frameworks for use in supercapacitors. *Advanced Materials* 27 (42): 6714–6721.

60 Huang, P., Lethien, C., Pinaud, S. et al. (2016). On-chip and freestanding elastic carbon films for micro-supercapacitors. *Science* 351 (6274): 691–695.

61 El-Kady, M.F. and Kaner, R.B. (2013). Scalable fabrication of high-power graphene micro-supercapacitors for flexible and on-chip energy storage. *Nature Communications* 4 (1): 1–9.

62 Wu, Z.S., Parvez, K., Feng, X. et al. (2013). Graphene-based in-plane micro-supercapacitors with high power and energy densities. *Nature Communications* 4 (1): 1–8.

63 Liu, Z., Liu, S., Dong, R. et al. (2017). High power in-plane micro-supercapacitors based on mesoporous polyaniline patterned graphene. *Small* 13 (14): 1603388.

64 Chmiola, J., Largeot, C., Taberna, P.-L. et al. (2010). Monolithic carbide-derived carbon films for micro-supercapacitors. *Science* 328 (5977): 480–483.

65 Wang, K., Zou, W., Quan, B. et al. (2011). An all-solid-state flexible micro-supercapacitor on a chip. *Advanced Energy Materials* 1 (6): 1068–1072.

66 Liu, W.W., Feng, Y.Q., Yan, X.B. et al. (2013). Superior micro-supercapacitors based on graphene quantum dots. *Advanced Functional Materials* 23 (33): 4111–4122.

67 Gao, W., Singh, N., Song, L. et al. (2011). Direct laser writing of micro-supercapacitors on hydrated graphite oxide films. *Nature Nanotechnology* 6 (8): 496–500.

68 Niu, Z., Zhang, L., Liu, L. et al. (2013). All-solid-state flexible ultrathin micro-supercapacitors based on graphene. *Advanced Materials* 25 (29): 4035–4042.

69 Wu, Z.-S., Tan, Y.-Z., Zheng, S. et al. (2017). Bottom-up fabrication of sulfur-doped graphene films derived from sulfur-annulated nanographene for ultrahigh volumetric capacitance micro-supercapacitors. *Journal of the American Chemical Society* 139 (12): 4506–4512.

70 Yuan, Y., Jiang, L., Li, X. et al. (2020). Laser photonic-reduction stamping for graphene-based micro-supercapacitors ultrafast fabrication. *Nature Communications* 11 (1): 1–11.

10

Graphene-Based Materials for Electrocatalysis

Lechen Diao[1] *and Chunnian He*[2,3]

[1] *Shandong University of Technology, School of Chemistry and Chemical Engineering, Zibo, Shandong, 255000, P.R. China*
[2] *Tianjin University, School of Materials Science and Engineering, Tianjin Key Laboratory of Composite and Functional Materials, Tianjin, 300350, P.R. China*
[3] *Joint School of National University of Singapore and Tianjin University, International Campus of Tianjin University, Binhai New City, Fuzhou, 350207, P.R. China*

10.1 Introduction

Due to the over consumption of energy, the traditional fossil fuel cannot satisfy the increasing energy need. To solve the environmental and energy issues, developing sustainable and renewable energy conversion technologies is imperatively demanded. Electrocatalysis is considered the most important electrochemical reaction to resolve the energy crisis for many energy conversion devices. Compared with the traditional fossil resources, electrocatalysis reactions toward energy conversion, such as hydrogen evolution reaction (HER), oxygen evolution reaction (OER), oxygen reduction reaction (ORR), carbon dioxide reduction reaction (CO_2RR), nitrogen reduction reaction (NRR), and electrocatalytic reaction of value-added chemicals are cleaner and more efficient to solve current environmental and energy crisis. Thus, designing efficient and inexpensive electrocatalysts is of great significance.

As one of the most plentiful elements on the earth, carbon is essential for living organisms, which makes it ideal to develop environment-friendly and renewable technologies, and it would replace potentially inefficient and high-cost catalysts. Therefore, carbon has been imperatively applied in the energy conversion field. In the past decades, various carbon materials, including carbon nanotubes, carbon dots, graphene, and organic derivative carbon, have attracted much attention as efficient catalysts for advanced energy conversion, including HER, ORR, OER, CO_2RR, NRR, and electrocatalytic reaction of value-added chemicals. As the research on graphene-based electrocatalysts goes deeper and deeper, all kinds of engineering strategies and design principles have been exploited, and ingenious modulation to the defect structure, doping configuration, and porous structure of graphene electrocatalysts have been developed for high electrocatalysis performance. Moreover, graphene serves as a superior conductive carrier, coupled with

Templated Fabrication of Graphene-Based Materials for Energy Applications, First Edition.
Edited by Chunnian He, Naiqin Zhao, and Junwei Sha.

metal or inorganic compounds to develop inorganic/graphene hybrid materials for advanced electrocatalysts.

This chapter focuses on graphene-based materials for electrocatalysis. We aim to ascertain the relation of the structure and activity origins by providing a comprehensive summary of the preparation and application of electrocatalysts. At first, the regulation strategies of electrocatalysts including dopants, edges, defects, and coupled with metal or inorganic compounds are discussed. Then, template-method synthesis strategies of graphene-based electrocatalysts are described. Afterward, the application of graphene-based electrocatalysts for water splitting (HER and OER), ORR, CO_2RR, and NRR are summarized. Finally, an outlook for the application prospect of graphene-based materials in the field of electrocatalysts is provided.

10.2 Preparation of Graphene-Based Materials for Electrocatalysis

Most of the electrocatalysis reactions, such as HER, OER, ORR, CO_2RR, and NRR, take place at the triple-phase interfacial zone of the gaseous products or reactants, liquid electrolyte, and solid catalysts, where the electrons and reactants contact with each other at the active sites. During an electrocatalytic reaction, several elementary steps are consecutively coupled together and determine the final reactivity of the electrocatalyst. As shown in Figure 10.1, this multistep reaction includes – (i) mass diffusion of the electrolyte on the electrocatalyst surface; (ii) electron transfer between the active sites and conducting support to participate in the electrochemical reaction; and (iii) the surface reaction originated from the adsorption of reactants, interfacial electron transfer, and the desorption of products at the active sites [1].

Graphene-based materials are extremely promising substitutes to noble metal-based electrocatalyst, due to the various advantages of graphene, such as excellent electronic conductivity, low cost, high specific surface area, structural flexibility on atomic scale, and long-term stability in alkaline/acidic media. However, unmodified pristine graphene shows poor electrocatalytic activity and cannot be directly used as an electrocatalyst. By heteroatoms doping, creating defects, or coupling with compounds, active sites could be created on the graphene for electrocatalysis reactions. Besides, the charge transfer between the graphene and heteroatoms or compounds can modulate the electron structure, thereby increasing the electrochemical activity. Until now, there are three strategies to turn pristine graphene into active electrocatalysts – (i) heteroatom doping, incorporating heteroatom into graphene to substitute sp^2 carbon; (ii) defects sites, creating zigzag edges and pentagon rings to increase structure defects; (iii) coupling with inorganic hybrid materials, metal compounds supported on the graphene. These processes provide various well-established preparation methods, usually resulting in various systems with excellent electrocatalytic activity.

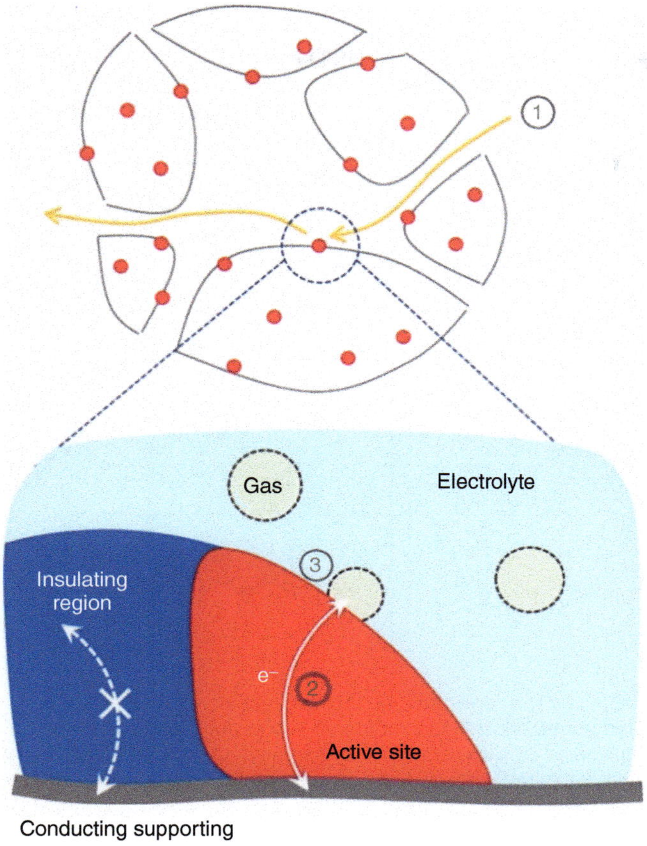

Figure 10.1 Schematic representation of the gas-involving electrocatalysis. Three critical steps are coupled with each other, including: (1) mass diffusion, (2) electron transfer, and (3) surface reaction. Source: Tang et al. [1]. Reproduced with permission of American Chemical Society.

10.2.1 Heteroatom Doping Graphene-Based Materials

So far, most of the reports for preparing metal-free graphene-based catalysts have been directed toward the doped graphene, which can be deemed as graphene structures with regulatable physicochemical properties arising after incorporating heteroatoms into the crystalline network. These graphene-based nanomaterials have been on the highlight of catalysis, endowed them appealing properties when compared with undoped graphene. Heteroatom-doped graphene can not only introduce more sites for electrocatalysis but also optimize the electrical properties of graphene. Experimental and theoretical results verify the doping heteroatoms in graphene sp^2 orbitals can efficaciously modulate the electronic distribution, create active sites, and enhance the electrocatalytic activity. Through displacing many carbon atoms by heteroatoms, the spin state and charge of the neighboring carbon atoms would restructure because of the different electronegativity between

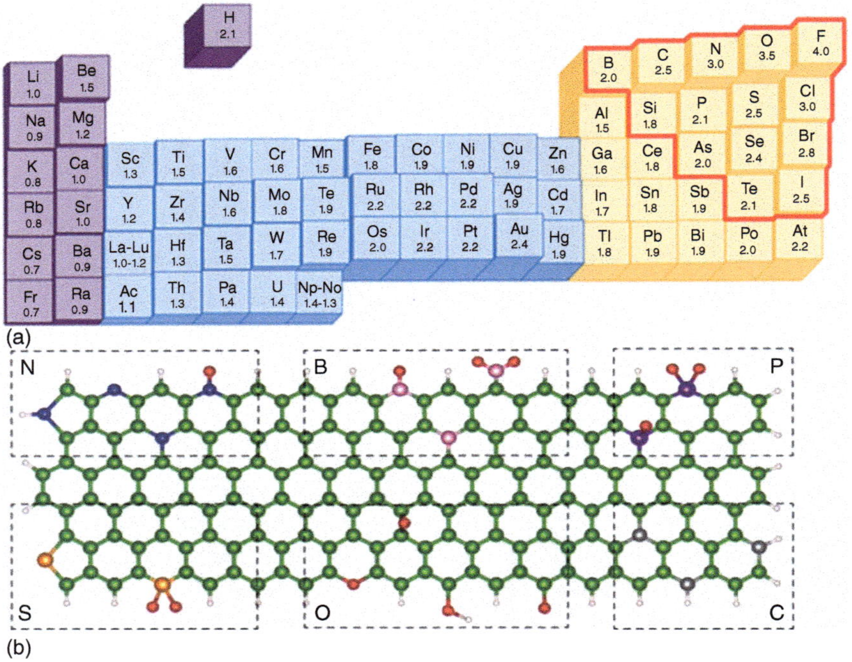

Figure 10.2 (a) Periodic table and the corresponding electronegativity of elements. The elements enclosed by the red border are the nonmetal ones successfully doped into sp^2 carbon Source: Dai et al. [2]. Reprinted with permission of American Chemical Society. (b) Schematic of the X-doped graphene, showing the possible positions of dopants Source: Jiao et al. [3]. Reproduced with permission of Springer.

carbon atoms and heteroatoms, and the spin of the carbon and change of charge can modulate its work function and the adsorption energy of reactive intermediates at active sites. Until now, more than ten heteroatoms have been successfully doped into graphene (Figure 10.2a). The doping positions and forms in each structure were different with respect to the heteroatoms (Figure 10.2b). The heteroatoms doping can lead to charge localization on the graphene surface, which can regulate the state of reaction intermediates. As a result, the electrocatalytic activity can be tuned by changing the adsorption energy and adsorption model. Moreover, theoretical calculations have also demonstrated that heteroatom doping induced extranuclear electrons redistribution and enhanced the electrocatalytic activity of electrocatalyst [1].

In 2010, nitrogen-doped graphene was first reported as metal-free electrocatalysts for ORR [4]. After that, the field of graphene-based catalysts developed very fast and various heteroatom-doped graphene-based electrocatalysts were prepared, such as sulfur-doped graphene [5], boron-doped graphene [6], iodine-doped graphene [7], phosphorous-doped graphite [8], and edge-halogenated (doped with chlorine, bromine, or iodine) graphene nanomaterials [9].

10.2.1.1 Single Doping Graphene

The conception of heteroatoms doping has been existed in the carbon family for many years, especially for N-doping, which has been most widely studied in recent years because it can improve the electrocatalysis activity of graphene. The larger electronegativity of N (3.04) compared with C (2.55) results in polarization of the carbon network and then influences the catalytic performance of N-doped graphene. When the N atoms were introduced into the sp^2 carbon structure, there are two advantages for electrocatalysis reaction – (i) N has one more electron than C atom, which is in favor of the reaction need electron; (ii) the similar atomic radius of N and C prevent lattice mismatch [10]. Indeed, N-doped graphene exhibits better activity than most of the other monodoped graphene prepared by similar technology. Furthermore, for co-doped graphene, N is the essential element for high activity [11, 12]. Therefore, N-doped graphene is the most widely explored metal-free graphene-based electrocatalysts.

A key point for N-doped graphene is regulating the configurations of N that usually coexist in multiple forms, i.e. pyrrolic N, graphitic N (quaternary N), pyridinic N, oxidized N, and the sp hybridized N (Figure 10.3). The configuration of N dopant lies on the surrounding chemical environment and can influence the charge distribution of neighboring carbon atoms, resulting in different catalytic activities. Therefore, to design efficient electrocatalysts, it is significant to comprehend the influence of N configurations on electrocatalytic performance. Graphitic N and pyridinic N have little influence on the structure of graphene owing to that the length of C—N bond is close to the C—C bond length [14]. Nevertheless, the planar structure of graphene could be destroyed by the sp^3-bonded pyrrolic N.

In the beginning, N-doped graphene was used as ORR catalyst, which is superior to Pt/C [4]. The incorporation locations and configurations of heteroatoms can notably affect the ORR performance. It is hard to identify the respective contribution to the ORR of different N configurations due to their coexisting in N-doped graphene. Generally speaking, pyridinic N is deemed as the active form to enhance the electrons donation capability and weaken the O—O bond due to that it owns a lone electron pair. Indeed, Ruoff and coworkers claimed controllable

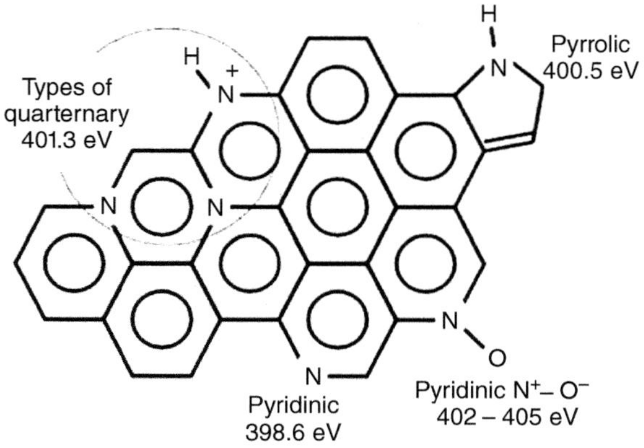

Figure 10.3 Typical nitrogen species in N-doped graphene and their binding energies of X-ray photoelectron spectroscopy (XPS) as marked. Source: Matter et al. [13]. Reproduced with permission of Elsevier.

preparation of N-doped graphene by changing the pyrolysis temperatures and appropriately selecting the N source. Their results indicate that pyridinic N reduces the overpotential and graphitic N raises the limiting current density [15]. Instead of controlling the annealing temperature, Xu et al. regulated the Ar gas-flow rate during the calcination of the precursors, which could change the local reaction conditions more subtle. By which they obtained a sequence of N-doped carbon with desired configuration. At last, they disclosed the electrocatalytic activity sequence of various N configurations – pyridinic N > pyrrolic N > graphitic N > oxidized N > C [16]. Although different ratios of N configurations were controlled, the contribution to ORR of every N-specie is still in argument. Thus, highly oriented pyrolytic graphite (HOPG) with definite content for the given N-specie was synthesized. Electrocatalysis performance tests in acidic electrolyte showed that the pyridinic-N dominated HOPG (5% for graphitic-N and 95% for pyridinic-N) possessed the better ORR activity than the graphitic-N dominated one (5% for pyridinic-N and 82% for graphitic-N) [17]. Similarly, Sun et al. prepared a model catalyst by hydrofluoric acid etching and subsequent NH_3 annealing method. The N concentration is as high as 12.6 at.%, and almost 100% belongs to pyridinic N, which exhibits improved performance [18].

However, researchers also illustrated that graphitic N is a contributor to the ORR activity. Samukawa and coworkers have studied the electrocatalytic performance of N-doped graphene for ORR by selecting the bonding state using a neutral beam, which suppressed defect emergence by the elimination of UV-photon irradiation and charge [19]. The performances of N-doped graphene were tested in alkaline electrolyte, and the results show that the graphitic N-doped graphene has superior activity to the pyridinic N-doped one. They deemed that the intrinsic graphene defects produced in the process of doping contribute most of the ORR activity of pyridinic N, and the graphitic N with less defect in the graphene contributes much more to the ORR proceeds than the pyridinic N. Thus, the excellent performance of N-doped graphene is attributed to different active sites in acidic and alkaline electrolytes, and the defects in the graphene may play a critical role in the ORR proceeds. Therefore, precisely controlling the specific N configurations in the doped graphene and eliminating the effect of the intrinsic defects are urgently demanded.

In addition to nitrogen, other heteroatoms, such as O, S, P, and B, have also been incorporated into the graphene skeleton as dopants. The abundant reactivity and precursor species expand the dimensions of regulation for exploring the new chemistry reaction and designing new multi-doped graphene-based electrocatalysts. These heteroatoms usually were doped in the plane or at the edge of graphene to substitute the carbon atoms and can break the uniform charge distribution, creating active sites. Total dopant content and doping form significantly influence the catalytic activity of doped graphene, which is tunable by precursor choice and doping procedure. As shown in Figure 10.4, the illustration of the proposed doping configurations of B-, S-, P-, P/N-, S/N-, B/N-, and B/O-doping show much more types than the single N-doping. Amini et al. [26] synthesized S-doped graphene nanosheets as a high-efficient electrocatalyst for HER by a simple, low cost, and

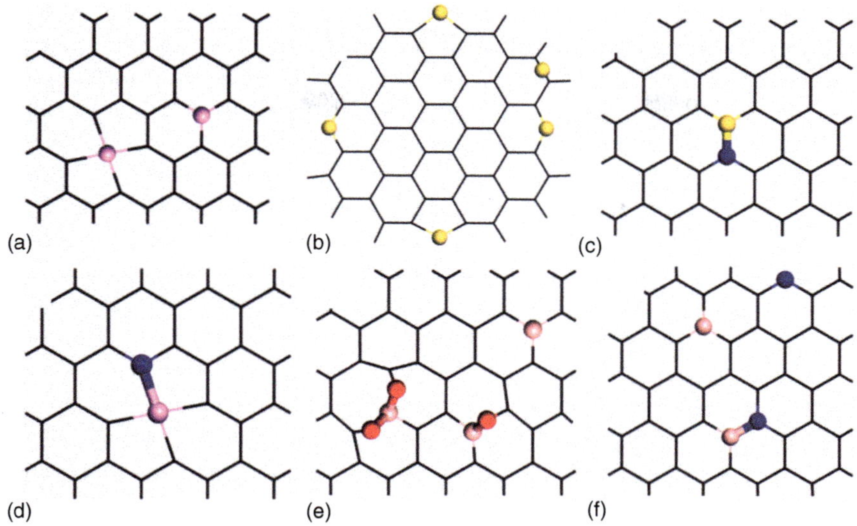

Figure 10.4 Schematic illustrations of the proposed P-doping [20] (a), S-doping [21, 22] (b), P/N co-doping [23] (c), S/N co-doping [24] (d), B and B/O co-doping [6] (e), and B/N codoping with separated and bonded B/N configurations [25] (f). Note that the dopant atoms are highlighted as the ball-and-stick model. Purple: P, yellow: S, blue: N, pink: B, red: O, black lines: C matrix. Source: Yang et al. [10]. Reproduced with permission of Wiley.

scalable method. The S doping created the structural defects in the graphene structure to form the S—C bonding, which contributes to the activity for HER.

The general method to introduce electron transfer effect in graphene is doping atoms with larger electronegative like N (electronegativity: 3.04) and S (electronegativity: 2.58) into sp^2 carbon structure and then introduce positive charge sites on the adjacent carbon atoms for electrochemical reactions. However, electron-deficient atoms with lower electronegative, like B (electronegativity: 2.04) and P (electronegativity: 2.19) can lead to electron transfer in sp^2 graphene network and change uniform charge distribution [27]. Xia and coworkers obtained B-doped graphene by calcining with boron oxide. B atoms can be successfully incorporated into the graphene lattice with a doping content of 3.2%. The resultant B-doped graphene exhibits excellent electrocatalytic performance toward ORR in alkaline electrolytes, which is comparable to the Pt catalysts [6]. Different from the N dopants, of which the active sites are the C atoms adjacent to the N atom [28], B atoms are considered to be the active sites in the B-doped graphene [29]. Hou et al. synthesized P-doped graphene via a low cost, facile, and scalable calcinating method by using triphenylphosphine and graphite oxide as phosphorus and carbon sources, respectively. The resultant P-doped graphene was used as an efficient ORR electrocatalyst, illustrating outstanding electrocatalytic activity, long-term stability, and tolerance to methanol crossover effect [30]. Huang and coworkers successfully fabricated iodine-doped graphene, which shows high activity, stability, and perfect methanol tolerance for ORR [7].

Moreover, halogen doped at the edge of graphene also causes charge imbalance of adjacent carbon atoms and then enhances the electrocatalytic activity of graphene. For instance, Jeon et al. prepared a series of graphene nanoplatelets with edge-selectively halogenated (Cl, Br, I) via ball-milling graphite with Cl_2, Br_2, and I_2. DFT calculations and experiment results revealed that the edge-halogenated graphene has remarkable electrocatalytic activity toward ORR [9]. Xing and coworkers synthesized F-doped carbon blacks for the first time as low cost and metal-free ORR catalysts and showed higher electrocatalytic performance and long-term operation stability than the commercial Pt/C catalysts. The power density that assembled methanol fuel cell with the F-doped carbon blacks as the air cathode electrocatalysts outperform the commercial Pt/C catalysts [31].

10.2.1.2 Multidoping Graphene

Beyond the above single heteroatom doping as mentioned, the recent reports have verified that the integration of multidoping with hierarchical structures is an effective method for performance improvement [32]. For example, dual-atom-doped graphene (N/B, N/S, N/P dopants) showed more excellent HER performance than the single-doped ones [3]. According to DFT calculations, the enhanced performance is originated from synergistic effects between the dopants. Xia and coworkers revealed an intrinsic descriptor that can precisely predict the OER/ORR electrocatalysis performance of dual-doped carbon materials. The OER and ORR performances are directly correlative to the descriptor in an inverted volcano shape (Figure 10.5) [33]. Tour and coworkers prepared B- and N-doped graphene that demonstrated excellent ORR activity due to synergism of N and B [25]. Multidoping may endow more functions to the graphene that are favorable to prepare multifunctional electrocatalysts. Furthermore, these predictions are demonstrated by experimental data. Qiao and coworkers synthesized high-efficient bifunctional electrocatalysts for ORR and OER by using PDA and graphene oxide to obtain the N/S co-doped graphene sheets [24]. Besides dual doping, triple-doping carbon

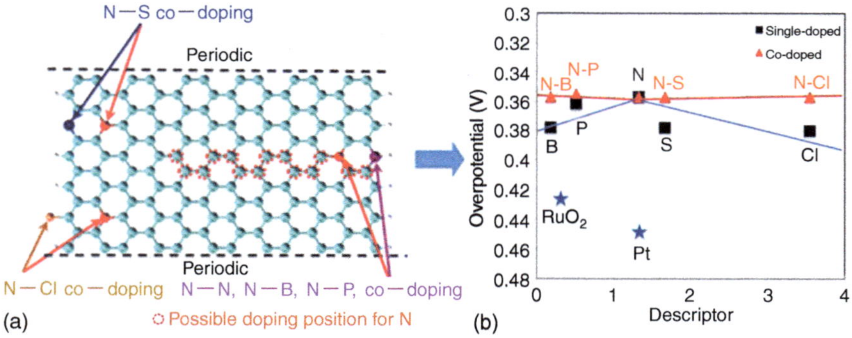

Figure 10.5 (a) Schematic of the N-X co-doped graphene nanoribbons (X = B, P, S, and Cl), showing the possible positions of dopants. (b) The lower limit of OER/ORR overpotentials for N-, and X-doped, and N-X co-doped graphene structures vs. descriptor Φ. Source: Zhao and Xia [33]. Reproduced with permission of American Chemical Society.

electrocatalyst was also prepared by Cao and coworkers [34]. The as-synthesized S, N, F co-doped carbon shows the same catalysis activity as Pt/C. Different configurations of the multi-dopants alter the spin and charge density, which is related to the catalytic properties [35].

In conclusion, a series of doped graphene has been successfully prepared. However, the accurate control of the doping concentrations and configurations within doped graphene is still a challenge and has been pursued intensively. After all, there are so many factors that influence the performance, such as the precursors, synthesis method, temperature, post-processing treatments, and type of catalyst used for synthesizing nanomaterial.

10.2.2 Edge and Defect Sites

As stated before, doping can lead to the internal charge redistribution in the graphene network and enhance the catalytic performance. The charge redistribution can enhance the ORR activity of the active sites, chemisorption, and bandgap. Apart from the heteroatom doping, the charge localization would be caused by the plane topological defects and the in-plane charge transfer. Furthermore, as is known to all, a certain number of defects or disorders are unavoidable in graphene according to the second law of thermodynamics [36]. There are various forms of defects in graphene, such as edge defects, Stone–Wales defects, pentagons or heptagons formations, vacancies, grain boundaries, and dislocation. Generally speaking, these defects preserve the graphene lattice without the dopants of heteroatoms that alter the charge balance. However, these structural changes adjust the atomic orbitals and can lead to the production of localized electronic states that promoted electrochemical activity. The importance of defects in many areas has been emphasized by many publications, but for electrocatalysis especially, it has been suggested that vacancies, high pentagon defects, curvature regions, and edges are commensurate in the electrocatalytic activity with hetero-doping of nanocarbons [37]. In Figure 10.6, we summarized some of the publications in this respect and depicted a universal defect classification of graphene materials.

To explore the contribution of intrinsic defects for the ORR activity, Jiang et al. reported the ORR performance of the defective carbon nanocages, which is comparable to that of the N-doped carbon nanostructures (Figure 10.7) [37]. DFT calculations and experiment results exhibit that zigzag edge defects and pentagon are in favor of the high catalytic activity, which is essential for understanding the influence of defects on ORR activity. Recently, Tang et al. also synthesized nitrogen-free graphene with some in-plane holes by direct annealing gelatinized sticky rice to understand the oxygen electrocatalytic mechanism. The theoretical and experimental results demonstrated that adjacent pentagons and heptagons play an important role to enhance the ORR and OER activity. Rybarczyk et al. reported that curved N-doped carbon synthesized with the assistance of $ZnCl_2$ and LiCl melt salts exhibited high curvature regions, which proposed that the curvatures of the graphitic domains improve the catalytic activity of these N sites [39].

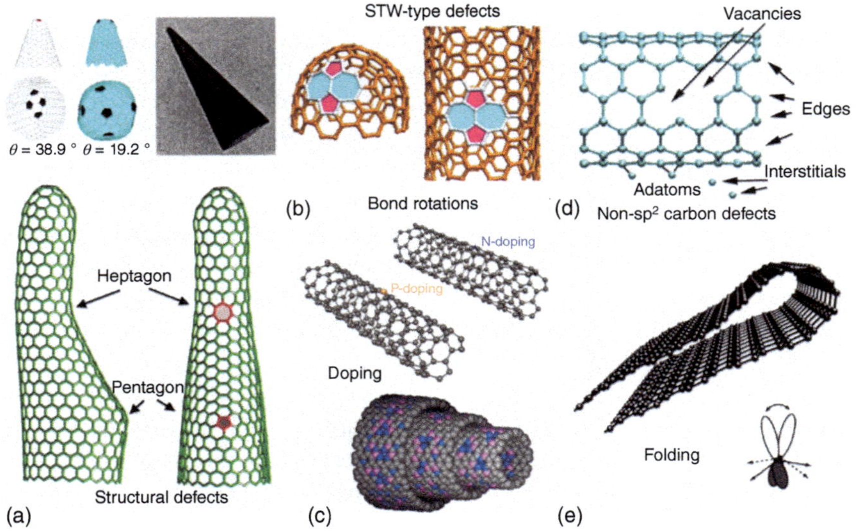

Figure 10.6 Defects in carbon nanostructures for catalysis. (a–e) Representation of different types of defects in nanocarbons. Source: Terrones et al. [38]. Reproduced with permission of Elsevier.

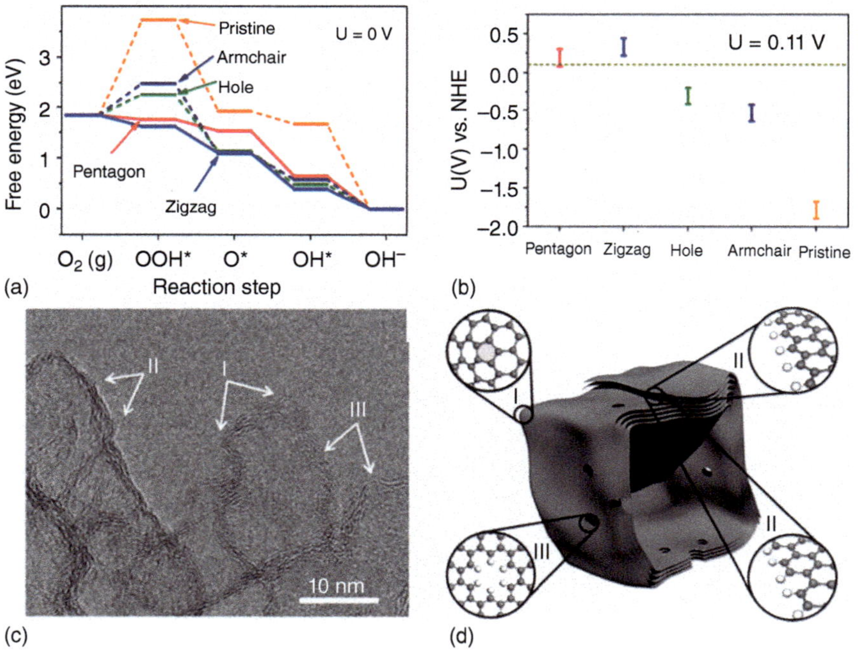

Figure 10.7 (a, b) The simulated free energy along the reaction step and the overpotential for each type of defect, respectively. (c) High resolution transmission electron microscope (HRTEM) image of carbon nanocages, where the characteristic features as folds and defects are highlighted and modeled in (d). Source: Jiang et al. [37]. Reproduced with permission of American Chemical Society.

Additional reports have also revealed and reviewed that defects would play the role of improving the performance of electrocatalytic reactions [27, 40]. Now, a popular strategy to prepare defective carbons is Zn evaporation. This new process can create the "pure" defective carbon structure without heteroatoms doping, which exhibited excellent ORR activity and confirmed role of defects [41]. Moreover, this strategy applies to other derived graphene materials, such as biomass and coal. Furthermore, defective graphene was synthesized by a simple N-doping and subsequent removal method, which creates various forms of defects for catalyses, such as ORR, HER, OER, CO_2RR, and NRR. In this way, Yao and coworkers fabricated defective activated carbon, which exhibits enhanced catalytic activity as a trifunctional catalyst toward the ORR, OER, and HER (Figure 10.3g) [42, 43]. For this sample, the defects are mainly located at the edge of the graphene without destruction of the graphene basal plane (Figure 10.8a–c). Figure 10.8d–f reveals that the defective carbon sites possess the lower energy barrier for electrocatalytic reactions. Commonly, the defective structures modify the distribution of the initial electron on the graphene network to introduce an electron localization region as the active site for the

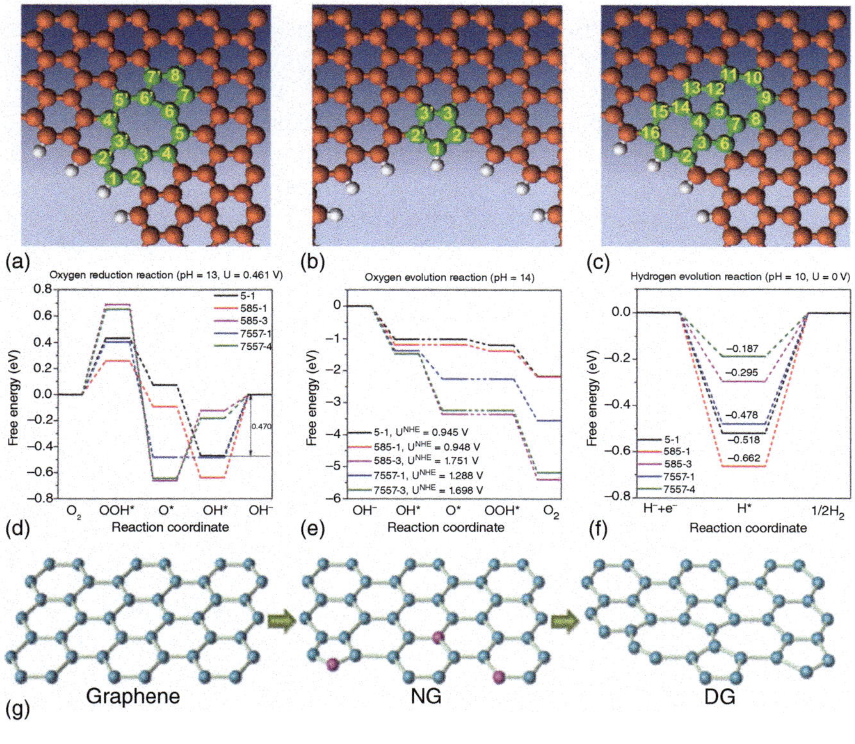

Figure 10.8 The schematics of different carbon lattice structures: (a) 5-8-5 defect; (b) Edge pentagon; (c) 7-55-7 defect. (d–f) Schematic energy profiles for the ORR pathway, the OER pathway, and the HER pathway on defective graphene in alkaline/acidic media. (g) The schematic of the formation of DG. Source: Jia et al. [44]. Reproduced with permission of Wiley-VCH.

electrocatalytic reactions. This mechanistic comprehending of the graphene-based defect-enhanced electrocatalytic reactions can be applied to all carbon structures.

Furthermore, several other ways have been explored and reported to treat graphene materials to create defect-rich graphene-based catalyst, such as Ar plasma etching technique. As is known, the plasma technique is usually used for etching and cleaning. Tao et al. produced defective graphite and carbon nanotubes by utilizing an Ar plasma technique, which obtained enhanced ORR activity when compared with the pristine counterparts [45]. This result confirms the significant role of edge defects in efficient ORR electrocatalysis. Similarly, Tian et al. utilized Ar plasma to treat N-doping graphene to develop a three-dimensional N-doped defective graphene as a highly efficient HER electrocatalyst. The plasma-induced abundant defects result in a lower overpotential compared with the sole N-doped counterpart, which indicates that the defects in graphene are efficient to improve electrocatalytic activity. Besides electrocatalysis reduction reaction, Wang and coworkers used Ar plasma treatment to in situ prepare edge-rich graphene from carbon fibers with the enhanced ORR/OER activities. Besides the virtues of carbon cloth, such as good intrinsic electron conductivity, high specific surface area, and excellent mass transport, the plasma etching endow carbon cloth with a raised amount of active sites for OER and ORR [46].

It is clear that a sizeable proportion of the reports on graphene-based electrocatalysis have been focusing on doping, with some changes in terms of doping sites and heteroatom species (co-doping) within the host graphene network. Moreover, the field associated with the above addressed intrinsic structural and topological defects, conduce to in-depth understanding of the mechanisms responsible for improved electrocatalytic performance. However, experimental and theoretical reports have illustrated that the effects of doping and intrinsic topological defects, and combining them together are undoubtedly exhibiting positive role on the electrocatalytic activities of graphene, making the graphene-based electrocatalyst a worthwhile field for further study and development.

10.2.3 Graphene as Supports

As mentioned above, heteroatom doping and defect engineering could cause the electron transfer to break the charge balance of graphene electrocatalysts to meliorate the electrocatalytic performance. In addition, composites with heterojunction structure may also occur to the electron transfer behavior to exhibit excellent activity. Owing to the different work functions of the individual component in the composite electrocatalyst, forming the heterojunction is an effective method to modulate the electron distribution to modify the electrocatalytic performance.

Inorganic nanomaterials suffer from poor electrons conductivity to hinder charge transport for electrocatalysis. Integrating the inorganic nanomaterials with conductive supports, such as graphene, carbon nanotube (CNT), and carbon materials, can solve this problem. Therefore, graphene has been widely used and is proved to be highly effective. The synthesis of inorganic/graphene hybrid materials represents another featured strategy to gain electrocatalysts with high stability and activity.

Carbon nitride (g-C_3N_4) has drawn much attention owing to its high N content, appealing chemical stability, and excellent electrocatalytic electronic structures. However, its low surface area and the inherent poor electronic mobility restrict it to electrocatalysis applications. Interestingly, with the composite structure of graphene and g-C_3N_4, the inhomogeneous planar g-C_3N_4 on the graphene could generate an electron–hole puddle, promoting hole-rich and electron-rich regions. This effect causes electron transfer between the g-C_3N_4 and graphene significantly, which will change the electron structure and modulate their catalytic activity. Qiao et al. fabricated a flexible three-dimensional HER electrode by integrating nitrogen-doped graphene with porous C_3N_4 nanosheets, which displayed an excellent HER performance [47]. Further study indicated that the performance improvement can be related to synergistic effect of abundance of exposed active sites in the in-plane pores of C_3N_4 synergistic, conductive graphene network, and coupling between graphene nanolayers and C_3N_4 sheets. Additionally, a g-C_3N_4@carbon ORR electrocatalyst was also prepared by incorporating g-C_3N_4 into mesoporous carbon. The coupling of them would facilitate charge transfer, increase the number of active sites, and decrease the reaction energy barrier for reducing the intermediates of OOH^-, which would lead to a more effective 4e$^-$ ORR pathway (Figure 10.9).

Transition metal with many merits such as low cost, rich reserve, and high catalytic performance made it promising alternative to replace the noble metals

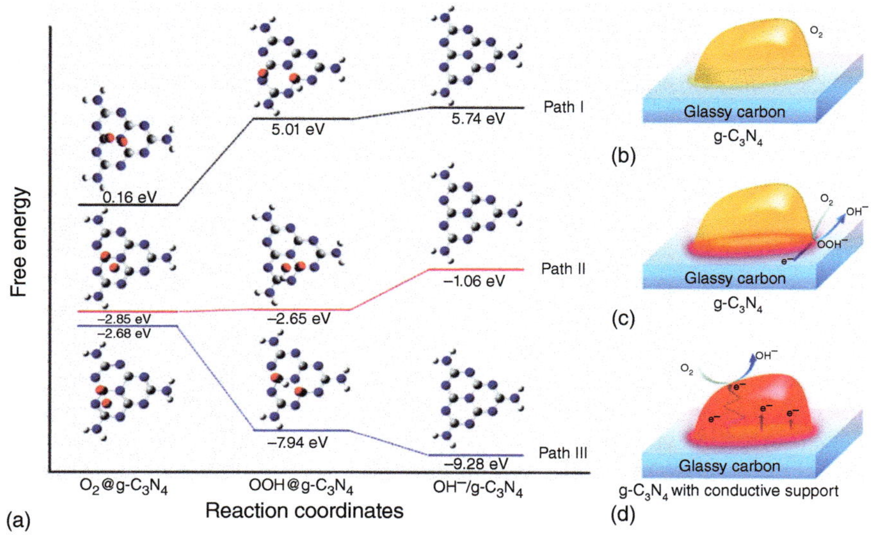

Figure 10.9 (a) Free energy plots of ORR and optimized configurations of the adsorbed species on zero-, two-, and four-electron participated g-C_3N_4 surface that are demonstrated as paths I, II, and III. (b–d) Schemes of ORR's pathway on pristine g-C_3N_4 without electron participation, pristine g-C_3N_4 with 2e− participation, and g-C_3N_4 and conductive support composite with 4e− participation, respectively (red areas represent the active sites facilitating ORR). Source: Zheng et al. [48]. Reproduced with permission of American Chemical Society.

catalysts. However, the nonprecious-metal catalysts confront poor stability in electrocatalytic reactions, especially under harsh situations, such as high potential and strong acidic/alkaline electrolyte. Encapsulating transition-metal catalysts in carbon materials could avoid corrosion of electrolytes and improve the stability during the electrocatalytic reactions. The interaction between the carbon materials and metal can adjust the electron structure of the composites and increase their electrocatalytic activity. Deng et al. reported a hierarchical structure that consists of a uniform CoNi alloy nanoparticle wrapped up by ultrathin graphene shells (1–3 layers) [49]. The resulted catalyst exhibits enhanced HER activity, which is comparable to that of commercial Pt/C. DFT results indicate that electron flow from the CoNi alloy to the surface of ultrathin N-doped graphene shells, which enhanced HER activity of the graphene shells. Moreover, increasing the N content and reducing the number of graphene layers can significantly promote the HER activity.

For metal compounds/graphene hybrid materials, graphene not only acts as the support for immobilizing but also plays the role of contributor to modify the electronic structure. The electron transfer between graphene and metal compounds would lead to electronic reconfiguration, and this would produce active sites at their interfaces. Moreover, the contact of metal compounds and graphene material would lead to the strain variation of the metal compounds, which is an efficient avenue to prepare high-activity electrocatalysts [50]. Charge transfer between metal species and carbon supports can adjust the Fermi energy level, which would influence their charge distribution. [51] Doping sites and defect sites supply nucleation sites for metal species with strong interaction, which will change the electron distribution, boost the activity, and enhance mass transfer and stability. As reported in previous work, Cu-doped Co_2P anchored on the N, P-doped graphene exhibits enhanced ORR performance. The N, P dual-doped graphene not only provides anchor sites for Co_2P nanoparticle but also increase electronic conductivity and active sites [52]. Xia et al. prepared Pt–TiO_2–rGO 3-phase junctions by synthesizing TiO_2 decorated graphene and subsequence depositing Pt nanoparticles on the interface of rGO and TiO_2. The obtained electrocatalyst exhibits enhanced catalytic activity owing to the distinctive structure and synergetic effect of the triple junctions [53]. Yang et al. developed an innovative method to synthesize $NiCo_2O_4$/N-doped graphene hybrids by low-intensity pulsed-laser irradiation. During which the relative contents of pyrrolic N and pyridinic N controllably and selectively regulate in N-graphene hybrids. According to experimental results and DFT calculation, they proposed that the pyridinic N—Co bonds at the composite interface play an important role to improve electrocatalysis effect, which promotes reversible oxygen electrocatalysis reactions [54].

Besides, transition metal sulfides composited with graphene also exhibit enhanced electrocatalysis activity. Dryfe and coworkers synthesized a series of dual-decorated graphene-based composites, including layered MoS_2 on the top side and metal nanoparticles on the bottom side. The obtained MoS_2/graphene/metal sandwich composites show an ideal platform for energy conversion and storage, such as photoelectrochemical hydrogen production and lithium intercalation [55]. Cobalt sulfides are common electrocatalysts in the transition metal chalcogenides. Dai and coworkers synthesized $Co_{1-x}S$/graphene hybrid by two-step hydrothermal method [56]. The $Co_{1-x}S$ particles with an average particle diameter of 10–20 nm were grown on the surface of graphene nanosheets. The rGO nanosheets coupling with the $Co_{1-x}S$ nanoparticles can provide an electron conductive path for the catalyst, control the size of the $Co_{1-x}S$ particles, and thus increase the electrocatalytic activity of the $Co_{1-x}S$.

10.2.4 Template Method Synthesis of Graphene-Based Electrocatalysts

Electrocatalysis reactions take place at the triple-phase interfacial zone of the gaseous products or reactants, liquid electrolyte, and solid catalysts. As a result, mass diffusion of the electrolyte on the electrocatalyst surface is crucial. Some special preparation strategies are required to create porous structures. Template method is one of the most commonly used methods, including hard templating and soft templating methods.

Hard templating method is the most widely used strategy to produce porous electrocatalyst, which is simple, straightforward, and effective in concept. In general, the synthesis procedure involves three elementary steps – (i) template synthesis, (ii) target material coating, and (iii) template removal. Coating the target material on the hard template is usually considered as the most pivotal step. The most utilized hard templates include oxide colloids, metal, and metal salts, which have the characteristics of uniform size, shape controllable, and ready availability in large amounts.

SiO_2 sphere is one of the most common hard templates, which is widely used as template for synthesizing porous graphene-based electrocatalysts. For example, Li et al. prepared N and S co-doped hierarchically porous graphene by using SiO_2 as templates, which was used for CO_2 reduction reaction and ORR [57]. The hierarchically porous structure largely facilitates electrolyte transport and exposes many catalysis sites to reaction species, thereby resulting in significantly increased catalytic activities. Moreover, MgO is also used as hard template to prepare graphene-based electrocatalysts [58]. Tang et al. synthesized defect-rich hierarchical graphene under the assistance of MgO templates. MgO templates and Co nanoparticles induced micro/mesosized nanopores, which facilitated the mass transfer during triple-phase reactions [59]. Although oxide colloid templates are widely used, template removal is always tedious, and strong acid or alkali are often used.

Metal template is also employed to prepare graphene-based electrocatalysts. Compared with other templates, metal template can catalyze graphene with higher quality. Especially, nanoporous metal has been an ideal template because of its high specific surface area, adjustable pore structure, and good conductivity. Chen and coworkers synthesized N and S co-doped nanoporous graphene by using nanoporous metal as template. The N and S co-doped 3D nanoporous graphene exhibits excellent HER activity [60]. Moreover, pyrolysis of organic carbon sources with metal powdery templates synthesis 3D graphene, which can be used as electrocatalyst. The acid etching is essential to remove the templates, which increases cost and causes the aqueous processing problem again.

Metal salts, with the advantages of solubility, large-scale production, and low cost, can tackle the above problems. As a result, they are always used as templates to synthesize graphene. Huang and coworkers fabricated monodispersed, densely populated cobalt atoms on nitrogen-doped graphene supported by a KCl salt-template method, and demonstrated excellent electrochemical activity for both the OER and ORR [61]. KCl plays an important role in preventing graphene shrinking and Co atoms aggregation. Na_2SO_4 is another salt template for producing graphene-based catalysts, and the in situ generated H_2S can result in the sulfur doping of graphene and the formation of metal sulfides [62]. Furthermore, Na_2SiO_3 is also usually used as template to create micropores in the graphene [63]. Compared with other templates, salt templates are easier to remove by washing.

Additionally, graphene can also be used as self-template to prepare doped graphene. Zhang et al. synthesized nitrogen- and phosphorus-doped carbon by pyrolysis of an aggregate of self-assembled phytic acid, melamine, and graphene oxide. Graphene plays the role of self-template to prepare doped carbon. The obtained catalyst exhibits excellent hydrogen evolution and oxygen reduction activity [64].

Soft templating method is another common strategy to produce porous electrocatalyst. Amphiphilic long-chain molecules, such as block copolymers and surfactants, can self-assemble into vesicles or micelles with different structures when their concentration in solution achieves the critical micelle concentration. Such micelles and vesicles play the role of soft templates directing the production of pore structures. The soft template can be easily removed by pyrolysis. As a result, template removal step can be omitted. The common soft templates include F127, polymethyl methacrylate, and polystyrene, etc.

10.3 Application of Graphene-Based Electrocatalysts

Energy conversion can be performed through electrochemical processes, especially those involving water splitting (include HER and OER), ORR, CO_2RR, and NRR. The reactions and electrocatalysts that used graphene-based materials are briefly summarized below.

10.3.1 Graphene-Based Electrocatalysts for Water Splitting

The water-splitting reaction includes HER and OER, occurring at the cathode and anode, respectively. The Gibbs free energy (ΔG) of the reaction is 237.2 kJ mol^{-1}, corresponding to a voltage contribution of 1.229 V based on the Nernst equation. In an acidic media, the HER occurs through the reduction of two protons, releasing a hydrogen molecule at the cathode (Eq. (10.2)), and the OER manifested in oxygen molecule by the oxidation of water molecules at the anode (Eq. (10.3)):

$$H_2O \rightarrow 1/2O_2 + H_2 \text{ (overall equation)} \tag{10.1}$$

$$2H + 2e^- \rightarrow H_2 \text{ (acidic HER)} \tag{10.2}$$

$$H_2O \rightarrow 1/2O_2 + 2H^+ + 2e^- \text{ (acidic OER)} \tag{10.3}$$

Despite the advantages of water splitting in acidic environment such as high hydrogen production rate, proton-exchange membrane that avoids mixing of the products, restricts the industrial application of the system. Therefore, alkaline water splitting has been proposed owing to its innate low cost resulting from the application of nonprecious metal catalysts. In an alkaline media, H_2O was reduced to H_2 (Eq. (10.4)), while OH^- was oxidized to O_2 (Eq. (10.5)):

$$2H_2O + 2e^- \rightarrow H_2 + 2OH^- \text{ (alkaline HER)} \tag{10.4}$$

$$4OH^- \rightarrow 2H_2O + O_2 + 4e^- \text{ (alkaline OER)} \tag{10.5}$$

Generally, studies of HER electrocatalysts focus mainly on acidic systems, while OER is alkaline system due to its high activities.

The HER is a multistep electrochemical reaction including H^+ adsorption, reduction, and H_2 desorption processes. The elementary steps of the HER in alkaline and acidic electrolytes are listed in Table 10.1. At first, hydrogen atoms from either a dissociated water molecule or hydronium ion (in acidic conditions) are adsorbed on the active site. Then, the adsorbed hydrogen atoms (H*) are reduced to form H_2 via either a Tafel or Heyrovsky step. Finally, the H_2 molecule desorbs from the active site. In the real electrocatalytic process, these pathways strongly rely upon the intrinsic chemical and electronic structure of the catalyst surface.

The overall HER reaction activity on the surface of electrocatalyst is largely dependent on the theoretical activity descriptor ΔG_{H^*}. If hydrogen adsorption is too weak ($\Delta G_{H^*} > 0$), the rate determining step is the Volmer step, and if hydrogen adsorption is too strong ($\Delta G_{H^*} < 0$), the rate determination step will be the desorption step (either the Tafel or Heyrovsky step). According to Sabatier principle, both will likely lead to poor HER activity. A volcano plot can be used to describe the HER trend of catalysts and prescribe parameters for the perfect HER catalyst. As shown in Figure 10.10, the perfect catalyst located at the top of the volcano plot, whose ΔG_{H^*} is 0 eV.

Recently, Chen et al. developed N and S co-doped porous graphene incorporated with high conductive graphene [66]. Heavy chemical doping modulates the electronic structure of graphene but reduces its electrical conductivity. However, the conductivity was remedied by coupling with a highly conductive pure graphene substrate. The combination of N, S doping, and high conductivity endowed hierarchical nanoporous graphene with high catalytic activity toward HER.

Table 10.1 HER mechanism in acid and alkaline electrolytes.[a)]

	Acid	**Alkaline**	**Tafel slope**
Overall	$* + 2H^+ + 2e^- \rightarrow H_2$	$* + 2H_2O + 2e^- \rightarrow H_2 + 2OH^-$	—
Volmer	$* + H + e^- \rightarrow H^*$	$* + H_2O + e^- \rightarrow H^* + OH^-$	120 mV dec^{-1}
Heyrovsky	$H^+ + e^- + H^* \rightarrow H_2 + *$	$H_2O + e^- + H^* \rightarrow H_2 + OH^- + *$	40 mV dec^{-1}
Tafel	$2H^* \rightarrow H_2 + 2*$	$2H^* \rightarrow H_2 + 2*$	30 mV dec^{-1}

a) Means an active site on the surface of an electrocatalyst.

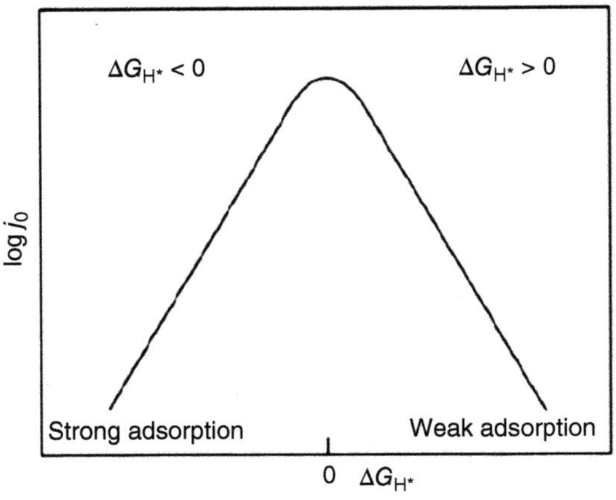

Figure 10.10 Schematic representation of a HER volcano plot. Source: Zheng et al. [65]. Reproduced with permission of Wiley.

The OER is a $4e^-$ transfer reaction which is reverse reaction with the ORR. It is an important reaction not only for water splitting but also for metal–air and fuel batteries. Many studies have shown that various graphene-based materials can efficiently catalyze the OER and have obtained significant development in recent years. Wang and coworkers synthesized edge selectively doped graphene with phosphorus atom through ball-milling graphite and red phosphorus, which exhibited competitive OER performance [67]. Although their individual roles remained unclear, the combination of P doping and defects resulted in enhanced OER performance. Moreover, Liu et al. reported N-doped three-dimensional graphene nanoribbon networks, which exhibited excellent bifunctional catalytic activities for both OER and ORR [68]. It is demonstrated that pyridinic N species are responsible for OER, while quaternary N sites serve as active sites for ORR for the first time in experiment. Besides, the special 3D nanostructure provide amount of OER active sites and promote the electron and electrolyte transports.

10.3.2 Graphene-Based Electrocatalysts for ORR

Besides the catalysis of the HER and OER, graphene-based catalysts are also applied for the ORR. ORR is among the most studied fields in electrocatalysis owing to its significance for electrochemical energy conversion applications. In the proton-exchange membrane fuel cell, the ORR exhibits relative sluggish kinetics compared with the anodic hydrogen oxidation reaction (HOR), which kinetically limits the overall device performance. Therefore, exploring a highly efficient and low-cost ORR catalyst is urgently needed to promote fuel cell technology development. Graphene and its derivatives show great application potential in the field of ORR electrocatalyst.

There are two possible processes for ORR in alkaline aqueous media, either direct 4-electron pathway (Eqs. (10.6)) or the 2-electron pathway (Eqs. (10.7) and (10.8)):

$$O_2 + 2H_2O + 4e^- \rightarrow 4OH^- \text{ (4 electron)} \tag{10.6}$$

$$O_2 + H_2O + 2e^- \rightarrow HO_2^- + OH^- \text{ (2 electron)} \tag{10.7}$$

$$HO_2^- + H_2O + 2e^- \rightarrow 3OH^- \text{ (2 electron)} \tag{10.8}$$

There are also two possible processes for ORR in acid aqueous media, either direct 4-electron pathway (Eqs. (10.7)) or the 2-electron pathway (Eqs. (10.10) and (10.11)):

$$O_2 + 4H^+ + 4e^- \rightarrow 2H_2O\,(4e^-) \tag{10.9}$$

$$O_2 + 2H^+ + 2e^- \rightarrow H_2O_2\,(2e^-) \tag{10.10}$$

$$H_2O_2 + 2H^+ + 2e^- \rightarrow 2H_2O\,(2e^-) \tag{10.11}$$

Between these two pathways, the direct 4-electron route as the most effective pathway is desirable for fuel cell and zinc–air battery applications. Due to the complex nature of ORR including multiple steps and intermediates, the ORR mechanism remains equivocal despite the abundance of studies conducted toward its clarification. According to Nørskov et al., the inherent principle behind the ORR catalytic activities of catalysts lies in their binding energy of oxygen intermediates bonding with the active sites [69]. Catalysts with too strong bonding energy result in slow desorption of products, while those with weaker bonding energy are faced with high energy barriers to electron transfer to oxygen and then the invalid dissociation of oxygen, leading to poor ORR catalytic performance. DFT calculation study exhibits that the carbon surfaces exhibit a high dissociation barrier, which is not conducive to $4e^-$ pathway [70]. Thus, the electron transfer number measured by experiment in almost all graphene materials is always between four and two.

Pure graphene exhibits poor activity for electrocatalytic reactions because of the absence of active sites. Therefore, various ways are widely applied to induce electronic transfer in the graphene frameworks and enhance catalytic activity, such as heteroatom doping, defects effect, and composite with other materials. Generally, the electrocatalysis activity originates from the special electron structure for reactant adsorption and various nanostructures that raise the amount of exposed catalytic active sites. For example, Dai et al. first used N-doped graphene as ORR electrocatalyst and showed superior activity to Pt/C through $4e^-$ pathway (Figure 10.11) [4]. Moreover, the excellent tolerance to methanol and CO largely prevent the crossover effect. Owing to its special chemical structure, N-doped graphene demonstrates excellent stability in alkaline electrolyte.

Some defects are unavoidably introduced into the graphene matrix during the synthesis processes or by posttreatment [36]. In general, these defects are regarded to be beneficial for electrocatalytic reactions. However, owing to the defects in graphene are always coexist with edge regions and heteroatom doping, distinguishing the real active sites is always complex in these materials. For ORR applications, in-plane topological defects (i.e. pentagons and heptagons) are the most widely studied cases. Jiang et al. revealed that the dopant-free carbon nanocages with extensive edges, holes, and positive topological disclinations exhibited enhanced ORR performance compared to doped counterparts [37]. A DFT calculation showed that both topological defects and edge effects have a greater influence on ORR, instead of on doping [71].

As mentioned above, accurately controlling intrinsic defects is difficult, because they always coexist with the edges of graphene and heteroatom dopants. From the point of electron structure, all of them are proved to have changed the spin/charge distribution in the graphene matrix and binding energy of the intermediates. More

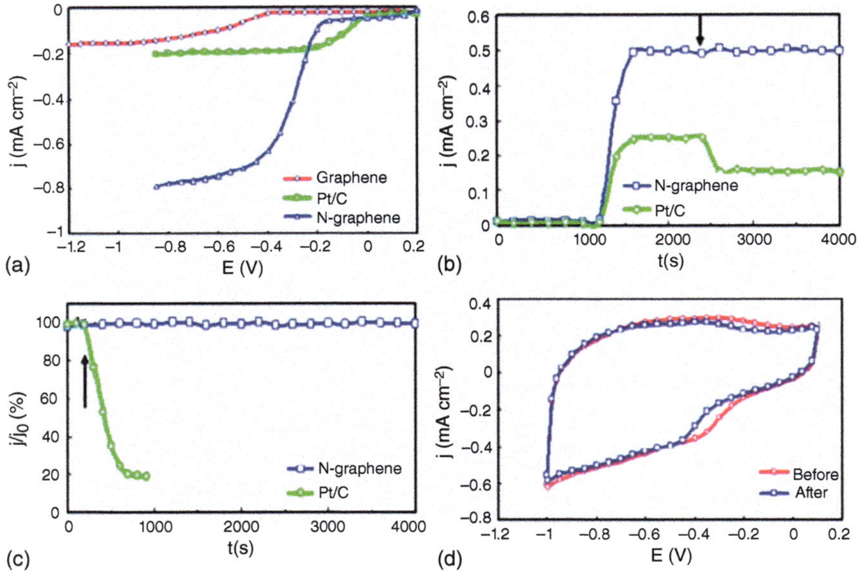

Figure 10.11 (a) Linear sweep voltammetry (LSV) for the ORR for graphene, N-graphene, and Pt/C electrocatalysts. (b, c) Chronoamperometric responses for N-graphene and Pt/C electrocatalysts (b) after adding 2% (w/w) methanol into the electrolytes (shown by the arrow) and (c) after adding 10% (v/v) CO into the electrolytes (shown by the arrow). (d) Cyclic voltammograms of N-graphene before and after a continuous potentiodynamic sweeping for 200 000 cycles Source: Qu et al. [4]. Reproduced with permission of American Chemical Society.

recently, extensive studies have been devoted to developing efficient ORR electrocatalysts in the whole pH range (acidic, neutral, and alkaline media), which are useful for various energy conversion technologies, including proton-exchange membrane fuel cells, microbial fuel cells, alkaline fuel cells, and metal–air batteries. Moreover, large-scale production of graphene-based electrocatalysts is still challenging, which deserves serious attention.

10.3.3 Graphene-Based Electrocatalysts for CO_2RR

Electrochemical CO_2 reduction (ECR) provides an important way for renewable energy conversion, fuels production, and mitigate CO_2 emissions. Recently, electrocatalytic reduction of CO_2 has received considerable attention. However, its slow kinetics and the poor selectivity to control the target products of the reaction limit its industrial applications. Despite the above difficulties, the field motivates researchers for the following reasons – (i) reducing the greenhouse influence by reducing the concentration of CO_2 in the atmosphere; (ii) generating fuels from CO_2; and (iii) transforming CO_2 into value-added chemicals. CO_2RR is a proton-coupled multielectron transfer reaction. Depending on the electrocatalysts, the transferred electron number differs from 2 to 18 leading to different end products (Table 10.2).

Table 10.2 Reaction and equilibrium potentials (E^0) of ECR pathways.

Reaction	E^0 (V vs. RHE)
$2H^+ + 2e^- \rightarrow H_2$	0
$CO_2 + 2H^+ + 2e^- \rightarrow CO + H_2O$	−0.11
$CO_2 + 2H^+ + 2e^- \rightarrow HCOOH$	−0.21
$CO_2 + 4H^+ + 4e^- \rightarrow HCHO + H_2O$	−0.10
$CO_2 + 6H^+ + 6e^- \rightarrow CH_3OH + H_2O$	0.03
$CO_2 + 8H^+ + 8e^- \rightarrow CH_4 + 2H_2O$	0.17
$2CO_2 + 8H^+ + 8e^- \rightarrow CH_3COOH + 2H_2O$	−0.26
$2CO_2 + 12H^+ + 12e^- \rightarrow C_2H_4 + 4H_2O$	0.07
$2CO_2 + 12H^+ + 12e^- \rightarrow C_2H_5OH + 3H_2O$	0.09
$3CO_2 + 18H^+ + 18e^- \rightarrow 1\text{-}C_3H_7OH + 5H_2O$	0.09

Electrochemical CO_2RR process with possible products is shown in Figure 10.12a [72]. Recently, extensive research vigor has been devoted to develop graphene-based catalysts for the CO_2RR and to control the reaction pathways for target products. By virtue of their excellent electrical conductivity, extremely high surface area, environmental friendliness, cost-effectiveness, and the adjustable doping of heteroatoms (e.g. B, N, etc.), heteroatom-doped graphene has been applied for ECR. Phani and coworkers prepared boron-doped graphene (B-graphene) that showed the electrocatalytic activity for CO_2RR in 0.1 m $KHCO_3$ aqueous solution [74]. CO_2 was reduced to formate as the product at low overpotentials during the electrolysis. DFT calculation result showed that boron-doping introduced asymmetric spin density, making it beneficial for CO_2 adsorbing on B doped G, and thereby undergoing reduction to formate. Besides B-doping, N-doping has also been demonstrated as an efficient means for promoting CO_2 reduction to formate and CO in aqueous electrolytes. For instance, Ajayan and coworkers reported N-doped three-dimensional graphene foam as CO_2RR catalyst with a faradaic efficiency for CO production up to 85% at a low overpotential of −0.47 V [75]. The dependence of catalytic activity on N-defect structures is unraveled by systematic experimental investigations. Indeed, the DFT calculations confirm pyridinic N as the most active site for CO_2 reduction, consistent with experimental results.

Although the doped graphene has been developed to produce formate or CO with low overpotentials and promising selectivity, the conversion of CO_2 into higher-order hydrocarbons (e.g. C_2H_4, C_2H_5OH, and 1-C_3H_7OH) on graphene-based electrocatalysts still remains challenging. Fortunately, changing the topological graphene nanostructure has been demonstrated as an effective way to design an electrocatalyst that could convert CO_2 into high-order products. Ajayan and coworkers first reported the N-doped graphene quantum dots as a catalyst for CO_2 reduction to multi-carbon hydrocarbons and oxygenate at low overpotentials, high current densities, and high faradaic efficiencies (up to 90% at

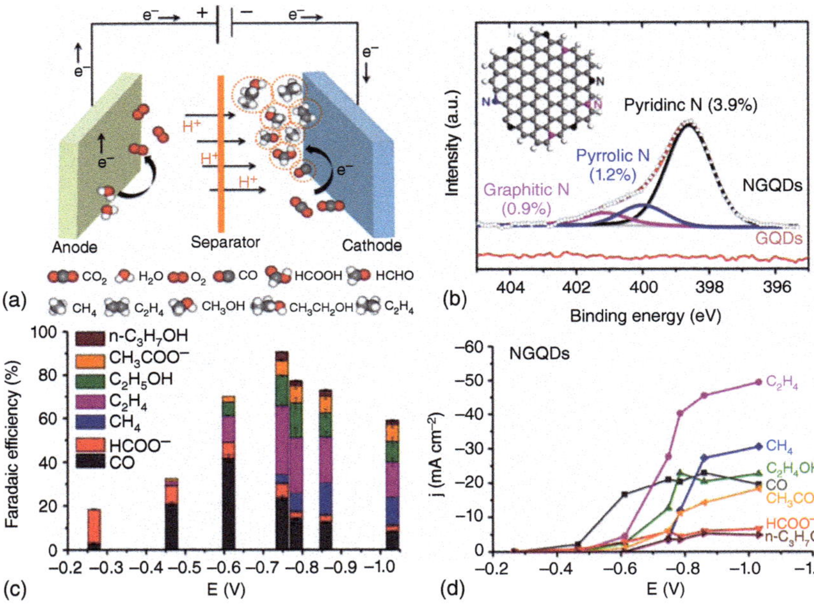

Figure 10.12 (a) Illustration of the electrochemical CO_2 reduction process and the possible products generated in an electrochemical reaction cell. Source: Zhang et al. [72]. Reproduced with permission of Wiley-VCH. (b) High-resolution N 1s spectrum of NGQDs and corresponding N atomic concentration. Inset is a schematic of specific N-bonding configuration. (c) A distribution of products for NGQDs as a function of cathode potential. (d) Partial current densities for different products. Source: Wu et al. [73]. Reproduced with permission of Springer Nature.

−0.74 V vs. RHE). Besides, the selectivity for C_2 and C_3 products reached up to 55% at −0.74 V vs. RHE. Detailed characterization reveals that the high performance attributes to the exposure of edge sites and pyridinic N dopant [73].

Although graphene-based materials have experienced considerable advancement in the field of CO_2RR electrocatalysis, it is still in their infancy. Its intrinsic feature of the active site and the reaction mechanism for heteroatom doped graphene catalysts are still unclear. To obtain a more comprehensive understanding of CO_2 reduction on graphene-based CO_2RR electrocatalysts, combining experiment and theory is necessary to obtain better designing principles for developing graphene-based CO_2RR catalysts with high product selectivity and catalytic activity.

10.3.4 Graphene-Based Electrocatalysts for NRR

Inspired by applications of graphene-based catalysts for the ORR and CO_2RR, the study of the graphene-based catalyst in electrocatalytic NRR has attracted great attention in recent years. However, owing to the differences in NRR, many critical problems need to be solved to realize the rational design of advanced graphene-based electrocatalysts to improve selectivity, activity, and stability for NRR. Heteroatoms

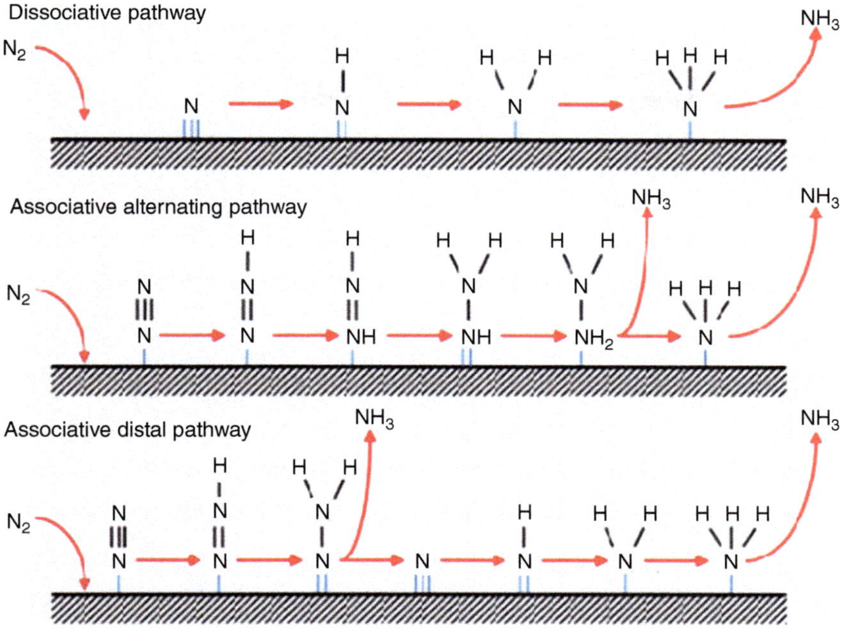

Figure 10.13 Schemes of the NRR via different pathways. Source: Shipman and Symes [76]. Reproduced with permission of Elsevier.

doping like N, B, and P could effectively change the energy bandgap, charge density, and spin density, and modulate the competition between HER and NRR through regulating charge-transfer interactions between the heteroatoms and surrounding carbon atoms.

NRR mechanisms can be classified into associative and dissociative mechanisms (Figure 10.13) [76]. In the associative mechanism, the two nitrogen atoms in N_2 keep bound to each other during the hydrogenation of N_2 molecule, with NH_3 being released when the N—N bond is broken. In the distal associative mechanism, the hydrogenation process could occur in two ways. Hydrogenation might prefer on the nitrogen far away from the catalyst surface, resulting in the release of NH_3. Then, the hydrogenation process occurs on the other N to produce the second NH_3. While in the alternating associative mechanism, the two nitrogen atoms undergo single hydrogenation process in turn, until the distal N is converted into NH_3 and the N—N bond is broken. However, in the dissociative mechanism, hydrogenation takes place after the N≡N bond cleaving, and the individual N atoms on the surface are converted into NH_3 respectively.

The B has a lower electronegativity of 2.04 than that of carbon (2.55), which can create Lewis acid catalytic sites to provide an empty orbital for binding N_2 [77, 78]. Recently, Zheng and coworkers synthesized B-doped graphene with different boron configurations, which enhanced the adsorption of N_2 [79]. B-doped graphene with a 6.2% doping level obtained an NH_3 production rate of $9.8\,\mu g\,h^{-1}\,cm^{-2}$ and high faradaic efficiency (10.8% at 0.5 V vs. RHE).

Although great progress has been made by controlling dopants, introducing surface defects, and constructing multidimensional architectures, deep understanding of the electrocatalytic mechanism to further enhance the activity of NRR catalysts is still very necessary for practical applications.

10.4 Outlook

Due to their large surface area, low density, suitable mechanical properties, and outstanding electrical conductivities, heteroatom-doped graphene-based materials have been promising catalysts not only for effective energy conversion but also for chemical productions and environmental protection. Recently, the availabilities of doped graphene-based multifunctional electrocatalysts with catalytic activity similar to or even higher than those of precious-metal catalysts open up a new avenue toward commercialization of energy conversion technologies from efficient and affordable graphene-based materials. Many graphene-based catalysts have already shown comparable or even superior HER, ORR, OER, CO_2RR, and NRR activities to precious-metal electrocatalysts' overall pH ranges. Their performances can be further enhanced through rational design of the physic-chemical structures on the carbon lattice and at the edges and the macroscopic structures. However, more efforts are needed to utilize graphene-based catalysts in integrated energy storage and conversion devices, chemical production processes, and environmental protection systems of practical significance.

References

1. Tang, C., Wang, H.F., and Zhang, Q. (2018). Multiscale principles to boost reactivity in gas-involving energy electrocatalysis. *Accounts of Chemical Research* 51 (4): 881–889.
2. Dai, L., Xue, Y., Qu, L. et al. (2015). Metal-free catalysts for oxygen reduction reaction. *Chemical Reviews* 115 (11): 4823–4892.
3. Jiao, Y., Zheng, Y., Davey, K. et al. (2016). Activity origin and catalyst design principles for electrocatalytic hydrogen evolution on heteroatom-doped graphene. *Nature Energy* 1 (10): 16130.
4. Qu, L., Liu, Y., Baek, J.B. et al. (2010). Nitrogen-doped graphene as efficient metal-free electrocatalyst for oxygen reduction in fuel cells. *ACS Nano* 4 (3): 1321–1326.
5. Jeon, I.Y., Zhang, S., Zhang, L. et al. (2013). Edge-selectively sulfurized graphene nanoplatelets as efficient metal-free electrocatalysts for oxygen reduction reaction: the electron spin effect. *Advanced Materials* 25 (42): 6138–6145.
6. Sheng, Z.-H., Gao, H.-L., Bao, W.-J. et al. (2012). Synthesis of boron doped graphene for oxygen reduction reaction in fuel cells. *Journal of Materials Chemistry* 22 (2): 390–395.

7 Yao, Z., Nie, H., Yang, Z. et al. (2012). Catalyst-free synthesis of iodine-doped graphene via a facile thermal annealing process and its use for electrocatalytic oxygen reduction in an alkaline medium. *Chemical Communications (Camb)* 48 (7): 1027–1029.
8 Liu, Z.W., Peng, F., Wang, H.J. et al. (2011). Phosphorus-doped graphite layers with high electrocatalytic activity for the O_2 reduction in an alkaline medium. *Angewandte Chemie International Edition* 50 (14): 3257–3261.
9 Jeon, I.Y., Choi, H.J., Choi, M. et al. (2013). Facile, scalable synthesis of edge-halogenated graphene nanoplatelets as efficient metal-free eletrocatalysts for oxygen reduction reaction. *Scientific Reports* 3: 1810.
10 Yang, L., Shui, J., Du, L. et al. (2019). Carbon-based metal-free ORR electrocatalysts for fuel cells: past, present, and future. *Advanced Materials* 31 (13): e1804799.
11 Zehtab Yazdi, A., Roberts, E.P.L., and Sundararaj, U. (2016). Nitrogen/sulfur co-doped helical graphene nanoribbons for efficient oxygen reduction in alkaline and acidic electrolytes. *Carbon* 100: 99–108.
12 Li, R., Wei, Z., and Gou, X. (2015). Nitrogen and phosphorus dual-doped graphene/carbon nanosheets as bifunctional electrocatalysts for oxygen reduction and evolution. *ACS Catalysis* 5 (7): 4133–4142.
13 Matter, P., Zhang, L., and Ozkan, U. (2006). The role of nanostructure in nitrogen-containing carbon catalysts for the oxygen reduction reaction. *Journal of Catalysis* 239 (1): 83–96.
14 Hou, Z., Wang, X., Ikeda, T. et al. (2013). Electronic structure of N-doped graphene with native point defects. *Physical Review B* 87 (16): 165401.
15 Lai, L., Potts, J.R., Zhan, D. et al. (2012). Exploration of the active center structure of nitrogen-doped graphene-based catalysts for oxygen reduction reaction. *Energy & Environmental Science* 5 (7): 7936.
16 Liu, J., Song, P., and Xu, W. (2017). Structure-activity relationship of doped-nitrogen (N)-based metal-free active sites on carbon for oxygen reduction reaction. *Carbon* 115: 763–772.
17 Guo, D., Shibuya, R., Akiba, C. et al. (2016). Active sites of nitrogen-doped carbon materials for oxygen reduction reaction clarified using model catalysts. *Science* 351 (6271): 361–365.
18 Sun, M., Wu, X., Xie, Z. et al. (2017). Tailoring platelet carbon nanofibers for high-purity pyridinic-N doping: a novel method for synthesizing oxygen reduction reaction catalysts. *Carbon* 125: 401–408.
19 Okada, T., Inoue, K.Y., Kalita, G. et al. (2016). Bonding state and defects of nitrogen-doped graphene in oxygen reduction reaction. *Chemical Physics Letters* 665: 117–120.
20 Zhang, X., Lu, Z., Fu, Z. et al. (2015). The mechanisms of oxygen reduction reaction on phosphorus doped graphene: a first-principles study. *Journal of Power Sources* 276: 222–229.
21 Yang, S., Zhi, L., Tang, K. et al. (2012). Efficient synthesis of heteroatom (N or S)-doped graphene based on ultrathin graphene oxide-porous silica

sheets for oxygen reduction reactions. *Advanced Functional Materials* 22 (17): 3634–3640.

22 El-Sawy, A.M., Mosa, I.M., Su, D. et al. (2016). Controlling the active sites of sulfur-doped carbon nanotube-graphene nanolobes for highly efficient oxygen evolution and reduction catalysis. *Advanced Energy Materials* 6 (5): 1501966.

23 Jintao, Z., Liangti, Q., Gaoquan, S. et al. (2016). N,P-codoped carbon networks as efficient metal-free bifunctional catalysts for oxygen reduction and hydrogen evolution reactions. *Angewandte Chemie International Edition* 55: 2230–2234.

24 Qu, K., Zheng, Y., Dai, S. et al. (2016). Graphene oxide-polydopamine derived N,S-codoped carbon nanosheets as superior bifunctional electrocatalysts for oxygen reduction and evolution. *Nano Energy* 19: 373–381.

25 Fei, H., Ye, R., Ye, G. et al. (2014). Boron- and nitrogen-doped graphene quantum dots/graphene hybrid nanoplatelets as efficient electrocatalysts for oxygen reduction. *ACS Nano* 8 (10): 10837–10843.

26 Shervedani, R.K. and Amini, A. (2015). Sulfur-doped graphene as a catalyst support: Influences of carbon black and ruthenium nanoparticles on the hydrogen evolution reaction performance. *Carbon* 93: 762–773.

27 Tao, L., Wang, Y., Zou, Y. et al. (2019). Charge transfer modulated activity of carbon-based electrocatalysts. *Advanced Energy Materials* 10 (11): 1901227.

28 Deng, D., Pan, X., Yu, L. et al. (2011). Toward N-doped graphene via solvothermal synthesis. *Chemistry of Materials* 23 (5): 1188–1193.

29 Jiao, Y., Zheng, Y., Jaroniec, M. et al. (2014). Origin of the electrocatalytic oxygen reduction activity of graphene-based catalysts: a roadmap to achieve the best performance. *Journal of the American Chemical Society* 136 (11): 4394–4403.

30 Zhang, C., Mahmood, N., Yin, H. et al. (2013). Synthesis of phosphorus-doped graphene and its multifunctional applications for oxygen reduction reaction and lithium ion batteries. *Advanced Materials* 25 (35): 4932–4937.

31 Sun, X., Zhang, Y., Song, P. et al. (2013). Fluorine-doped carbon blacks: highly efficient metal-free electrocatalysts for oxygen reduction reaction. *ACS Catalysis* 3 (8): 1726–1729.

32 Zheng, Y., Jiao, Y., Li, L.H. et al. (2014). Toward design of synergistically active carbon-based catalysts for electrocatalytic hydrogen evolution. *ACS Nano* 8 (5): 5290–5296.

33 Zhao, Z. and Xia, Z. (2016). Design principles for dual-element-doped carbon nanomaterials as efficient bifunctional catalysts for oxygen reduction and evolution reactions. *ACS Catalysis* 6 (3): 1553–1558.

34 Lv, Y., Yang, L., and Cao, D. (2018). Sulfur, nitrogen and fluorine triple-doped metal-free carbon electrocatalysts for the oxygen reduction reaction. *ChemElectroChem* 6 (3): 741–747.

35 Yang, S., Peng, L., Huang, P. et al. (2016). Nitrogen, phosphorus, and sulfur co-doped hollow carbon shell as superior metal-free catalyst for selective oxidation of aromatic alkanes. *Angewandte Chemie International Edition* 55 (12): 4016–4020.

36 Banhart, F., Kotakoski, J., and Krasheninnikov, A.V. (2011). Structural defects in graphene. *ACS Nano* 5 (1): 26–41.

37 Jiang, Y., Yang, L., Sun, T. et al. (2015). Significant contribution of intrinsic carbon defects to oxygen reduction activity. *ACS Catalysis* 5 (11): 6707–6712.

38 Terrones, M., Botello-Méndez, A.R., Campos-Delgado, J. et al. (2010). Graphene and graphite nanoribbons: morphology, properties, synthesis, defects and applications. *Nano Today* 5 (4): 351–372.

39 Rybarczyk, M.K., Gontarek, E., Lieder, M. et al. (2018). Salt melt synthesis of curved nitrogen-doped carbon nanostructures: ORR kinetics boost. *Applied Surface Science* 435: 543–551.

40 Ortiz-Medina, J., Wang, Z., Cruz-Silva, R. et al. (2019). Defect engineering and surface functionalization of nanocarbons for metal-free catalysis. *Advanced Materials* 31 (13): e1805717.

41 Zhao, X., Zou, X., Yan, X. et al. (2016). Defect-driven oxygen reduction reaction (ORR) of carbon without any element doping. *Inorganic Chemistry Frontiers* 3 (3): 417–421.

42 Yan, X., Jia, Y., Odedairo, T. et al. (2016). Activated carbon becomes active for oxygen reduction and hydrogen evolution reactions. *Chemical Communications (Camb)* 52 (52): 8156–8159.

43 Paul, R., Zhu, L., Chen, H. et al. (2019). Recent advances in carbon-based metal-free electrocatalysts. *Advanced Materials* 31 (31): e1806403.

44 Jia, Y., Zhang, L., Du, A. et al. (2016). Defect graphene as a trifunctional catalyst for electrochemical reactions. *Advanced Materials* 28 (43): 9532–9538.

45 Tao, L., Wang, Q., Dou, S. et al. (2016). Edge-rich and dopant-free graphene as a highly efficient metal-free electrocatalyst for the oxygen reduction reaction. *Chemical Communications (Camb)* 52 (13): 2764–2767.

46 Liu, Z., Zhao, Z., Wang, Y. et al. (2017). In situ exfoliated, edge-rich, oxygen-functionalized graphene from carbon fibers for oxygen electrocatalysis. *Advanced Materials* 29 (18): 1606207.

47 Duan, J., Chen, S., Jaroniec, M. et al. (2015). Porous C_3N_4 nanolayers@N-graphene films as catalyst electrodes for highly efficient hydrogen evolution. *ACS Nano* 9 (1): 931–940.

48 Zheng, Y., Jiao, Y., Chen, J. et al. (2011). Nanoporous graphitic-C_3N_4@carbon metal-free electrocatalysts for highly efficient oxygen reduction. *Journal of the American Chemical Society* 133 (50): 20116–20119.

49 Deng, J., Ren, P., Deng, D. et al. (2015). Enhanced electron penetration through an ultrathin graphene layer for highly efficient catalysis of the hydrogen evolution reaction. *Angewandte Chemie International Edition* 54 (7): 2100–2104.

50 Zhou, M., Zhang, A., Dai, Z. et al. (2010). Strain-enhanced stabilization and catalytic activity of metal nanoclusters on graphene. *The Journal of Physical Chemistry C* 114 (39): 16541–16546.
51 Zhu, Y.P., Guo, C., Zheng, Y. et al. (2017). Surface and interface engineering of noble-metal-free electrocatalysts for efficient energy conversion processes. *Accounts of Chemical Research* 50 (4): 915–923.
52 He, C., Diao, L., Yang, T. et al. (2019). Electronic reconfiguration of Co2P induced by Cu doping enhancing oxygen reduction reaction activity in zinc–air batteries. *Journal of Materials Chemistry A* 7 (37): 21232–21243.
53 Xia, B.Y., Wu, H.B., Chen, J.S. et al. (2012). Formation of Pt-TiO_2-rGO 3-phase junctions with significantly enhanced electro-activity for methanol oxidation. *Physical Chemistry Chemical Physics* 14 (2): 473–476.
54 Wang, X.R., Liu, J.Y., Liu, Z.W. et al. (2018). Identifying the key role of pyridinic-N-Co bonding in synergistic electrocatalysis for reversible ORR/OER. *Advanced Materials* 30 (23): e1800005.
55 Toth, P.S., Velický, M., Bissett, M.A. et al. (2016). Asymmetric MoS_2/graphene/metal sandwiches: preparation, characterization, and application. *Advanced Materials* 28 (37): 8256–8264.
56 Wang, H., Liang, Y., Li, Y. et al. (2011). Co(1-x)S-graphene hybrid: a high-performance metal chalcogenide electrocatalyst for oxygen reduction. *Angewandte Chemie International Edition* 50 (46): 10969–10972.
57 Li, R., Liu, F., Zhang, Y. et al. (2020). Nitrogen, sulfur co-doped hierarchically porous carbon as a metal-free electrocatalyst for oxygen reduction and carbon dioxide reduction reaction. *ACS Applied Materials & Interfaces* 12 (40): 44578–44587.
58 Wang, H.-F., Tang, C., and Zhang, Q. (2018). Template growth of nitrogen-doped mesoporous graphene on metal oxides and its use as a metal-free bifunctional electrocatalyst for oxygen reduction and evolution reactions. *Catalysis Today* 301: 25–31.
59 Tang, C., Wang, B., Wang, H.F. et al. (2017). Defect Engineering toward atomic Co-Nx -C in hierarchical graphene for rechargeable flexible solid Zn-air batteries. *Advanced Materials* 29 (37): 1703185.
60 Ito, Y., Cong, W., Fujita, T. et al. (2015). High catalytic activity of nitrogen and sulfur co-doped nanoporous graphene in the hydrogen evolution reaction. *Angewandte Chemie International Edition* 54 (7): 2131–2136.
61 Wu, J., Zhou, H., Li, Q. et al. (2019). Densely populated isolated single Co-N site for efficient oxygen electrocatalysis. *Advanced Energy Materials* 9 (22).
62 Al-Mamun, M., Yin, H., Liu, P. et al. (2017). Carbon-encapsulated heazlewoodite nanoparticles as highly efficient and durable electrocatalysts for oxygen evolution reactions. *Nano Research* 10 (10): 3522–3533.
63 Ma, G., Xie, X., Sun, K. et al. (2020). Facile preparation of Fe_3C decorate three-dimensional N-doped porous carbon for efficient oxygen reduction reaction. *International Journal of Hydrogen Energy* 45 (24): 13272–13281.
64 Zhang, J., Qu, L., Shi, G. et al. (2016). N,P-codoped carbon networks as efficient metal-free bifunctional catalysts for oxygen reduction and hydrogen

evolution reactions. *Angewandte Chemie International Edition* 55 (6): 2230–2234.

65 Zheng, Y., Jiao, Y., Jaroniec, M. et al. (2015). Advancing the electrochemistry of the hydrogen-evolution reaction through combining experiment and theory. *Angewandte Chemie International Edition* 54 (1): 52–65.

66 Chen, L., Han, J., Ito, Y. et al. (2018). Heavily doped and highly conductive hierarchical nanoporous graphene for electrochemical hydrogen production. *Angewandte Chemie International Edition* 57 (40): 13302–13307.

67 Xiao, Z., Huang, X., Xu, L. et al. (2016). Edge-selectively phosphorus-doped few-layer graphene as an efficient metal-free electrocatalyst for the oxygen evolution reaction. *Chemical Communications (Camb)* 52 (88): 13008–13011.

68 Yang, H.B., Miao, J., Hung, S.F. et al. (2016). Identification of catalytic sites for oxygen reduction and oxygen evolution in N-doped graphene materials: Development of highly efficient metal-free bifunctional electrocatalyst. *Science Advances* 2 (4): e1501122.

69 Nørskov, J.K., Rossmeisl, J., Logadottir, A. et al. (2004). Origin of the overpotential for oxygen reduction at a fuel-cell cathode. *The Journal of Physical Chemistry B* 108 (46): 17886–17892.

70 Yan, H.J., Xu, B., Shi, S.Q. et al. (2012). First-principles study of the oxygen adsorption and dissociation on graphene and nitrogen doped graphene for Li-air batteries. *Journal of Applied Physics* 112 (10): 104316.

71 Tang, C., Wang, H.F., Chen, X. et al. (2016). Topological defects in metal-free nanocarbon for oxygen electrocatalysis. *Advanced Materials* 28 (32): 6845–6851.

72 Zhang, W., Hu, Y., Ma, L. et al. (2018). Progress and perspective of electrocatalytic CO_2 reduction for renewable carbonaceous fuels and chemicals. *Advanced Science (Weinh)* 5 (1): 1700275.

73 Wu, J., Ma, S., Sun, J. et al. (2016). A metal-free electrocatalyst for carbon dioxide reduction to multi-carbon hydrocarbons and oxygenates. *Nature Communications* 7: 13869.

74 Sreekanth, N., Nazrulla, M.A., Vineesh, T.V. et al. (2015). Metal-free boron-doped graphene for selective electroreduction of carbon dioxide to formic acid/formate. *Chemical Communications (Camb)* 51 (89): 16061–16064.

75 Wu, J., Liu, M., Sharma, P.P. et al. (2016). Incorporation of nitrogen defects for efficient reduction of CO_2 via two-electron pathway on three-dimensional graphene foam. *Nano Letters* 16 (1): 466–470.

76 Shipman, M.A. and Symes, M.D. (2017). Recent progress towards the electrosynthesis of ammonia from sustainable resources. *Catalysis Today* 286: 57–68.

77 Hering-Junghans, C. (2018). Metal-free nitrogen fixation at boron. *Angewandte Chemie International Edition* 57 (23): 6738–6740.

78 Yang, L., Jiang, S., Zhao, Y. et al. (2011). Boron-doped carbon nanotubes as metal-free electrocatalysts for the oxygen reduction reaction. *Angewandte Chemie International Edition* 50 (31): 7132–7135.

79 Yu, X., Han, P., Wei, Z. et al. (2018). Boron-doped graphene for electrocatalytic N_2 reduction. *Joule* 2 (8): 1610–1622.

Index

Templated Fabrication of Graphene-Based Materials for Energy Applications, First Edition.
Edited by Chunnian He, Naiqin Zhao, and Junwei Sha.

l

m

n

o

t